# GIS: A Computing Perspective

MICHAEL F. WORBOYS

*Department of Computer Science,*
*University of Keele, Keele, UK*

D1494096

Taylor & Francis
*Publishers since 1798*

UK    Taylor & Francis Ltd, 4 John Street, London WC1N 2ET
USA   Taylor & Francis Inc., 1900 Frost Road, Suite 101, Bristol, PA 19007

**British Library Cataloguing in Publication Data**

A catalogue record for this book is available from the British Library.

ISBN 0-7484-0064-8 (cased)
ISBN 0-7484-0065-6 (paper)

**Library of Congress Cataloging in Publication data are available**

Cover design by Amanda Barragry

Typeset in Times 10/12pt by Santype International Limited, Salisbury, Wilts.

Printed in Great Britain by Burgess Science Press, Basingstoke, on paper which has a specified pH value on final paper manufacture of not less than 7.5 and is therefore 'acid free'.

*This book is dedicated to Moira*
*and to*
*Bethan, Colin, Ruth, Sophie, Tom and William*

# Contents

# Preface

Truth lies within a little and certain compass, but error is immense.

Viscount Bolingbroke

A geographic information system (GIS) is a computer-based information system that enables capture, modelling, manipulation, retrieval, analysis and presentation of geographically referenced data. The study of GIS has emerged in the last decade as an exciting multi-disciplinary endeavour, spanning such areas as geography, cartography, remote sensing, image processing, the environmental sciences and computer science. The treatment in this text is unashamedly biased towards the computational aspects of GIS. Within computing science, GIS is a special interest of fields such as databases, graphics, systems engineering and computational geometry, being not only a challenging application area but also providing foundational questions for these disciplines. An underlying question throughout is 'What is special about spatial information?', and answers to this question will be attempted at several different levels, from conceptual and formal models, through representations and data structures that will support adequate performance, to the special-purpose interfaces and architectures required.

The book is intended for readers from any background who wish to learn something about the issues that GIS engenders for computing technology, and for the more advanced researcher. The reader does not have to be a specialist computing scientist: the text develops the necessary background in specialist areas, such as databases, as it progresses. However, some knowledge of the basic components and functionality of a digital computer is essential for understanding the importance of certain key issues in GIS. Where some aspect of general computing bears a direct relevance to our development, the background is given in the text.

This book inevitably reflects the interests and biases of its author, in particular emphasizing spatial information modelling and representation, as well as developing some of the more formal themes of the object. I have tried to avoid detailed discussion of particular currently fashionable systems, but to concentrate upon the foundations and general principles of the subject area. I have also tried to give an overview of the field from the perspective of computing science. The task of computing practitioners in the field of GIS is to provide the application experts, whether geographers, planners, utility engineers or environmental scientists, with a set of tools based around digital computer technology that will aid them in solving problems in their domains. These tools will include modelling constructs, data structures that will allow efficient storage and retrieval of data, and generic interfaces that may be customized for particular application domains.

GIS are not just systems, but also constitute an interdisciplinary endeavour involving many people in industry and academia. The aim of this endeavour is to provide better solutions to geo-spatial problems. Contributions vary from constructing purpose-built computer hardware to conceptual modelling for spatial problems and novel approaches to spatial analysis. The work in many cases moves beyond the single or multi-disciplinary arenas to truly interdisciplinary collaboration: therein lies its excitement and its strength.

Current systems are almost all based on a static two-dimensional view of the world, and this then is the focus of the main body of the text. However, as hardware and software develops, the challenging prospect of handling information referenced to the three spatial dimensions is emerging. Also, there is a need in many applications to work with dynamic information, modelling phenomena as they evolve through time. These issues are given special attention in the final chapter.

Anybody familiar with GIS will know that one of the biggest problems is integration of different systems, each one with its own models and representations of data and processes. Allowing systems to combine to solve common problems is no trivial task, but standards are emerging that will help to alleviate these difficulties. Standards exist for systems and communications (e.g. the 'open system' philosophy), information systems (e.g. the SQL standards) and data interchange (e.g. SDTS and NTF). Appendix 2 lists the names and main features of some of the most important and relevant standards.

This book can be used as a teaching text, taking readers through the main concepts by means of definitions, explications and examples. However, the more advanced researcher is not neglected. Each chapter has at its close a bibliography that provides pointers to further reading, including in many cases recent research findings. The final chapter is devoted to the development of three themes that are very much part of the current research agenda. Also included is Appendix 3, which lists some of the conferences and journals in which the latest research findings are reported.

Not every topic can be covered, and I have deliberately neglected two areas, leaving these to people expert in those domains. The first is the historical background. The development of GIS has an interesting history, stretching back to the 1950s, and readers who wish to pursue this further will find an excellent introduction in (Coppock and Rhind, 1992). The other area that is given scant treatment is spatial analysis, which is judged to be specifically the province of the domain experts, having specialist statistical techniques. An introduction to spatial analysis is provided by Unwin (1981) and a more recent survey by Openshaw (1991). Openshaw (1994) discusses some recent trends in the identification and analysis of patterns in space.

Michael Goodchild in his keynote address to the Fourth International Symposium on Spatial Data Handling, held at Zurich in 1990, used what he called the 'S' word in GIS: not system, but science. He identified a need to do science in this field, both by tackling the scientific questions that geographic information handling raises and by pursuing scientific goals using the technology that the systems provide. I hope that this book contributes to those high ideals.

## BOOK STRUCTURE

The overall plan of the book is shown below and in Figure 0.1. The difference between model, representation and structure can be unclear. It was sometimes difficult to decide where to place particular items. For example, a triangulated irregular network can have roles as model, representation and structure.

Chapter 1: Motivation and introduction to GIS; preparatory material on general computing.

Chapters 2–3: Background material on general databases and formalisms for spatial concepts.

Chapters 4–6: (Models, representations, structures and access methods): Exposition of the core material, forming a progression from high-level conceptual

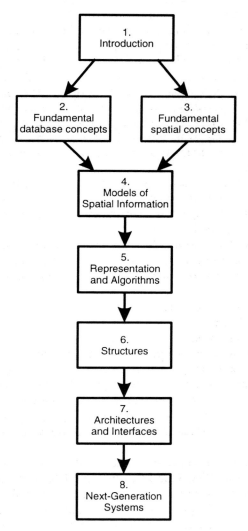

**Figure 0.1**  Relationships between chapters.

models, through representations, to indexes and access methods that allow acceptable performance.

Chapter 7:   Overview of architectures and interfaces for GIS.

Chapter 8:   Selected topics of current research interest.

### ACKNOWLEDGEMENTS

Many people have contributed to this book, either explicitly by performing tasks such as reading a draft or contributing a diagram, or implicitly by being part of the lively GIS research community and so encouraging the author to develop this material. Waldo Tobler, of Santa Barbara, California, contributed a diagram of an interesting travel-time structure for Chapter 3 from his doctoral thesis, and cheerful encouragement. Also, from the National Center for Geographic Information (NCGIA) at the University of California Santa Barbara, Mike Goodchild, Helen Couclelis and Terry Smith provided a congenial environment for developing some of the ideas leading to this book. Max Egenhofer, from another NCGIA site at the University of Maine, provided hospitality and a useful discussion about the structure of the book. At Keele University, Keith Mason and Andrew Lawrence provided invaluable help with some of the example applications introduced in Chapter 1. Dick Newell of Smallworld has generously provided Smallworld GIS as a research and teaching tool, enabling me to develop and implement many ideas using a state-of-the-art system. Tigran Andjelic and Sonal Davda checked some of the technical material. Richard Steele from Taylor & Francis has been encouraging and patient with my failure to meet deadlines. Mostly, I am indebted to my wife Moira for her invaluable assistance with the final draft, and for her moral and practical support throughout this work.

MICHAEL WORBOYS

# Introduction

I could be bounded in a nut-shell, and count myself a king of infinite space . . .

William Shakespeare, *Hamlet*

This chapter introduces the elements of a geographic information system (GIS), set against a general background of computing. The scene is set in section 1.1, where GIS is defined, briefly described and distinguished from other related information systems. Section 1.2 provides motivation for GIS by describing some scenarios in which GIS would be useful and by producing a list of some functional requirements. Two of the most important technologies underlying GIS are database systems and computer graphics, and some of the main considerations for these areas are outlined in sections 1.3 and 1.4. Section 1.5 considers some of the hardware supporting GIS and the final section discusses the computer communication technology that is already having a very large impact in our field. The chapter concludes with pointers to further reading on these topics.

## 1.1 WHAT IS A GIS?

Any kind of computer-based information system aims to use computer technology to assist people to manage the information resources of organizations. 'Organization' is used here in a wide sense that includes corporations and governments, but also more diffuse forms such as global colleges of scientists with common interests or a collection of people looking at the environmental impact of a proposed new rail-link. For example, an information system manages the information resources of the human genome project and contains at its heart not only a database of genetic information, but also details of contributing scientists, experiments, and so on. An information system (assumed to be computer-based) will have the capability to collect, manage, and distribute information of importance to the organization. It is supported by a database and its management software, systems developed for the specific needs of the organization (often called *applications software*), personnel, the data themselves, capturing devices, and presentation and dissemination media. GIS adds its own special flavour to that of a generic information system (see Figure 1.1).

A *geographic information system* (*GIS*) is a computer-based information system that enables capture, modelling, manipulation, retrieval, analysis and presentation of geographically referenced data.

1

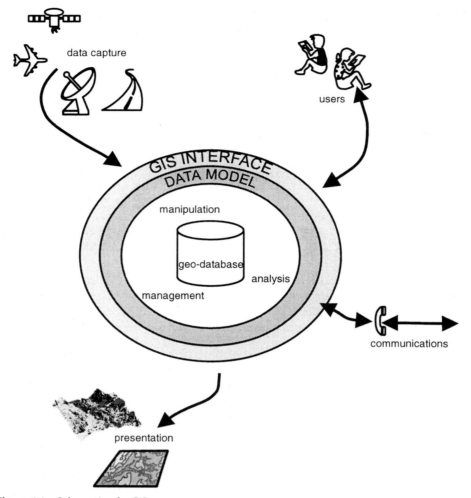

**Figure 1.1** Schematic of a GIS.

One way of understanding GIS is to look at its application. We all exist within a spatial and temporal world and have a need for information that has spatial and temporal dimensions. The environment is a dynamic system that is sampled, monitored, modelled and simulated. Future decisions that affect us all, such as planning new roads or cities, formulating agricultural strategies and locating mineral extraction sites, rely upon properly collected, managed, analysed and presented spatio-temporal information. To get the most from such complex information, we should know its limitations (accuracy, legacy) and have tools to visualize it effectively.

At the heart of any GIS is the database. The data that are held in the database are given context and meaning by the application domain(s) and then take on an informational value. Such data must first be captured in a form that the system can handle and input into the system. Database systems provide facilities that allow their population of data to be input in appropriate forms. Data are only useful (taking on value as information) when they are part of a structure of inter-relationships that forms the context of the data. Such a context is partly provided by the *data model* (Chapter 2). Some applications require a relatively simple data model: for example, in a library system, data about books, users, reservations and loans are structured in a clear way. Many applications, including most GIS applications, demand a more complex data model (consider for example a global climatological system) and much of the concern in this book is the provision of facilities for handling such models. This book does not discuss the appropriateness of particular models in an application area, but it is concerned with how to provide computational facilities for such models once they are determined by the application domain experts. For example, our first concern is not that a gravitational model of transportation flows between cities is suitable for a particular application, but to provide facilities for the representation of such a model in an information system. In the case of simple models, the technology for the provision of these facilities is well understood. Handling more complex models will take us to the research frontiers.

Along with the model come the manipulation tools. Many modern models allow the representation of not only the structural relationships within the data but also the operations and processing requirements that the data will require. It is important that a GIS has a sufficiently complete functionality to provide the higher-level analysis and decision support required within the application domain. For example, a GIS that is used for utility applications will require network processing operations (optimal routing between nodes, connectivity checks, etc.), while a GIS for geological applications will require three-dimensional operations. It is possible to make some generalizations and lay down a set of primitive operations required by a generic GIS; this will be a major concern of the book. However, a discussion of particular approaches to spatial data analysis are within the province of the application experts and outside the scope of this text.

Efficient retrieval of data depends not only on properly structured data in the database to provide satisfactory performance (for example, fast data retrieval times) but also on appropriate human computer interaction (HCI), including well designed interfaces and query languages for the information system. The former optimizes structures, representations and algorithms for the data and operations, while the latter provides an environment that optimizes human interaction with the information system. Retrieval and performance raise many interesting questions for spatial data and will be a further main theme of this text. Related to these issues is the presentation of the results of computation. Most traditional information systems require only limited presentational forms, usually based around tables of data, numerical computation and textual commentary. While these forms are required to support decisions with GIS, the spatial nature of the information allows a whole new range of possibilities, including cartographic forms and visualization techniques. Presentation of the results of retrievals and analyses using GIS therefore takes us beyond the scope of most traditional databases and will be a further concern of this text.

There are several terms that are in common use and are more or less synonymous with GIS. A *spatial information system* (*SIS*) has the same functional components as a GIS but may handle data that are referenced at a wider range of scales than geographical, e.g. data on molecular configurations. We shall assume that a *spatial database* comprises all the database functionality within a spatial information system. A *geographic database* (*geo-database*) provides the database functionality of a GIS. An *image database* is fundamentally different from a spatial database in that the images are 'unintelligent'. That is, there are no explicit data on the structural inter-relationships or topological features of the objects in the database. For example, a database of brain-scan images may be termed an image database.

Computer graphics are designed to display information of a visual nature but a GIS provides more than purely graphics processing. While geographic information has a very important visual (graphical) component, there is an equally important collection of non-visual components to geodata. As Burrough (1986, p. 7) observed, 'Good computer graphics are essential for a modern geographic information system but a graphics package is by itself not sufficient for performing the tasks expected, nor are such drawing packages necessarily a good basis for developing such a system.'

Computer-aided design (CAD) is a further area that has some elements in common with GIS. Indeed, some current proprietary GIS stem from CAD systems. CAD provides interactive graphics as an aid to the design of a variety of artificial structures, such as buildings, vehicles, very large-scale integrated (VLSI) chips and computer networks. The emphasis here is on the interaction between the designer and computer-based model, so as to iterate towards an acceptable design. It should be possible to simulate tests of the properties of the users' designs (for example, structural properties of a building). Three-dimensional modelling tools are also a feature of such systems. While there are many similarities between CAD systems and GIS, particularly when CAD begins to support non-graphical properties of objects for simulations, there are enough differences to make the beasts quite distinct. These range from the types of data underlying the systems (GIS has a wide range captured from natural and artificial sources) to the functional requirements. However, as with graphics systems, it often benefits us to see connections and take advantage of research undertaken in these fields. Future simulations and visualizations will use virtual reality technology, which again has implications for GIS.

We should note finally that geographic reference implies not only spatial reference at geographic scales, but also temporal reference. Geographic phenomena are usually highly dynamic, and this fact is ignored by present systems only because of technological limitations. Handling temporality with geographic information will be pursued as a topic later in the book.

## 1.2 SOME FUNCTIONAL REQUIREMENTS OF A GIS

A GIS is a computer system that has special characteristics. To demonstrate the range of functionality contained within a GIS, a particular area of England has been chosen. The region, familiarly called 'The Potteries', due to its dominant eponymous industry, comprises the six pottery towns of Burslem, Fenton, Hanley, Longton, Stoke and Tunstall, along with the neighbouring town of Newcastle-under-Lyme (Figure 1.2). The region developed rapidly during the English industrial revolution, local communities producing ware of the highest standard (for example from the potteries of Wedgwood and Spode) from conditions of poverty and cramp, but then declined during this century. The landscape is scarred by the extraction of coal, ironstone and clay, but the area is currently in a phase of regeneration.

We provide six example applications to give the reader a feel for the capabilities of a GIS. We emphasize that the examples are here merely to show functionality, rather than serious descriptions of application areas. For application-driven texts,

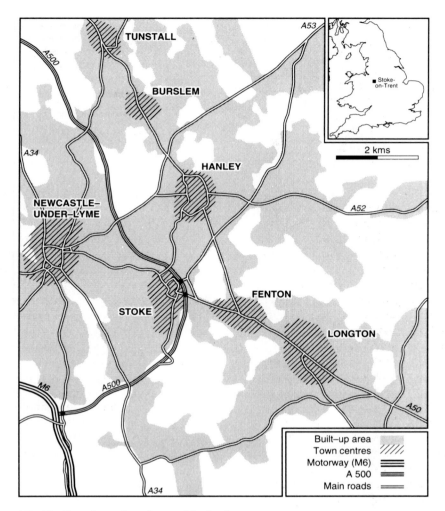

**Figure 1.2**   The Potteries region of central England.

see the bibliography at the end of the chapter. The final part of the section summarizes typical GIS functionality.

### 1.2.1   Resources inventory: a tourist information system

The Potteries, because of its past, has a locally important tourist industry based upon the industrial heritage of the area. A GIS may be used to support this by drawing together information on cultural and recreational facilities within the region and combining these appropriately with details of local transport infrastructure and hotel accommodation. Such an application gives an example of a simple resource inventory. The power of almost any information system lies in its ability to relate and combine information gathered from disparate sources. This power is increased dramatically when the data are spatially referenced, provided that the

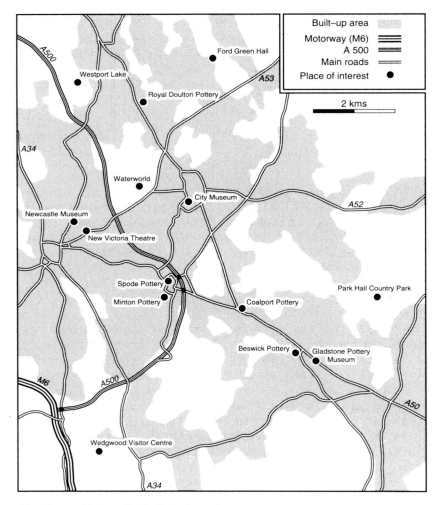

**Figure 1.3** Places of interest in the Potteries region.

different sources of spatial data are made compatible through some common spatial unit or transformation process. Figure 1.3 shows the beginnings of such a system, including some of the local tourist attractions, the major road network and built-up areas in the region. Multimedia systems could play an important role here, allowing the retrieval of video clips and audio commentary.

### 1.2.2   Network analysis: a tour of the Potteries

Network analysis is one of the cornerstones of GIS functionality and applications may be found in many areas, from transportation networks to the utilities. To provide a single simple example, we remain with the tourist information example above. The major potteries in the area are famous worldwide. Many of the potteries include factory tours and shops. The problem is to provide a route using the major road network, visiting each pottery (and the City Museum) once, and once only, and

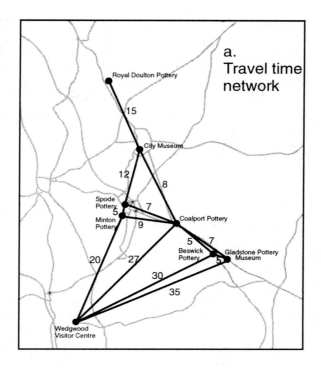

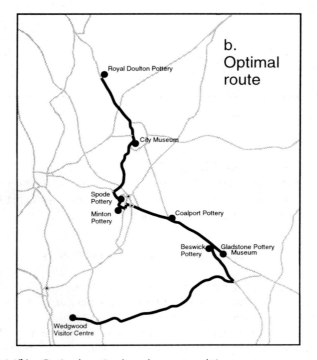

**Figure 1.4(a) and 1.4(b)** Optimal routing based upon travel times.

minimizing the travelling time. The dataset required is a travel–time network between the potteries: a partial example is given in Figure 1.4(a). This travel–time network was derived from average times on the main roads shown on the map. The network may then be used to generate the required optimal route between Royal Doulton and the Wedgwood Visitor Centre, shown in Figure 1.4(b). The route visits the potteries in sequence: Royal Doulton, City Museum, Spode, Minton, Coalport, Gladstone, Beswick, Wedgwood. The specific network analysis technique required here is the 'travelling salesperson algorithm', which constructs a minimal weight route through a network that visits each node at least once. The analysis could be dynamic, assigning weights to the edges of the network and calculating optimal routes depending upon changeable road conditions.

### 1.2.3   Terrain analysis: siting an opencast coal mine

Terrain analysis is usually based upon datasets giving topographical elevations at point locations. Examples of information to be directly derived from such datasets are degree and direction of slope, leading to the path of least resistance down a slope. More complex is the analysis of visibility between locations and the generation of viewsheds (points visible from a given point under given conditions). Applications of terrain analysis are diverse. For example, in the Potteries conurbation the search for new areas of opencast coal mining has resulted in much interest from local communities who might be concerned about the effects of such operations. One factor in this complex question is the visual impact of proposed opencast sites: visibility analysis can be used to evaluate this, for example by measuring the sizes of local populations within given viewsheds. Sites that minimize population may be considered desirable. The left-hand part of Figure 1.5 shows a contour map of topographical elevations for the area around Biddulph Moor: the cross indicating the proposed sight of the opencast mine. The right-hand of Figure 1.5 shows an isometric projection of the same surface. Such projections allow a powerful visualization of the terrain. Figure 1.6 shows the same surface again, this time draped by the viewshed. The shaded regions give the area visible from the proposed mine and therefore from which the mine would be visible. Of course, a real case would take into account much more than just visibility, but the principle remains.

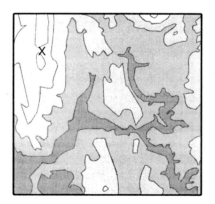 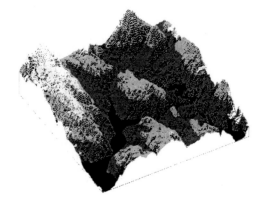

**Figure 1.5**   Contour map and isometric projection of the surface of Biddulph Moor.

**Figure 1.6**   Isometric projection of the surface of Biddulph Moor draped with the viewshed of the projected opencast coal mine.

### 1.2.4   Layer-based analysis: the potential of extraction sites for mineral ore deposits

The Potteries area is rich in occurrences of superficial and bedrock sand and gravel, although few such sites have been worked in recent times. This application exemplifies the drawing together and analysis of data from disparate sources in order to measure the potential of different locations for sand and gravel extraction. Geological data describing the location of appropriate deposits are of course key here. Other important considerations are local urban structure (e.g. urban overbidding), water-table level, transportation network, and land prices and restrictions. A sample of the data available, showing built-up area, sand and gravel deposits and the major road network, overlaid on a single sheet, is shown in Figure 1.7.

Layer-based analysis results from asking a question such as: 'Find all locations that are within 0.5 km of a major road, not in a built-up area and on a sand/gravel deposit.' In Figure 1.8, the grey areas show: (a) the area within 0.5 km of a major road (not the motorway) by placing a buffer around the road centre-lines; (b) areas of sand and gravel deposits; and (c) areas that are not built up. Figure 1.8(d) shows the overlay of all three of these areas, and thus the areas that satisfy our query. The analysis here is simplistic. A more realistic exercise would take into account grading of the deposit, land-prices, regional legislation, etc. However, the example does show some of the main functionality required of a GIS engaged in layer-based analysis, including:

**Buffering**:   The formation of areas containing locations within a given range of a given set of features. Buffers are commonly circular or rectangular around points, and corridors of constant width about lines and areas.

**Boolean overlay**:   The combination of one or more layers into a single layer that is the union, intersection, difference, or other Boolean operation applied to the input layers.

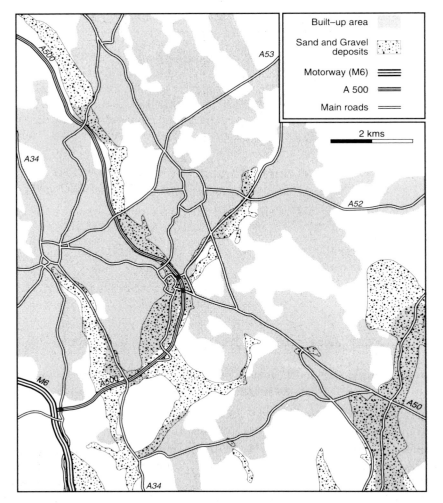

**Figure 1.7**   Locations of sand and gravel deposits in the Potteries region (redrawn from Wilson *et al.*, 1992).

Layer-based functionality is explored further in the context of field-based models and structures later in this book.

### 1.2.5   Location analysis: locating a clinic in the Potteries

Location problems have been solved in previous examples using terrain models (opencast mine example) and layer-based analysis (estimating the potential of sites for extracting sand and gravel). Our next example is the location of clinics in the Potteries area. A critical factor in the decision to use a particular clinic is the time it takes to travel to it. We will construct the 'neighbourhood' of a clinic, based upon positions of nearby clinics and travel times to the clinic. With this evidence, we can then support decisions to relocate, close or create a new clinic. Figure 1.9 shows the

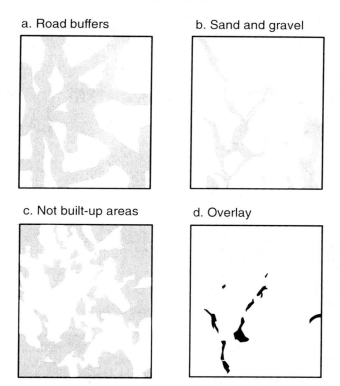

**Figure 1.8**   Layer-based analysis to site a mineral ore extraction facility.

positions (idealized) of clinics in the Potteries region. Assuming an 'as the crow flies' travel–time metric between points (i.e. the time is directly related to the Euclidean distance between points), Figure 1.9(a) shows the 'isochrones' surrounding a given clinic. It is then possible to partition the region into areas (proximal polygons), each containing a single clinic, such that each area contains all the points that are nearest (in travel time) to its clinic (see Figure 1.9(b)). Of course, we are making a big assumption about travel time. If the road network is taken account of in the travel–time analysis, the isochrones will no longer be circular and the areal partition will no longer be polygonal. This more general situation is discussed later in the book.

### 1.2.6   Spatio-temporal information: 30 years in the Potteries

Geographic information sometimes becomes equated with purely static spatial information, thus neglecting the importance of change and of the temporal dimension(s). In a temporal GIS, data are referenced to three kinds of dimensions: space, time and attribute. Our last example suggests possibilities for a future dynamic GIS functionality, just beyond the reach of present systems. We return to the problem of adding temporal dimensions in the final chapter.

   The main period of industrial activity in the Potteries has long since past and the history of the region in the latter half of the twentieth century has been one of

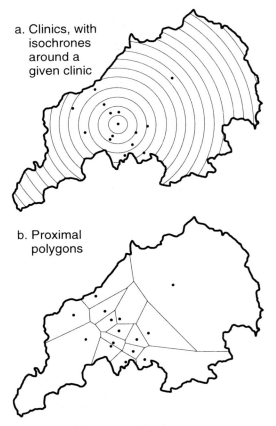

a. Clinics, with
   isochrones
   around a
   given clinic

b. Proximal
   polygons

**Figure 1.9**   The Potteries' clinics and their proximal polygons.

industrial decline. The maps below show the changing scene within one small area of the Potteries over a period of just over 30 years. Figure 1.10 shows the area around the Gladstone Pottery, recorded in snapshot at three times: 1937, 1959 and 1969. The potbanks used for firing the pottery are shown on the maps as circles. It is clear that as time passed so the quantity of potbanks changed and ultimately decreased to almost zero. Examples of questions that we may wish to ask of our spatio-temporal system are:

Retrieve the potbanks closest to Chadwick Street in 1937, 1957 and 1969, respectively.

Identify those streets that have changed name during the period 1937–1969.

Identify those streets that have changed spatial reference during the period 1937–1969.

At what year is the existence of Park Works last recorded in the system?

Did the areas occupied by Park Works and Sylvan Works at some times (not necessarily the same times) ever overlap?

Present in a vivid way the change in the spatial pattern of this region between 1937 and 1969.

**Figure 1.10** History of an area around the Gladstone Pottery, recorded in snapshot in 1937, 1959 and 1969 (based upon old Ordnance Survey maps).

### 1.2.7   Summary of processing requirements

The examples of applications reviewed in this section demonstrate some of the specialized requirements for GIS technology. At the heart of the GIS is the data store (cf., section 1.2.1), and this must be managed appropriately to provide fast, secure, reliable, correct information. Much of the data have a spatial connection and will be presented using computer graphics. However, we have seen that a GIS is not just a graphics database. A graphical system is concerned with the manipulation and presentation of screen-based graphical objects, whereas a GIS handles phenomena embedded in the geographic world and having not only spatial dimensions, but also structural placement in multi-dimensional geographic models. We summarize some of the analytical processing requirements that give GIS its special flavour.

*Geometric/topological/set-oriented analyses*:   Most if not all geographically referenced phenomena have geometric, topological or set-oriented properties. Set-oriented properties include membership conditions, Boolean relationships between collections of elements and handling of hierarchies (e.g. administrative areas). Topological operations include adjacency and connectivity relationships. A geometric analysis of the clinics (section 1.2.5) produced their proximal polygons. All these properties are key to a GIS and form one of the main themes of this book.

*Fields, surfaces and layers*:   Many applications involve spatial fields, that is, variations of attributes over a region. The terrain in section 1.2.3 is a variation of topographical elevation over an area. The gravel-sand deposits and built-up area of section 1.2.4 were variations of other attributes. Fields may be discrete (a location is either built-up or not) or continuous (e.g. topographical elevation). Fields may be scalar (variations of a scalar quantity and represented as a surface) or vector (variations of a vector quantity such as wind-velocity). Field operations include overlay (sections 1.2.4), slope/aspect analysis, path finding, flow analysis, viewshed analysis (section 1.2.3). Fields are discussed further throughout the text.

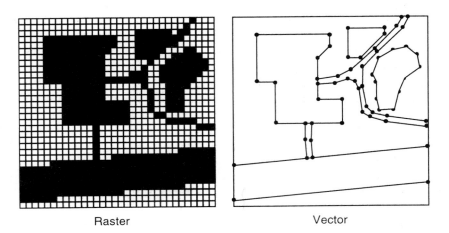

Raster                                             Vector

**Figure 1.11**   Raster and vector data.

*Network analysis*:  A network is a configuration of connections between nodes. The maps of most metro systems are in network form, nodes representing stations and edges being direct connections between stations. Networks may be directed in the case where edges are assigned directions (e.g. in a representation of a one-way street system) or labelled, when edges are assigned numeric or non-numeric attributes (e.g. travel–time along a rail link). Network operations include connectivity analysis, path finding (trace-out from a single node and shortest path between two nodes), flow analysis and proximity tracing (the network equivalent of proximal polygons). The Potteries tourist information system application (section 1.2.2) gives an example of a network traversal. Networks are considered further in Chapter 3.

Beyond the analytical processing required of a GIS, there are several other classes of requirements, from data capture to presentation and visualization of results. These will emerge as the chapter progresses.

## 1.3    DATA AND DATABASES

Data are at the heart of a GIS and any system is only as good as its population of data. This section introduces some of the main themes associated with data and their management. GIS data are handled in a database or databases, which will have special functional requirements as well as the general characteristics of any standard database. To support the database, there is a need for appropriate capability for data capture, modelling, retrieval, presentation and dissemination. An issue that arises throughout the book is the functionality that a geo-database system needs to provide beyond the standard functions of a general purpose database system. This section provides an overview, but the topic is taken further in Chapter 2 and then throughout the book.

### 1.3.1    Spatial data

First, a reminder of some data basics. Data are stored in a computer system in *bits*. Each bit records one of two possible states: 0 (off, false) and 1 (on, true). Bits are amalgamated into *bytes*, usually of 8 bits, each byte representing a single character. Typically a character may be represented by 7 bits with an extra bit used as an error check, thus each byte allows for $2^7 = 128$ possible character combinations. This is the basis of the ASCII code (see section 1.6.1).

Spatial data are traditionally divided into two great classes, *raster* and *vector*, and in the past, systems have tended to specialize in one or other of these classes. Translation between then (*rasterization* for vector-to-raster and *vectorization* for raster-to-vector) has been a thorny problem, and will be discussed later in this book. Figure 1.11 shows raster and vector data representing the same situation of a house, outbuilding and pond next to a road.

Raster data are structured as an array or grid of cells, often referred to as *pixels*. (The three-dimensional equivalent is a three-dimensional array of cubic cells, called *voxels*). Each cell in a raster is addressed by its position in the array (row and column number). Rasters are able to represent a large range of computable spatial objects. Thus, a point may be represented by a single cell, an arc by a sequence of neighbouring cells and a connected area by a collection of contiguous cells. Rasters are handled naturally in computers, because all commonly used programming languages support array handling and operations. However, a raster when stored in a raw state with no compression can be extremely inefficient in terms of usage of computer storage. For example, a large uniform area with no special characteristics may be stored as a large collection of cells, each holding the same value. We will consider efficient computational methods of raster handling in later chapters.

The other common paradigm for spatial data is the vector format. A *vector* is a finite straight line segment defined by its end-points. The locations of end-points are given with respect to some coordinatization of the plane or higher-dimensional space. A discretization of space is not explicit as it is with the raster structure; however it must exist implicitly in some form, because of the discrete nature of computer arithmetic. Vectors are an appropriate representation for a wide range of spatial data. Thus, a point is simply given as a coordinate point. An arc is discretized as a sequence of straight-line segments, each represented by a vector, and an area is defined in terms of its boundary, represented by a collection of vectors. The vector data representation is inherently more efficient in its use of computer storage than raster, because only points of interest need be stored. A disadvantage is that, at least in their crude form, vectors assume a hard-edged boundary model of the world that does not accord with our observations. (The vector approach does not rule out a less hard-edged approach to models of the world. For example, we could associate fuzzy coordinates to end-points of vectors.)

Before leaving data, we touch briefly upon data quality and usability issues. The value of any database lies ultimately in the quality and usability of its data. Data quality considerations include correctness (accuracy), timeliness, appropriateness, relevance and usability, and has been the subject of considerable work, especially in the GIS community where geodata raise their own special problems. It is now well recognized that data should come with *lineage*, which describes their quality. There follow some notes on the main issues. The bibliography at the close of the chapter contains some further references.

*Correctness and accuracy*:   Accuracy concerns the consistency and completeness of the link between the data and the original source about which the data are collected. Accuracy of data has several components, including accuracy of attribute values and, for geodata, accuracy of spatial and temporal references. Data quality is an issue when observations and measurements are taken, measurements are input to the computer, data are processed and results are presented.

*Timeliness, relevance and cost*:   Clearly the age, appropriateness and cost of the data is relevant to their quality.

*Usability and accessibility*:   There are several considerations here. Accessibility is dependent upon physical links to databases through networks. Even if a database is accessible it may not be the case that the data are usable. Data must be in a form that can be handled with the tools that users have available. At a deeper level, users need to know what data are in the database and what are their characteristics. Particularly in very large databases this is a considerable problem. *Metadata*, data stored in the database with and about the original data, can provide help here as can high-level models (maps?) of the structure of the database that can be used as navigational aids.

### 1.3.2 Database as data store

A database is a store of interrelated data to be shared by a group of users. Items of data may be numbers, character strings, text, images, sounds, spatial configurations; in fact anything that can be measured, recorded and to which it is possible to assign meaning. Data that have been given a meaning due to their context and relationship with other pieces of data become information. For example, 'Mike' is an item of data, and the fact that Mike is the usual name by which the author of this book is addressed is information. Information may not only come in the form of facts, but also as rules, like 'all counties in England are connected areas'. Rules are more general than facts and apply over an entire data domain or domains. Traditional databases are able to handle the storage of facts. General rules may then only be handled by storing all instances of the facts to which they apply. Future databases (*knowledge bases* or *deductive databases*) allowing the handling and manipulation of facts and rules are described in Chapter 8.

A database is a repository of data, logically unified but possibly physically distributed over several sites, and required to be accessed by many users. It is created and maintained using a general purpose piece of software called a *database management system* (DBMS). For a database to be useful it must be:

*Secure*:  Preventing unauthorized access and allowing different levels of authorized access. For example, we may be allowed to read our bank balances but no one else's (*read access*); we may not be allowed to change our bank balances (*write access*).

*Reliable*:  Usually up and running when required by users and taking precautions against unforeseen events such as power failures. We would want to avoid the situation, for example, where we deposit money into an automatic banking device and the power fails before our balance has been updated.

*Correct and consistent*:  Data in the system should be correct and consistent with each other. This has been a problem with older file systems, where data held by one department contradict data from another. While it is not possible to screen out all incorrect data, it is possible to control the problem to some extent. Clearly, the more we know about the kinds of data that we expect, the more errors can be detected.

*Technology proof*:  Not changing with each new technological development. Both hardware and software continue to develop rapidly, as new chips, storage devices, software and modelling techniques, etc., are invented. When we check our balance at the automatic banking device we do not really want to know that on the previous night, new storage devices were introduced to support the database. Database users should be insulated from the inner workings of the database system.

### 1.3.3 Data capture

The data capture requirement is two-fold. First, to provide the physical devices for capturing data external to the system and inputting to the database. Second, to provide software for converting data to structures that are compatible with the data model of the database and checking for correctness and integrity of data before entry into the system. The cost of data capture for GIS is usually regarded as a major problem, and as systems become cheaper these costs will dominate more and more the overall cost of GIS technology.

With regard to the first requirement, geographic databases have a much wider variety of sources and types of data than traditional databases. These sources may be primary or secondary (stemming from other sources). Primary data may be captured by remote sensing or direct surveying (possibly employing positioning tools that satellite technology provides). Among the secondary data sources are existing paper-based maps. The hardware to handle such capture is primarily digitizer and scanner technology. The software often comes in the form of drivers to convert from one data format to another.

As system hardware and software becomes cheaper and provide more functionality, the cost of spatial data capture increasingly dominates. An earlier survey (Dickinson and Calkins, 1988) showed that data capture costs may account for as much as 70 per cent of total GIS system costs. The digitizing process is slow and expensive, but it does deliver a well-structured and information-rich vector data source. Scanning is fast and very much cheaper than digitizing, but the raster end-product is limited in structure and in the ability to add value by attaching attributes to spatial references. Scanning technology has improved dramatically in the last decade and therefore two independent possibilities for improved data capture arise:

Developments in raster-vector conversion, with the use of semi-automatic and automatic raster-to-vector conversion methods.

Developments in raster systems to provide equal or enhanced functionality with vector systems.

On the former possibility, there is much current interest in new methods for semi-automatic and automatic raster-to-vector conversion approaches. At the time of writing, fully automatic approaches have yet to be realized in practice, even though they are the subject of a considerable research effort (Musavi et al., 1988). The problem here is similar to that in several areas of computing where intelligence, knowledge and handling a suitable rule base are critical factors. However, semi-automatic approaches are now feasible and some of the principles behind these are discussed in Chapter 5.

### 1.3.4   Data modelling

Choice of appropriate data model can be the critical factor for the success or failure of an information system; in fact the data model is the key to the database idea. Data models function at all levels of the information system. High-level *conceptual data models* provide a means of communication between the user and the system: they are independent of the details of the implementation and express the details of the application in a way that is comprehensible to the user (ideally, being the way that the user perceives the application), and also provide a link with the computational model of the database. On the other hand, low-level *physical data models* provide a description of the physical organization of the data in the machine, for example the configuration of the data in secondary storage. Physical data models are more useful to system developers than domain experts. To summarize, data models describe the database at some level (whether application-oriented or system-dependent) and act as mediators between humans and machines.

### 1.3.5 Data retrieval

The database must be accessible to users, whether they be domain experts, casual enquirers, or system programmers. Accessibility is accomplished by the provision of interfaces through which interactions can take place. This can be problematic, because the user group may be diverse, including: the casual user, maybe accessing a GIS from a library; the regular user, proficient with computer technology; and the system developer, creating, maintaining, tuning and revising the database. The latter two groups may be best served with a database query language that can be embedded in other programming languages that they use. Other groups may be happier with a simple menu interface. Interfaces for GIS are discussed in Chapter 7.

Most interactions with a database are attempts to retrieve data: traditional databases support retrieval of text and numerical data. To express to the database what data are required, we may apply a filter, usually in the form of a logical expression. For example

Retrieve names and addresses of all opencast coal mines in Staffordshire.

Retrieve names and addresses of all employees of Wedgwood Pottery who earn more than half the sum earned by the managing director.

Retrieve the mean population of administrative districts in the Potteries area.

Retrieve the names of all hospital patients at Stoke City General Hospital, who are over the age of 60 years and have been admitted on a previous occasion.

Assuming that the first query accesses a national database and that the county in which a mine is located is given as part of its address, the data may be retrieved by means of a simple look-up and match. For the second query, each employee's salary would be retrieved along with the salary of the managing director and a simple numerical comparison made. For the third, populations are retrieved and then a numerical calculation is required. For the last, a more complex filter is required, including a check for multiple records for the same person.

There are spatial operators in the above queries. Thus, in the first query, whether a mine is 'in' Staffordshire has a spatial flavour. However, it is assumed that the processing needs no special spatial component. In our example, there was no requirement to check whether a mine was within the spatial boundary of the county of Staffordshire, because this information was given explicitly by textual means in the database. However, GIS allow real spatial processing to take place. Examples of explicit spatial query types are:

*What is at a particular location?*

A mouse may be clicked to indicate that the cursor is at a particular location on the screen, or coordinates may be given. In either case, a location is specified spatially that forms an integral part of the query. For example, 'What is the name, address and value of this (*click*) land area?'

*What is the position of this object?*

Consider the query: 'Bring to the screen the appropriate map and highlight the land owned by Gladstone Pottery. In this case, the database retrieval is based upon a non-spatial condition (name of a pottery), but this name is required to be linked with a land area, whose boundary and interior are to be highlighted.

*What locations satisfy these requirements?*

A retrieval with a more complex filter is the following: Find the names of all land areas in the Potteries that satisfy the following set of conditions:

- Less than the average price for land in the area.
- Within 15 minutes drive of the M6 motorway.
- Having suitable geology for gravel extraction.
- Not subject to planning restrictions.

Some of these conditions are spatial and some are non-spatial. The query requires the retrieval of land areas satisfying the logical conjunction of all the conditions.

*What is the spatial relationship between these objects?*

Here, we specify a set of objects and ask for a relationships between the objects. Examples are:

> Find the adjacency relationship between the English counties of Staffordshire, Cheshire and Shropshire. (Each pair shares a boundary.)

> Is there any correlation between the spatial references of automobile accidents (as recorded on the hospital database) and designated 'accident black spots' for the area?

The first of these examples is rather more prescriptive than the second, specifying the type of relationship required. The second will require a more open-ended application of spatial analysis techniques. Recent work (e.g. Openshaw, 1994) discusses approaches to handling exploratory situations, where it is unknown what type of pattern (if any) amongst spatial data may exist. In general database research, the topic of data mining addresses questions of the kind 'Here is a lot of data, is there anything significant or interesting about it?'

All of the above functionality is attractive and desirable, but will be useless if not matched by commensurate performance. Computational performance is usually measured in terms of storage required and processing time. As hardware technology improves by leaps and bounds storage becomes cheaper, yet requirements (for example, the development of satellite data capture and consequent huge volumes of data) continue to outstrip technology. Processing time is easily understood and is quite critical for successful systems. It matters whether we have to wait six seconds or six minutes for the balance of our account at an automatic banking machine; it matters whether a map is presented on the screen in half a second or half a minute. Response times are a key factor for the usability of databases.

Performance is an even bigger problem for a geographic database than a general purpose database. Spatial data are notoriously voluminous. Also, the nature of the data is that they are often hierarchically structured (a point being part of an arc, which is part of a polygon). We shall see that such structures present problems for traditional database technology. Also, geodata are often embedded in the Euclidean plane and special storage structures and access methods are required. This problem has exercised computer scientists interested in GIS more than any other. We shall be looking at some of their work, much of which is elegant and ingenious, later in the book.

### 1.3.6 Data presentation

Traditional general purpose databases provide output in the form of text and numbers, usually in tabular form. A report generator is a standard feature of a DBMS and allows the embedding of database output in a high-level report, laid out flexibly and with due regard to the audience at whom the report is aimed. As inter-active access to databases has increased and hardware such as the PC and Mac-intosh have come into widespread use, presentations have been enhanced with so-called 'business graphics', allowing charts and graphical displays, commonly using colour as a means of communication. GIS requires a further step to the presentation of maps and map-based material. Such presentations should enable the user to visualize the results of their analysis, which may be multi-dimensional. Information will be required at varying scales, and appropriate detail should be presented at each scale. This is closely related to the issue of *cartographic generalization*. Presen-tational functionality is a highly distinctive feature of a GIS compared with a general purpose database.

### 1.3.7 Data distribution

A centralized database system has the property that data, DBMS and hardware are all gathered together in a single computer system. Although the database is located at a single site, that does not preclude remote access from terminals away from the computer system. Indeed, this is one of the guiding principles of a database: facili-tating shared access to a centralized system.

The trend in recent technology is to move from centralization towards distrib-uted computer systems communicating with each other through a network. The logical consequence of this trend for database technology is that data and even DBMS should be distributed throughout a computer network. There are natural reasons for this move. Data themselves may be more appropriately associated with one site rather than another: for example, details of local weather conditions may be more usefully held at a local site where local control and integrity checks may be maintained. Commonly occurring accesses to the local site from local users will be more efficient. Performance will be weaker for those accesses to remote sites, but such accesses may be fewer. Another advantage of a distributed network is increased reliability, where failure at one site will not mean failure of the entire system. The down-side is that distributed databases have a more intricate structure to support, for example, handling queries where the data are fragmented across sites, or main-taining consistency of commonly accessed data items that are replicated across sites. The distributed database management system (DDBMS) is correspondingly complex. Distributed geographic databases will be important in supporting the next generation of GIS.

### 1.3.8   GIS design

A GIS, like any information system, is a multi-faceted entity, consisting of hardware, software and data, and going through several processes in its life-cycle, from conception, through design and implementation, to use. The life-cycle begins with *analysis*, when the issue of what the system is required to do is clarified. Following analysis comes *design*, when the problem of how the system will satisfy its requirements is tackled. The translation of the design into something that really works is the *implementation* stage, which may be the *customization* of a proprietary system. Following that comes continued *usage* and *maintenance*. This sequencing of activities in system development is not meant to suggest a strict temporal dependency between stages. In fact, in most cases the dependency is logical but not temporal: for example, in an evolutionary approach to system development, a new round of analysis and design might follow a pilot implementation and this process might continue to loop, iterating to a possibly acceptable solution.

For GIS, these processes may be elaborated a little. Figure 1.12 shows a refinement of the system life-cycle to account for a further stage at the beginning, undertaken by domain experts. The single analysis phase above has been decomposed

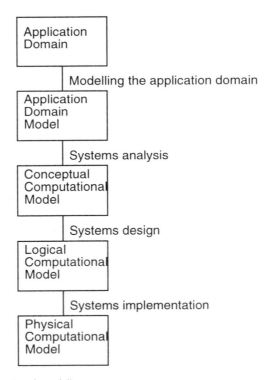

**Figure 1.12**   Computational modelling.

into two stages. First, phenomena in the application domain are represented as processes in an appropriate abstract *application domain model*; then this model is transformed into the *conceptual computational model* that is computationally tractable but independent of any specific computational paradigm. The design phase moves from the specification provided within the conceptual computational model to the design of the *logical computational model*, embedded in a particular computational paradigm, whether relational database, object-oriented, deductive, or some other. The *physical computational model* corresponds to implementation and is the eventual representation of the real-world phenomena as computational processes inside a physical computer. Chapter 2 will give a more concrete picture of how this process works for information system construction.

### 1.4  COMPUTER GRAPHICS AND MULTIMEDIA

Graphics is a key component of GIS and is defined by Foley *et al.* (1990) to include 'the creation, storage, and manipulation of models and images of objects'. These objects may exist in the physical world as natural, engineered or architectural objects, or they may be abstract, conceptual or mathematical structures. This section examines some of the graphics fundamentals, such as presentation, storage and transmission. Later in the book, mathematical models, graphics data structures, graphics algorithms and graphics database storage will all be taken up as important issues for GIS. As hardware and communications improve, so it is possible to extend the range of presentation media beyond basic text, number and still graphics to animated graphics, video and audio. This multimedia approach to presentation is briefly discussed as the conclusion of the section. Appendix 2 describes some common image transfer formats.

#### 1.4.1  Basic graphics concepts

Graphics software has progressed from low-level routines, dependent on manufacturer and computing device, to general purpose device-independent software, usable in, and portable between, a wide variety of applications. Until this decade, almost all graphics were two-dimensional, but during the 1990s 3D software became widely available. Graphics packages typically allow operations upon a collection of graphics primitives, such as point, line segment, polyline, B-spline, rectangle, circle and ellipse. These operations range from presentation upon the screen (positioning and rendering) to transformations (such as rotation, translation, scaling and reflection). Graphics primitives have attributes such as line widths, pen patterns, colours and whether shapes are outlined or filled, including the style of the fill. Text may also be part of an image. As with GIS in general, graphics packages handle either or both vector and raster data.

Common standards for graphics packages are the Graphics Kernel System (GKS), established in 1985 and extended to three dimensions in 1988, and the Programmer's Hierarchical Interactive Graphics System (PHIGS) (1988). PHIGS supports a full 3D capability and allows the definition of hierarchies of graphics primitives (*structures*). The third dimension is of great interest in graphics, but is only just becoming available in GIS. Renderings of 3D objects may be *pseudorealis-*

*tic*, simulating the basic optical laws, or *photorealistic*, more closely approximating to the way that the object would appear if photographed but requiring more computation. The ability to define the shape of three-dimensional objects on a two-dimensional screen requires an understanding of parallel and perspective transformations, and hidden-surface detection and elimination. PHIGS allows 3D primitives to be defined in device-independent 3D floating-point coordinate space.

Graphics algorithms cover a wide spectrum, from the rendering of a simple shape to complex perspective and surface removal operations necessary when presenting a three-dimensional image on the screen. We provide a brief list below.

*Scan conversion algorithms* provide a set of pixels to display geometric objects such as a straight-line segment, circle, ellipse, rectangle, filled rectangle, disc or filled polygon.

*Buffering algorithms* that allow arcs to be made thicker or thinner.

*Clipping* intersects the graphical objects of interest with a rectangular, circular, or other shaped window.

*Antialiazing* that diminishes blocky edges (staircase effects) from graphical objects.

Basic *geometrical transformations* such as scaling, translation, reflection and rotation of two-dimensional and three-dimensional graphical objects.

*Projections*, so that three-dimensional objects may be presented with perspective in two dimensions. Eye-position and horizon may be specified as parameters.

*Visible surface determination* for the presentation of a three-dimensional object in two dimensions.

*Illumination, shading and colouring* of a three-dimensional object presented in two dimensions.

### 1.4.2    Coloured and achromatic images

Three of the forms in which images may be presented are bilevel (e.g. black and white), grey-scale or colour. Images that are without colour (black-and-white or grey-scale) are termed *achromatic* and each element of them is attributed an *intensity*. An achromatic monitor is usually capable of producing a large number of intensities at each pixel position. The characteristics of elements of coloured images are often measured with respect to three orthogonal dimensions: hue, saturation and lightness/brightness. *Hue* distinguishes the spectral frequency or tint of the colour as, for example, red, yellow, green or blue. *Saturation* measures how much spectral colour is in the element or how far it is from greyness. Thus, a crimson red will be highly saturated whereas a washed-out pink would have low saturation. *Lightness* corresponds to the achromatic level of intensity of a cell of reflected light (e.g. on a paper sheet), while *brightness* embodies the same idea for emitted light (e.g. from a computer monitor).

There are several computer models of colour. The RGB colour model is hardware-oriented, based upon the red, green and blue primary colours that may be emitted from the phosphor dots of a cathode ray tube. The model represents the colour component of each element of the image as a triple $(x,y,z)$ where each of $x,y,z$ range between 0 and 1, and each measures the degree of red, green or blue in the element. The model is additive, for example red is (1,0,0), green is (0,1,0) and yellow is (1,1,0). The corresponding subtractive model for reflected light, useful for hard copy, is the CMY (cyan, magenta and yellow) colour model. A model that is oriented towards the user rather than the display device is the HSV (hue, shade and brightness) colour model. Another user-oriented model is the HLS (hue, lightness and saturation) model.

Many display devices, especially hard-copy devices are bilevel, in that for each output cell there is the possibility of recording one of two states (black or white, for example). However, it is possible for a cluster of neighbouring cells, by acting as a single logical unit, to produce a much wider range of output levels. This is the principle of *dithering*. For example, a bilevel output device can be used to output a multi-level grey-scale image. Figure 1.13 shows ten intensity levels produced by nine neighbouring cells. As the intensity increases, cells register black in the order shown by the *dither matrix*, also shown in Figure 1.13.

Dithering is a component operation in colour-to-achromatic image conversion. The lightness of the colour at a particular cell is calculated (there is usually a function that returns the luminance for each colour). The lightness is then converted to a grey-scale intensity, and if hard-copy output is required, dithering is used to provide a grey-scale image. There are also colour dithering techniques (Luse, 1993).

*Thresholding* is, in a sense, a special case of dithering, where the dither matrix has a single row and column. Thus, thresholding converts the intensity of each element of an image to one of two levels, to produce a bilevel image. We shall see later (Chapter 5) that thresholding is an important operation in automatic raster-vector conversion.

### 1.4.3  Image compression

Images, particularly those in raster or bit-mapped form are notoriously greedy in their space requirements. Therefore, compression is a key concept in bit-mapped

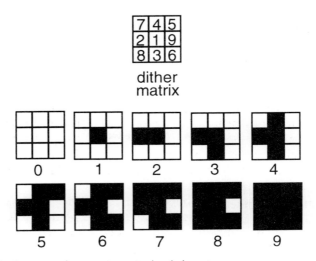

**Figure 1.13**   Dithering to produce ten intensity levels from two.

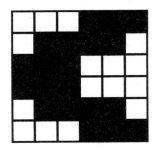

**Figure 1.14**   Simple raster with a run-length encoding as [3,3,1,4,1,3,3,3,4,4,4,3].

image storage. Compression encodes the data representing the image so that (we hope) less storage space is required. Compression may be *lossy* (we cannot reconstruct precisely the original information after compression) or *lossless*. An example of a compression technique is *run-length encoding (RLE)*, which counts the runs of equal values in cells of the image and stores the counts. Thus, the simple image in Figure 1.14 may be run-length encoded as [3,3,1,4,1,3,3,3,4,4,4,3], counting along the rows from top-left to bottom-right. RLE is lossless and works well for simple blocky patterns, but not so well for continuous-tone photographic images.

Two other image compression schemes in common use are:

*Huffman compression*:   Encodes a string of image intensities into a sequence of variable-length bit codes. The nice idea here is that the more common intensities are given the shorter bit codes. However, the image will need to be examined to determine the order of frequency of intensities. This lossless compression technique is commonly used in facsimile images.

*LZW (Lempel-Ziv-Welch) compression*:   Similar to Huffman compression in that codes are generated for intensities. However, the codes are generated as the image is read, thus only a single pass through the image is required. LZW compression is a general file compression technique and is used in such archiving packages as **zip**. As an image compression technique, it is used in modified form in the GIF format (discussed below) and is the copyright of UniSys and CompuServe. LZW compression is lossless.

A collection of standard structures for image compression are described in Appendix 2. They include GIF (lossless) and JPEG (lossy).

### 1.4.4   Multimedia

Multimedia extends the range of computational data types beyond numbers and basic text to:

- still and animated computer graphics
- still and animated images (video)
- audio
- structured text (e.g. hypertext)

Multimedia computing presents exciting new opportunities for the GIS community, but raises several difficulties. A single graphic or image is itself usually voluminous: animated graphics, video and audio raise this level of data volume to new heights. Storage, compression and transmission of such data is a non-trivial task and solving these problems is a major research effort for computer scientists and engineers. Standards are emerging (e.g. MPEG for video) and database systems are beginning to handle these new types, although support for new types such as binary large objects (BLOBs) is limited in most systems to retrieval of the object as a whole based on its global properties but not its internals. For example, while we may easily retrieve all satellite images relating to cyclones if 'cyclone' is a textual attribute associated with the image, it is much more problematic to retrieve images of cyclones based upon properties of the image itself. The bibliography provides some entries into this fast-changing field.

## 1.5   HARDWARE SUPPORT

This section discusses the structure and function of computer hardware, to the degree necessary for understanding of the role that it plays in supporting computer-based GIS. Most of the important components will be described, particularly those directly relevant to GIS, although we assume a broad familiarity with the architecture of a computer. This is not a specialized text on hardware, so detail will be kept within bounds. Specialized texts are referenced in the bibliography.

There is a large range of computers, from single-chip microcomputers affordable for personal and private business use, to supercomputers available for governmental and large corporate applications. Hardware technology has a track record of extremely rapid advance: this has certainly been the case up to the present, but whether it will continue at such a pace is questionable. GIS have always made great demands on processor power, storage capacity and graphics interface, previously restricting for serious use the possible platforms to workstations and above. However, today it is possible to buy a medium-range microcomputer capable of running many of the proprietary GIS available.

### 1.5.1   Overall computer architecture

Despite the wide range of tasks required of computers, it is possible to describe the functionality of a general purpose machine in simple terms. The functionality of a computer can be divided into four parts, which are described below, and their broad relationships to each other are indicated in Figure 1.15.

*Processing*:   Data are processed in the sense that operations are carried out to combine and transform them. Complex data processing functions may be reduced to a small set of primitive operations, performed by the *Central Processing Unit* (*CPU*).

*Storage*:   Data are held in storage so that they may be processed. This storage may range from short-term (held only long enough for the processing to take place) to long-term (held in case of future processing needs). Short-term storage is accomplished by *main memory*; long-term storage by *storage devices*.

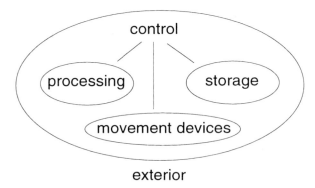

**Figure 1.15**    The four major functional components of a computer.

*Data movement*:   Data are moved to and fro between the computer and its external environment. *Computer peripherals* (*input/output devices*) are directly connected to the computer and allow the movement in and out of the computer of data. *Communications devices* permit movement of data to and from other remote computers.

*Control*:   The above functions are the subject of control. The *control unit* within the CPU of the computer performs this function, managing and allocating resources for the processing, storage and movement of data.

In terms of components, rather than functionality, amongst a computer system's modules are CPU, memory, storage, input/output devices, and interconnections between these parts. The architecture of almost all contemporary computers (although see later for variations such as parallel architectures) conforms to that originated by the mathematician John von Neumann, who developed the basic concepts at the University of Princeton Institute for Advanced Studies during the 1940s and 1950s. The three key concepts of the von Neumann architecture, to be elaborated in the following sections, are:

Storage of data and instructions in a single module, called *main memory*. Main memory may be written into and read from.

Main memory constituted as a set of locations, each of which can store an item of data, coded as a binary number. The locations are assigned addresses sequentially and an item of data is accessed by specifying the address of the location in which it is held.

Execution of operations upon the data by calling items of data from main memory into the CPU and executing the operations sequentially (unless explicitly modified) from one instruction to the next.

### 1.5.2   Processing and control

Processing of data in the computer hardware is handled by the Central Processing Unit (CPU). The CPU is itself made up of several subcomponents: Arithmetic-Logic Unit (ALU), control unit and register file. The CPU's main function is to execute machine instructions, each of which performs a primitive computational operation. The CPU executes machine instructions by fetching data into special registers and then performing computer arithmetic upon them. Computer arithmetic is commonly performed on two types of numbers (fixed point and floating point) and is handled by the ALU.

The various components of the CPU are synchronized by the control unit, with the aid of the system clock. Each *instruction cycle* (performing the processing required for a single machine instruction) is actually a sequence of micro-operations that generate control signals. These signals originate in the control unit and are received by the ALU, registers and other infrastructure of the CPU. One instruction cycle is divided into one or more *machine cycles*, defined to be the time required for bus access (see next section). Thus, the number of machine cycles that make up an instruction cycle depends upon the number of communications that the CPU makes with external devices to execute that instruction.

### 1.5.3 Computer system infrastructure

By infrastructure we mean the connections and communication channels between internal components of a computer system. Such connectivity is provided by the *bus*, which is a communication channel between two or more internal modules. A bus may be shared by several devices, and a message sent on the bus by one module may be received by any of the other modules attached to it. However, only one module may send a message along the bus at any one time. The principal buses providing connectivity between the CPU, memory and input/output devices are called *system buses*, and each may contain in the order of one hundred component lines holding data, specifying location and control. In order to synchronize events in a computer system, a system bus may contain a clock line that transmits a regular sequence of signals based upon an electronic clock.

### 1.5.4 Input devices

Data from the application domain have to be captured and got inside the computer system. As the computer is essentially a digital machine, the implication is that the data must be converted to strings of bits in an appropriate format and then directed to the correct places in the computer system. The performance of this task is accomplished by means of data capture, transfer and input devices. Input devices can feed data directly into the CPU. We may distinguish between *primary data* (data captured directly from the application world, for example remotely sensed data or river pollutant data directly from source), and *secondary data* (data to be captured from devices that are already storing the data in another form, for example paper-based maps). Another division is between numerical/textual data, of the kind traditionally held in a database and spatially referenced data. Table 1.1 shows examples of data capture devices, categorized according to data type and whether primary or secondary.

The numerical and textual data capture and input devices are standard for all information systems. A more detailed description of two primary spatial data capture devices follows:

**Table 1.1**  Classification of input devices

|                    | Primary                   | Secondary                             |
|--------------------|---------------------------|---------------------------------------|
| Textual/numerical  | Keyboard                  | Optical character reader              |
|                    | Voice recognition system  | Scanner/character recognition device  |
|                    | Data logger               |                                       |
| Spatial            | Remote sensor             | Scanner                               |
|                    | GPS                       | Digitizer                             |
|                    |                           | Stereoplotter (3D)                    |

*Remote sensing*:  Captures digital data by means of sensors on satellite or air-craft that provide measurements of reflectances or images of portions of the Earth. Remotely sensed data are usually raster in structure. They are characterized by high data volumes, high resolution but low-level of abstraction in the data model. Thus, a river (high-level concept) is captured as a set of pixels within a range of reflectances (low-level).

*Global Positioning Systems (GPS)*:  Allow capture of terrestrial position and vehicle tracking, using a network of navigation satellites. Data are captured as a set of point position readings and may be combined with other attributes of the object by means of textual/numerical devices such as dataloggers. The data are structured as sequences of points, that is in vector format.

For secondary data capture, usually from existing paper-based maps, either scanners or digitizers are the traditional means of capturing two-dimensional data and stereoplotters for three dimensions.

*Scanners*:  Convert an analogue data source (usually a printed map, in the case of GIS) into a digital dataset. The data are structured as a raster. The process is to measure the characteristics of light that is transmitted or reflected from the gridded surface of the data source. The principle is precisely the same as for remote sensing. In this case, the sensor is much closer to the data source. However, the characteristic of a large volume of low-level data is still operative.

*Digitizers*:  Provide a means of converting an analogue spatial data source to a digital dataset with a vector structure. Digitizing is a manual process (although it is possible to provide semi-automatic or even automatic tools) using a digitizing tablet upon which the source is placed. A cursor is moved about the surface of the data source and its position is recorded as it goes, along with non-spatial attributes. For example, the cursor may be moved along the boundary of an administrative district on a paper-based map of political regions, a sequence of points recorded, and the whole point sequence labelled with a boundary identifier. In this way a vector structure is established, containing nodes, arcs, and polygonal areas enclosed by collections of arcs.

When the data are captured, they are transferred to the database within the GIS by file transfer devices along computer communication channels. The main problem is to get the data into a format that is acceptable, even understandable, to the particular proprietary GIS that is the recipient. To this end, data transfer standards have been agreed (see Appendix 2).

### 1.5.5   Storage devices

All digital data in a GIS, whether spatial or non-spatial, must be physically kept somewhere in the computer system. From an application perspective, any storage device is measured in three dimensions: capacity (how much data can be stored), performance (how quickly the data can be accessed) and price. Persistent data, which are to be retained beyond immediate processing requirements, must also be stored on media that are non-volatile, where data are not lost by natural decay or when the power is removed. Devices for holding data are called *storage media* (Figure 1.16) and are divided into three categories:

*Primary storage*:   Storage that can be directly manipulated by the CPU. The main example of primary storage is the computer's core memory, but also includes local CPU memory and associated cache memories. Primary storage is currently based on semiconductor technology: it is fast but relatively expensive.

*Secondary storage*:   Storage devices, usually based upon magnetic and optical technology, that are accessed only indirectly by the CPU (via input/output controllers) and are slower but cheaper than primary storage. Examples include magnetic and optical disks.

*Tertiary storage*:   Storage devices that are off-line to the computer in normal activity, and used to archive very large amounts of information that could not be efficiently handled by secondary storage. An example is a tape library of satellite imagery.

Primary storage may be divided into CPU local register memory and other main memory (*core*). The CPU requires its own local memory in the form of registers so that data may be operated on by the ALU. Also the control unit will require some internal memory. Such register memory is the fastest and most expensive of all

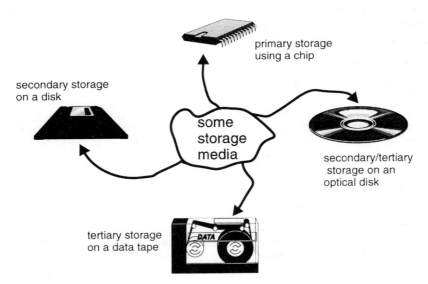

**Figure 1.16**   Some types of storage media.

types, the speed of which must be comparable with machine cycle times. Primary storage is generally *volatile*, data being lost when the system is in a power-off state.

Main memory originally consisted of an array of ferrite cores, and the term *core memory* is still in use. However, the technology is now based on semiconductor chips. Table 1.2 gives some of the main types of semiconductor memory. The first and most common is *random access memory* (*RAM*), a misnomer because all main memory is randomly accessible (i.e. directly addressable). RAM is volatile, thus requiring a continuous power supply, and capable of both read and write *access*. RAM is used for holding data that are coming and going between the CPU and peripheral devices.

ROM (read-only memory) is non-volatile and contains data that are 'hard-wired', in the sense that once written then they cannot be changed. This hard-wiring takes place during the ROM chip manufacture. ROM is used to hold data that are fixed and often required, such as systems programs and code for often-used functions (for example, graphics functions in a GIS). Data from ROM are thus permanently in main memory.

ROM has the disadvantage (apart from only allowing read access) of being expensive to manufacture and therefore only feasible for manufacture in large quantities. PROM (programmable ROM) allows the writing of data or 'programming' to be done after chip manufacture by a cheaper, electrical process and consequently are more flexible. Thus they are suitable for specialized chips required in small quantities.

There are other variants of non-volatile ROM technology, such as EPROM (erasable PROM) that allows multiple erasure of data to take place by optical means (ultraviolet light). Writing of new data is performed electrically, as with PROM. EEPROM (electrically EPROM) allows electrical write and read access. Unlike the EPROM, where the whole chip is rewritten at each new write, writing to an EEPROM can be done to just part of the chip (even a single byte). EEPROM is more volatile and more expensive than other variants of ROM. However, its flexibility may give it the advantage in the future.

Critical to the day-to-day performance of a GIS is the balance between primary and secondary storage. With current technology and dataset sizes, it is impractical to handle all the data required by a GIS in primary storage. Therefore, the structuring of data files on secondary storage devices has been the key factor. Much effort and ingenuity has been spent by computer scientists devising suitable data structures to ensure good performance with spatial data. To understand the issues that arise with such data structures, it is worth examining in some detail the properties of the typical and prevalent secondary storage technology, the magnetic disk.

A magnetic disk has a *capacity*, which is the number of bytes that may be stored on it at any one time. Current floppy disks for microcomputers may hold about a megabyte (Mbyte, $10^6$ bytes) of data. Hard disks for micros usually have a capacity

**Table 1.2**  Main memory types

| Type | | Access allowed | Volatile |
|------|------|------|------|
| RAM | Random-access memory | Read and write | Yes |
| ROM | Read-only memory | Read | No |
| PROM | Programmable ROM | Read | No |
| EPROM | Erasable PROM | Read and rare write | No |

of a few hundred megabytes. Large disks packs for standard computers typically have capacities measured in gigabytes (Gbyte, $10^9$ bytes). The technology changes rapidly and disk capacities are growing all the time.

A magnetic disk is literally a disk coated with magnetic material, having a central hole so that it may be rotated on a spindle. Disks may be *single-sided* (holding data on only one side), *double-sided* (holding data on both sides) and may come in *disk packs* for storage of larger data volumes. Figure 1.17 shows the logical arrangement of a disk. The disk rotates and so all data held on single track may be accessed without moving the read/write head. The magnetic disk drive is often cited as the prime example of a direct access device (that is, storage where the time to access a data item is independent of its position on the storage medium). However, this is not quite the case. Although access times are tiny compared with a device such as a tape drive (sequential access), there is still some dependence upon position. The time taken to move the head (*seek-time*) is the over-riding factor in access to data on a disk. Thus there is an advantage in physically structuring the data on the disk (or disk pack) so as to minimize this time as far as possible. The physical structure of spatial data on a disk is a question that has received attention from the computer science community. The *block* is the unit of data access from secondary storage. Data are retrieved a block (or a cluster of blocks) at a time and placed in an I/O buffer within primary storage. The total time that is taken to transfer the block to the I/O buffer is the *block transfer* time.

The magnetic disk is important because it can store and provide direct access to large amounts of data. Another non-primary storage device is the magnetic tape medium. Although tapes can store large amounts of data more cheaply than a disk, they only allow sequential read and write access. Thus, tapes are more useful as tertiary storage devices. Other storage media based upon optical technology are becoming more important: for example, CD-ROM allows direct read-only access.

Access methods provide the means of getting data items into and out of computer storage. They come in four main types: sequential, direct, random and associative. These types are distinguished as follows:

*Sequential access*:   Made in a specific sequence. The magnetic tape device is an example of a sequential access device. For example, if census records are stored

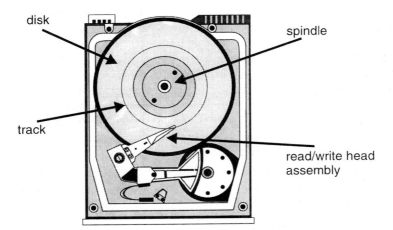

**Figure 1.17**   Schematic of a single magnetic disk assembly.

on a magnetic tape and it is required to access all the records for Stoke, then each block of data would be read in sequence, records within the block examined and retrieved in the case that they were Stoke records. As with playing an audio cassette, to access a track, we have to start from where the tape is after the previous use and wind to the required position(s). Access times vary with the amount of wind required.

*Direct access*: To a specific neighbourhood, based upon a unique physical address. The magnetic disk drive is an example of a direct access device. After reaching the neighbourhood, a sequential search or other mechanism leads to the precise block. When locating a bar of music on a compact disk, we may go directly to the track and then scan through the track for the bar we want. Access times vary (but are much less variable than for sequential access).

*Random access*: To the precise location, based upon a unique physical address. Main memory is randomly accessible. The access time is invariant of position of data on the medium.

*Associative access*: To the precise location, based upon the contents of the data cell. All cells are accessed simultaneously and their data matched against a given pattern. Data items in cells for which there is a match are retrieved. Some cache memories use associative access.

### 1.5.6    Output devices

There is a variety of devices that may be used to present the information resulting from computational analysis to the user. Textual and numerical data may be output using devices such as printers and visual display units (VDUs). Output from a spatial information system often comes in the form of maps. Maps may be output using plotters and printers. If permanent records are not required, then VDUs with high-powered graphics functionality may be used to view output.

### 1.5.7    Hardware developments

*RISC*

The reduced instruction set computer architecture (RISC) offers a substantial change from traditional CPU design. As hardware becomes cheaper, the general movement has been to provide high-level programming languages that have more features, are more complex and have greater power (more processing per line of program). However, the traditional CPU has supported low-level processing operations. In order to reduce the so-called 'semantic gap' between high-level languages and the level of processing in the CPU, the trend has been towards increasing functionality in the CPU (more operation types, addressing modes, etc.). The RISC philosophy is the reverse, in that it aims to simplify the CPU architecture. The main features of RISC architecture are:

- Fewer and simpler instruction formats
- One instruction per machine cycle

- Large number of general purpose registers
- Simple address modes

A well-known RISC project was undertaken during the 1980s at the University of California at Berkeley. Two prototype computers, RISC I and RISC II, were built, leading to the proprietary Pyramid computer. Today, there are several proprietary machines available that are based upon RISC architecture.

### Parallel processing

The traditional, so-called von Neumann architecture of a computer assumes sequential processing, where each machine instruction is executed in turn before control passes to the next instruction. As processors have become smaller and cheaper, the possibility of allowing the simultaneous, parallel execution of machine instructions on separate processors within the same computer is being explored. The goal is to improve performance (by the simultaneous execution of independent operations) and reliability (by the simultaneous execution of the same operation). These developments may be particularly beneficial to GIS, where the processing demands are extreme and the opportunities for parallel execution natural, for example in some raster operations where each pixel (or group of pixels) may be processed independently. There is a growing body of work in the application of parallel architectures to GIS, and the bibliography at the end of the chapter provides some starting points.

### 1.6  COMMUNICATIONS

Of all areas of information technology, the one set to have the most dramatic impact on all our lives in the next few years is that of communications and distributed computing. The next generation of systems will provide an integrated service to users, who may be physically distant from each other, giving them the opportunity to interact with a large number of services provided by different computers but perceived as a seamless whole. A *distributed system* is a collection of autonomous computers linked in a network, together with software that will support integration. Such systems began to emerge in the early 1970s following the construction of high-speed local area networks. Examples of distributed computing are:

Communication between computer users via electronic mail.

A remote login to a distant computer for a specialized application (e.g. accessing national census records or satellite imagery).

A transfer of files between computers using the file transfer protocol (FTP).

A hypertext document with embedded links to services around the globe in a distributed network.

To understand GIS in this distributed context, it is important to take stock of some of the key general notions in this area. This section provides a general background, beginning with an emphasis on hardware and the principles of information transmission, continuing with a consideration of network technology and concluding with some applications of distributed computing.

### 1.6.1   Information transmission

Data are transmitted through a data network by means of electromagnetic signals. Such signals may be either continuous (*analog*) where the signal strength varies smoothly over time, or discrete (*digital*) where the signal jumps between a given finite number (almost always two) of levels. Examples of these signal types are shown in Figure 1.18. Some signals may be periodic, where the pattern of the signal repeats cyclically. A particularly simple and important signal type is the sine wave, shown approximately in Figure 1.19.

A binary digital signal (or just digital signal) transmits bit-streams of data in the form of a wave with higher intensities corresponding to 1 and lower intensities corresponding to 0. Thus 010010 may be represented by the idealized digital signal shown in Figure 1.20.

Any electromagnetic signal may be expressed as a combination of sine waves. The rate of repetition of a sine wave is called its *frequency* and is measured in hertz (Hz) (cycles per second). Thus, any periodic wave may be expressed as a combination of frequencies of its component sine waves. The *spectrum* of a signal is the range of frequencies that it contains and the *bandwidth* of the signal is the width of its spectrum. There is a clear relationship between the bandwidth of a signal and the amount of information that it can carry. The higher the bandwidth, the greater the amount of information.

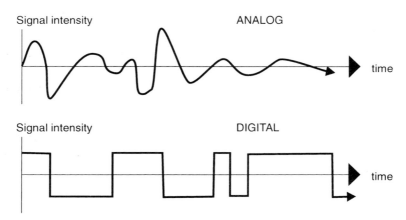

**Figure 1.18**   Analog and digital electromagnetic signals.

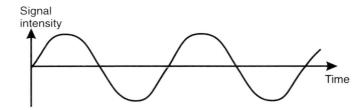

**Figure 1.19**   Sine wave.

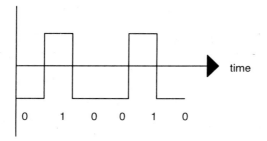

**Figure 1.20** Idealized digital signal carrying a bit-stream.

*Signal encoding: text and images*

Data that are to be communicated, whether text, numbers or images, must be encoded into a bit-stream. Numbers, whether integer or rational, may be represented as a bit-stream using base two arithmetic. Numbers and text may be represented as a string of characters, which are then encoded as binary sequences. The most common code is the seven-bit ASCII code. ASCII is an acronym for American Standard Code for Information Interchange and provides codes for up to $2^7 = 128$ numeric, alphabetic and control characters. Precise details of the code are not needed for this text and may be found in many introductory texts on computing. Seven-bit ASCII characters are normally transmitted with an extra (parity) bit for error detection.

Images may be encoded in several different ways for the purpose of communications, amongst which are:

*Analog Facsimile (FAX)*: An image is coded as a sequence of horizontal lines, each of which has a continuously varying intensity.

*Digital Facsimile and Raster-scan Graphics*: An image is coded as an array of pixels, each pixel representing either black or white (or a colour code if using colour graphics).

*Vector Graphics*: An image is decomposed into a set of points, lines and other geometric objects (rectangles, circles, etc). Each object is specified by giving its position, size and other determining characteristics.

*Transmission terminology and media*

Digital data are transmitted as a bit-stream using an electromagnetic signal as a carrier. We have noted that the information carrying capacity of a signal is determined by its bandwidth (the width of the spectrum of frequencies that it contains). Data transmission is a communication between transmitter and receiver. Only in the ideal world is it the case that the signal transmitted is equal to the signal received: in the real world degradations occur:

*Attenuation*: The weakening of a signal as it travels through the transmission medium. Attenuation is usually greater for higher frequencies ($\omega$ and $\omega t$) and this variation causes signal distortion. It is important that attenuation does not reduce the strength of the signal so that background noise becomes significant: a solution is to provide *amplification* during transmission.

*Delay*: Caused by transmission along cables. Delay of a component of a signal depends upon the frequency. As with attenuation, this variation may cause distortion. Delay places a limit upon data transmission rates though cables.

*Noise*: Extraneous electromagnetic energy in and around the medium during transmission. Noise has a variety of causes, some related to the properties of the transmission media and others to environmental effects such as other signals and electromagnetic storms. As with attenuation and delay, noise can distort the signal, particularly if the signal strength is of the same order of magnitude as the noise. The solution is to provide a strong signal and error detection and correction mechanisms.

Attenuation, delay and noise place limitations upon the information carrying capacity of a transmission channel. These limitations are measured using bandwidth, data rate (in bits per second), noise and error rate. Transmission media come in a variety of types and have varying channel characteristics. Media types may be broadly separated into guided and unguided. Guided media restrict the signal to a predetermined physical path, such as a cable or fibre. Unguided media place no such restriction: an example is transmission through the atmosphere of the earth. With a guided medium, both transmitter and receiver are directly connected to the medium. Examples are twisted-pair (of two insulated copper wires), coaxial cable (hollow outer cylindrical conductor surrounding single wire conductor) and optical fibre (flexible and thin glass or plastic optical conductor). Optical fibre (fibre-optic) is the newest and will surely become the dominant guided transmission medium. Table 1.3 below outlines some of the characteristics and applications of these technologies.

With unguided media, both transmission and reception are by means of antennae. A useful frequency band for unguided transmission is microwave (2–40 GHz). Terrestrial microwave is an alternative to coaxial cable for long-distance telecommunications. It requires line-of sight transmission and therefore long distances require intermediate stations (but fewer than coaxial cable). Other applications include short-range data links (maybe between buildings). Satellite microwave uses earth-orbital satellites to reflect signals, thus avoiding the problem of line-of-sight in terrestrial microwave. Satellite microwave has a huge and growing number of applications, from commercial television to business data communications.

**Table 1.3**  Guided transmission media

| Guided medium | Characteristics | Applications |
| --- | --- | --- |
| Twisted-pair | Cheap | Telephone networks |
|  | Limited distance, bandwidth, data rate | Small PC LANs |
| Coaxial cable | More expensive than twisted-pair | Cable TV |
|  | Higher bandwidth and data rate | LANs |
|  |  | Long-distance telecommunications |
| Optical fibre | Greatest bandwidth and data rate | Long-distance telecommunications |
|  | Much less attenuation | LANs |

### 1.6.2 Networks

As communications technology develops, so the possibilities for cooperation between persons/machine and machine/machine have expanded. This cooperation is made possible through the use of computer networks. A *computer network* is a collection of computers, called *stations*, each attached to a *node*. The nodes are communications devices acting as transmitters and receivers, capable of sending data to and receiving data from one or more other nodes in the network. Computer networks are customarily divided into two types, wide-area networks and local-area networks.

A *wide-area network (WAN)* is a network of computers spread over a large area. Such an area may be as small as a city or as large as the globe. WANs transmit data at relatively low speeds between stations linked together by dedicated node computers called *packet switches*. The packet switches route messages (packets) between stations. This routing process makes transmission between stations slower than local-area networks, where direct transmission takes place. The wide-area telecommunication network Integrated Services Digital Network (ISDN) allows communication of data between stations. Broadband ISDN (B-ISDN) provides a sufficiently high bandwidth to carry video, as well as the more traditional forms of information.

A *local-area network (LAN)* is a network of computers spread over a small area. Such an area may be a single building or group of buildings. LANs transmit data between nodes at relatively high speeds using a transmission medium such as optical fibre or coaxial cable. Their main feature is direct communications between all nodes in the network with the consequence that no routing of messages is required. Most computerized work environments are connected as a local-area network. Examples of current LAN technologies are Ethernet and IBM token ring.

*High-speed networks*

It is almost impossible to keep up-to-date in this fast moving area of computing. At present, high-speed networks are developing that take advantage of transmission technologies such as ATM (Asynchronous Transfer Mode) using optical fibre and satellite microwave transmission media. ATM currently allows data transfer at rates approaching 1 gigabit per second. The optical fibre cabling of cities will allow the transmission of video, voice and bulk data city-wide, with typical transmission times (latencies) of less than one millisecond. By the time that you read this paragraph, it may already be yesterday's news!

### 1.6.3 Applications of communications technology

*Open systems*

The Open Systems Interconnection (OSI) model attempts to solve problems that arise from a distributed system, where the components are from different vendors using different data formats and exchange protocols. The International Standards Organization (ISO), the world-wide standards body, has developed OSI as a common framework for systems communication. OSI is a layered model, each of the seven layers dealing with a separate component or service required in distributed computing. The seven layers are given in Table 1.4.

**Table 1.4**  OSI layers

| Level | Layer | Notes |
|-------|-------|-------|
| 1 | Physical | Physical properties of the link |
| 2 | Data link | Reliable transmission of data (frame) along transmission medium |
| 3 | Network | Communications network protocols. Provides independence of upper levels from physical details. |
| 4 | Transport | Ensures interchange is error-free. |
| 5 | Session | Control structure, including dialogue discipline and recovery protocols. |
| 6 | Presentation | Provides application programs with data transformations such as compression and encryption. |
| 7 | Application | Management services such as transaction server, file transfer protocol (FTP) and email. |

To see how OSI works in practice, consider the configuration of a hypothetical computer system as an open system within a network. Suppose that the computer is connected to a single network, for example a local-area network. A specific protocol, depending upon the nature of the LAN and setting out the logic of the network connection, will be required at each of the three lowest OSI layers. Session and transport layer protocols will be provided to give reliable data transfer and dialogue control between our computer and others on the network. File transfer, electronic mail and other application facilities are provided by the top layer.

Apart from the detail, the important advance that OSI provides is the standardization of interfaces for inter-computer communication. This makes possible the introduction of new software and hardware into a networked system, provided that there is conformance to the OSI standards. OSI allows us to select hardware, software and services for a networked system from a variety of vendors.

### Client-server computing

Client-server computing is a form of distributed computing in which some stations act as servers and some as clients. A *server* holds data and services available for transmission to other stations, *clients*, in the network. The servers are usually higher specification machines, capable of running complex system software. Clients typically may be PCs or workstations. Client stations request information from servers, which they then subject to their own specialized processing requirements. Thus, the network may have servers holding national census data, administrative boundary polygons and medical statistics. A client may then request census information for a particular region from a server, combine this with medical statistics obtained from a second server and use the boundary data from a third server to conduct a spatial analysis, code for which may be obtained from yet another server. For GIS applications, client-server computing holds many possibilities. A GIS at present combines both information system, spatial analysis and presentation components: maybe, with client-server computing, these and other components will be physically separated.

A particularly important and very widely used application of the client-server model is the X Window System. The *X-server* controls a *display*, which comprises keyboard, mouse and one or more screens. The X-server usually starts by presenting to the user a login window. The user logs in to begin the X-session and to initiate the X-clients. The X-server provides the user with a *root window*, within which other client windows are arranged. The *window manager* is a special client that provides functionality to arrange and manipulate client windows on the root window. Figure 1.21 shows an example of a display of X windows upon a screen.

Communications between clients and server take place using the *X protocol*. Connection between client and server is established with a *connection setup packet* message sent by the client, followed by a reply from the server. During connection, the types of messages possible include:

- *requests* from client to server.
- *replies* from server to client.
- *events* from server to client (e.g. the click of a mouse button).
- *errors*.

The clients act upon the server in terms of *resources*. Resources are abstract objects, examples of which are windows, fonts, cursors and graphics contexts. Information about each client may therefore be expressed in terms of attributes associated with its resources, collectively known as the *state* of the client.

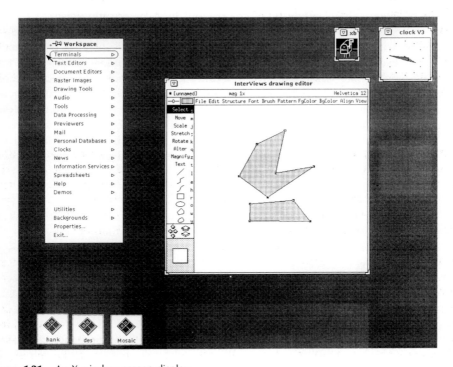

**Figure 1.21**   An X window screen display.

*Internetworking, the Internet and World-Wide Web*

Distributed computing has arisen out of the need of computer users to access not only their immediate information resources but those of others. Thus, a network allows all connected users to share information, subject to constraints imposed by the network architecture. The next logical step is to allow networks to share information. This is the concept of an internet. An *internet* is an interconnected collection of networks (called *subnetworks*), each subnetwork retaining its own individual characteristics but the entire collection appearing as a single network to a user. The technology of an internet requires a further set of components:

*Bridge*: Used to connect to subnetwork LANs with identical protocols; operates in the OSI data link layer.

*Router*: Used to connect subnetworks (usually LANs) with different protocols; operates in the OSI network layer.

*Gateway*: Used to connect networks with different protocols; operates in the OSI application layer.

A special and important example of internetworking is the Internet, which originated from a research initiative under the auspices of the US Government. The Internet is a fast-growing global collection of networks. At the time of writing (summer 1994), the Internet comprises approaching 50 000 constituent networks, a new network being added to the system about every 10 minutes, and about 20 million people have access to Internet resources such as electronic mail and file transfer.

The *World-Wide Web* originated from an approach to communication and common problem-solving made by engineers and high-energy physicists at CERN (European Particle Physics Laboratory) in Geneva, Switzerland. The Web runs on the Internet with client programs at each member station displaying hypermedia objects. For example, a document with buttons for sounds, images, video and text may be displayed, and pressing a button invokes a call for those data to a different server on the Internet. Thus, the user can trace routes through to new documents and links. It is the hypermedia format that gives the Web its particular flavour. While menus and directory structures are available, hypermedia provides powerful tools for communicating and gaining information. The Web user interface is the same regardless of client or server, thus insulating users from unnecessary protocol details. In summary, the Web provides:

A network with vast and interconnected sources of information on every conceivable topic.

A system of addresses (URI) and a network protocol (HTTP).

A simple hypertext mark-up language (HTML) that is used to construct the constituent hypertext documents.

### BIBLIOGRAPHIC NOTES

Further reading suggested here will expand on GIS functionality and computational support for GIS technology. GIS functionality is nicely summarized in (Dangermond, 1983) and (Maguire and Dangermond, 1991), with a short update in

(Maguire and Dangermond, 1994). Laurini and Thompson (1992), Burrough (1986) and Maguire *et al.* (1991) are general texts providing overviews of GIS technology. Peuquet and Marble (1990) provide an edited collection of papers covering a range of issues for GIS. Goodchild (1993) summarizes GIS functionality for environmental applications. The inspiration for some of the example Potteries applications was provided by Phillips (1993).

Overview books on general database technology are (Date, 1995; Elmasri and Navathe, 1994). Ozsu and Valduriez (1991) and Bell and Grimson (1992) specialize in distributed databases. Distributed GIS are the subject of an overview by Laurini (1994). Software design in the context of GIS is discussed in (Marble, 1994). Mayo (1994) and Flanagan *et al.* (1994) discuss research on data capture for GIS. Generalization is covered in the book by McMaster and Shea (1992) and the chapter by Muller (1991). The classic book on computer graphics is by Foley *et al.* (1990). Luse (1993) discusses image representations, colour models, compression techniques and standard formats. A collection of research papers on multimedia is edited by Herzner and Kappe (1994). Visualization is treated in the edited collection of Hearnshaw and Unwin (1994) and the chapter by Buttenfield (1991).

A general overview of hardware is provided by Stallings (1987), and performance considerations for GIS with an emphasis on parallel architectures is discussed by Gittings *et al.* (1994). Hopkins *et al.* (1992) give an example of a parallel algorithm for GIS. Computer communications are described in the texts by Stallings and van Slyke (1994) and Coulouris *et al.* (1994). ATM Networks are discussed from several angles in a collection of articles in the *Communications of the ACM* (Vetter and Du, 1995). The X Window System is comprehensively discussed by Schleifler and Gettys (1992).

# Fundamental database concepts

It is a capital mistake to theorize before one has data

Sherlock Holmes, from Arthur Conan Doyle, *Scandal in Bohemia*

The database is the foundation of a GIS. An understanding of GIS technology cannot be achieved without some knowledge of the principles of databases. Many existing GIS are built upon general purpose relational databases: certainly all GIS will connect with such systems in a distributed environment. This chapter introduces the reader to the main principles of databases. The general database approach is discussed in section 2.1 along with the high-level architecture of a database management system. At present, almost all databases are relational, therefore the relational model and its major interaction language SQL are described in detail in section 2.2. The reader is introduced to database design principles in section 2.3. Object-oriented databases provide extensions of relational functionality, particularly at the modelling level, that could be very useful for GIS and these are described in section 2.4.

## 2.1 INTRODUCTION TO DATABASES

To introduce the database approach, it is useful to contrast the computational flavour of databases with that of more traditional computing paradigms. In the bad old days, the almost universal use of a computer was to convert one dataset into another by means of a large and complex transformation process (typically a FORTRAN program). For example, we might wish to apply a transportation model to datasets of city population sizes and road distances between cities in order to predict average annual traffic flows. The inputs are datasets of city populations and

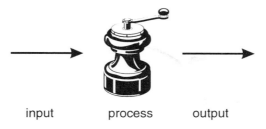

input      process      output

**Figure 2.1** The 'computer as giant calculator' paradigm.

**Figure 2.2**   The 'computer as data repository' paradigm.

road distances; the transformation is provided by a procedural program based upon the transportation model; the output is the traffic flow prediction dataset. This approach to computation is shown in Figure 2.1. We may call this the 'computer as giant calculator' paradigm. The alternative offered by the database approach is shown in Figure 2.2. In this case, the computer acts as a useful repository of data, by facilitating its deposit, storage and retrieval. It also allows the data to be modified and analysed while it is in the store.

### 2.1.1   Review of database requirements (databases in a nutshell)

In order to act effectively as a data store, a computer system must have the confidence of its users. Data owners and depositors must have confidence that the data will not be used in unauthorized ways (*security*) and that the system has fail-safe mechanisms in case of unforeseen events such as power failure (*reliability*). Both depositors and data users must be assured that as far as possible the data are correct (*integrity*). There should be sufficient flexibility to give different classes of users different types of access to the store (*user views*). For example, in a bank database, a customer should not be allowed to have read access to other customers' accounts nor have write access to her own account! Not all users will be concerned how the database works and should not be exposed to low-level database mechanisms (*independence*). Data retrievers will need a flexible method for finding out what is in the store (*metadata support*) and for retrieving it according to their requirements and skills (*human-database interaction*). The database interface should be sufficiently flexible to respond differently to both single-time users with unpredictable and varied requirements, and regular users with little variation in their requirements. Data should be retrieved as quickly as possible (*performance*). It should be possible for users to link pieces of information together in the database to get added value (*relational database*). Many users may wish to use the store, maybe even the same data, at the same time (*concurrency*) and this needs to be controlled. Data stores may need to communicate with other stores for access to pieces of information not in their local holding (*distributed systems*). All this needs to be managed by a complex piece of software (*database management system*).

### 2.1.2   The database approach

The 'computer as giant calculator' and 'computer as data repository' are extreme positions. Most applications require calculation (*process*) and proper treatment of data upon which the processes are to act. So a synthesis must be achieved. In the early days of computers the emphasis was on the processes in computations but, particularly for corporate applications, weaknesses resulting from the imbalance between data and process became apparent. We illustrate this with a fictitious example.

'Nutty Nuggets' is a vegetarian fast-food restaurant, established in the 1970s. The owner-manager, a computer hobbyist, decided to apply computer technology to help with some aspects of the management of the restaurant. She began by writing a suite of programs to handle the menu. Menu items were stored as records in a menu file, programs were written to allow the file to be modified (items deleted, inserted and updated) and for the menu to be printed each day. Figure 2.3 shows, on the left, the menu file being operated on by the two programs and, on the right, the constitution of a menu item record of the menu file. The menu file is held in the operating system and accessed when required by the programs.

Time passed, the menu system was successful and the owner gained the confidence to extend the system to stock management. A similarly structured stock system was set up, consisting of stock file, and programs to modify the stock file and periodically print a stock report. Once stock and menu details were in the system, it became apparent that changes in stock costs influenced menu prices. A program was written to use stock costs to price the menu. Stage two of the Nutty Nuggets system is shown in Figure 2.4.

The system continued to grow with new files for supplier and customer details added. But as the system became enlarged, some problems began to emerge, including:

*Loss of integrity*: Linkages between programs and files became complex. The programs made the relationships between the data in the files: if the relationships changed then the programs had to be changed. The development of software was becoming complex and costly and errors crept in: for example, the only supplier of a crucial raw material went out of business and was deleted from the supplier file but the material supplied was not deleted from the stock file.

*Loss of independence*: A too close linkage between program and data caused high software maintenance costs. For example, a change in secondary storage medium required a partial rewriting of many of the programs.

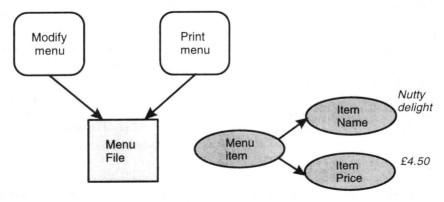

**Figure 2.3**   Nutty Nuggets stage one: menu system.

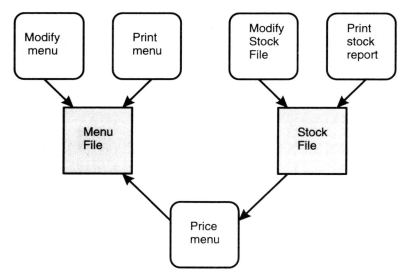

**Figure 2.4**   Nutty Nuggets stage two: menu, stock and menu pricing system.

*Loss of security*:   A personnel file was added without regard to security and it was discovered that a member of staff had been accessing personal information about colleagues.

The database philosophy is an attempt to solve these and other problems that occur in a traditional file systems approach. Figure 2.5 shows a reorganization of the system so that all data files are isolated from the rest of the system and accessible to the processes only through a controlled channel. The idea is to place as much of the structure of the information into the database as possible. For example, if there is a relationship between suppliers and the goods they supply, then this relationship should be stored with the data. The means of expressing the structure and relationships in the data is provided by the database *data model*. The data model also allows the user to enter properties of the data in the database that are always true (*integrity constraints*): for example, that the price of any menu item must be greater than zero but less than £10. Integrity constraints are an aid to maintaining correctness of the data in the database, because they only allow modifications to the database that conform to them.

The data are collected in one logical centralized location. The *database management system* (DBMS) is a complex piece of software that manages the database by insulating the data from uncontrolled access, allowing definition of the data model, supporting manipulation of the data and providing appropriate two-way access channels between the exterior and the database. It allows the designer to define the structure of the data in the database, providing levels of authorization that allow different groups of users access to appropriate data and managing transactions with the database that may be occurring concurrently. The DBMS also provides *data independence*, so that the data in the database are accessible without precise knowledge of implementation details.

A *database* is a unified computer-based collection of data, shared by authorized users, with the capability for controlled definition, access, retrieval, manipulation

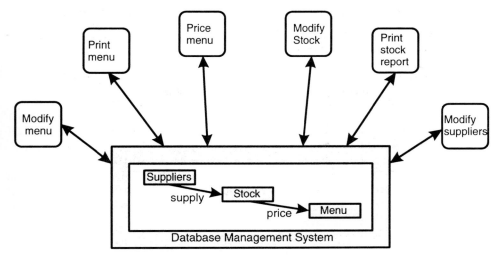

**Figure 2.5**    Nutty Nuggets stage three: database approach.

and presentation of data within it. Databases have applications that give them particular flavours.

*Single-user small database*:    Often runs on PC or work-station; does not need concurrent access so software is much simpler and cheaper; an example is the Nutty Nuggets system.

*Corporate database*:    The standard from which other applications deviate; runs on many platforms; key issues are security, reliability, integrity, performance, concurrent access; different groups of users (maybe in different departments) require specialized views; transactions are usually simple; data structures are also usually quite simple (tables of numbers and character strings).

*Office information system*:    Characterized by mainly textual data; management of multiple versions of documents is required.

*Engineering database, CAD/CAM database*:    Characterized by mainly graphics-based data; management of multiple versions of designs is usually required; data are often structured hierarchically (sub-assemblies of assemblies), transactions with the database are often long and complex (e.g. when modifying a design).

*Bibliographic database*: Characterized by mainly textual data; specialized searches required (e.g. by combinations of keywords).

*Scientific database*: Often complex data structures (e.g. human genome project); complex analyses are performed on the data.

*Image and multimedia database*: BLOB (binary large object) data; specialized techniques for data retrieval; range of data types such as image, video and audio need support; very large volumes of data (e.g. video data).

*Geographic database*: Combination of spatial and non-spatial data; complex data structures and analyses; discussed in this book.

*Sharing data: the view concept and layered ANSI-SPARC architecture*

A key element of database philosophy is data-sharing and almost all but the smallest DBMS support multi-user access to the data in a database. Not all users have the same requirements of the database: each user group may require a particular window onto the data. This is the concept of *view*. Views provide users with their own customizable data model, a subset of the entire data model, and the authorization to access the sectors of the database that fall within their domain. There is a distinction between the data model for the entire database (*global conceptual scheme*) and a data model for a subclass of users (*local conceptual scheme*, or *external view*).

The schemes so far discussed are called *conceptual* to distinguish them from the schemes that structure the physical nature of the database (addresses on physical storage media, etc.). Data independence is provided by separating the implementation details, handled in the *internal scheme* from the higher-level, user-oriented considerations provided by the *conceptual scheme*. This layered model of a generic database architecture is embodied in the so-called ANSI-SPARC (Tsichritzis and Klug, 1978) database architecture shown in Figure 2.6. The DBMS provides mappings between each layer. Users may only interact with the database by means of an external view. They may interact directly through an interactive query language, such as SQL, or via an application program in which are embedded one or more calls to the database.

### 2.1.3   Elements of a database management system

This section describes some of the components of a DBMS that allow it to perform the complex functions described above. Imagine that someone at Nutty Nuggets

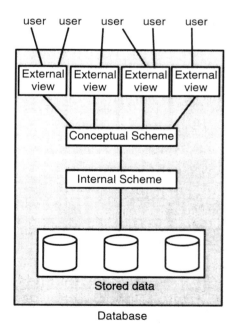

**Figure 2.6**   The layered ANSI-SPARC architecture.

wants to know the quantity of mung beans available in the stock room. This task will require retrieval of data from the database, in particular from the stock file. The retrieval request must be communicated to the database, maybe using a database interaction language (usually called a *query language*) or through a menu/forms interface. Let us suppose that a database language expression is used. The DBMS has a *query compiler* that will parse and analyse the query and, if all is correct, generate execution code that is passed to the *runtime database processor*. Along the way, the compiler may call the *query optimizer* to optimize the code so that performance on the retrieval is as good as possible (usually there are several possible *execution strategies*): it may also use the *constraint enforcer* to check that any modifications to the database (none in this case) satisfy any integrity constraints in force. If the database language expression had been embedded in a general purpose computer language such as C++, then an earlier pre-compiler stage would be needed.

To retrieve the required data from the database, mappings must be made between the high-level objects in the query language statement (e.g. the quantity available field of the stock file) and the physical location of the data on the storage device. These mappings are made using the *system catalogue*, which stores the external views, conceptual and internal schemes described in the previous section, as well as the mappings from one to the other. Access to DBMS data is handled by the *stored data manager*, which calls the operating system for control of the physical access to storage devices.

Figure 2.7 shows schematically the place of these components in the processing of an interactive query, or an application program that contains within the host general purpose programming language some database access commands. The

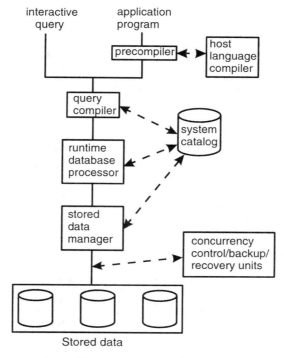

**Figure 2.7** DBMS components used to process user queries.

figure also shows units that handle transaction management, concurrency control, back-up and recovery in the case of system failure. Further DBMS elements, not considered here, are utilities for loading large files into the database and for monitoring performance so that database tuning can take place.

### 2.1.4   Interaction with a database

A database usually supports a wide variety of usage. For example:

1. *Regular, day-to-day routine usage*, where the access is limited to a few retrieval patterns. For example, a librarian at a check-out desk may interact with the database to check the status of users and register books as loaned to users. Other examples include ticket reservation for air travel and car rental.

2. *Regular, expert usage*, performed by expert database users to access the database in any way required by applications and process the results accordingly. This usage would include system development.

3. *Irregular but predictable usage*, for example where a library user consults the book catalogue to find whether a book is owned by the library.

4. *Irregular, flexible, casual usage*, for example browsing through an information resource to identify relevant information. The interactions may be unpredictable.

For cases, such as 1 and 3, where the usage pattern can be predicted to be composed of a small number of so-called *canned transactions*, interaction facilities can be set up in advance for these transactions. Such interactions may be facilitated by the presentation on screen of a form for filling or a set of menus. The database may be optimized to perform well for its main canned transactions. Case 2, involving more flexible and varied usage by professionals, may not be supportable by a forms interface. A query language would be necessary so that the user could formulate her query. Case 4 is the most challenging because the requests are flexible, varied, *ad hoc*, maybe complex, and unknown in advance by the system, but the users are not necessarily computer experts. It may be possible to provide a database interaction language simple enough to be used with just a little practice. (This was an initial goal of the database interaction language SQL.) More realistically, other forms of interaction, such as menu-based systems, must be considered. Human interaction with information systems, particularly GIS, is discussed further in Chapter 7.

### 2.1.5   Sharing information: transaction management

We have already emphasized the importance of DBMS support for database access. Although the layered ANSI-SPARC architecture facilitates multi-user access by providing multiple views, there are further lower-level features that must be supported for data-sharing to work properly. These are considered under the heading of transaction management, a *transaction* being an atomic unit of interaction between user and database. Transactions are primarily of the following types:

- insertion of data into the database
- modification of data in the database
- deletion of data in the database

- retrieval of data from the database

All transactions simplify finally to two basic operations: read an item from store (usually disk) into main memory, and write an item from main memory into store. An item from store will usually be a field of a record, although it may be useful to read the entire record or disk block.

Transaction management is about recovery control and support for concurrency. The problem with which concurrent access confronts us is that if the same data item is involved in more than one concurrent transaction, the result may not be predictable and may result in loss of database integrity. Problems can occur if the atomicity and independence of each concurrent is not preserved. For example, suppose that I add £500 to my bank account at roughly the same time as my bank transfers the full balance of my account to a creditor. If, prior to these transactions I had £1000, then after the transactions two possible states may obtain: I will have either nothing or £500 in my account. The first case results if the system adds the £500 to my account and then transfers £1500 to my creditor; the second if the bank transfers £1000 to my creditor and then adds my £500. It depends upon the order in which the transactions are executed by the DBMS.

A similar problem (*lost update*) can occur as in the following example. As before, my bank balance is £1000. Two transactions are in progress, $T_1$ to credit me with £300 and $T_2$ to debit me with £400. Table 2.1 shows a particular *interleaving* of the constituent operations of each transaction. Transaction $T_1$ begins by reading the contents of my balance from the database into a program variable $X$ and increasing $X$ to £1300. Transaction $T_2$ then starts by reading the same balance from the database into $Y$ and decreases $Y$ to £600. $T_1$ then concludes by writing $X$ to the database as the new balance of £1300 and $T_2$ writes $Y$ to the database as the new balance of £600. It is as if transaction $T_1$ never occurred, hence 'lost update'.

The problems discussed so far concern database integrity. The situation now described involves *recovery management*, which may affect the reliability of the database if not properly handled. (There are other problems with transactions, such as 'dirty read' and 'incorrect summary'. These are fully discussed by Elmasri and Navathe (1994)). Imagine that you are able to update the balance of your bank account by inserting a cheque at an automatic teller machine. You insert the cheque, the system logs the cheque as accepted and then goes down due to an unforeseen failure. Has the database been updated with your new balance or not? This needs to be very clear. The DBMS must ensure that either the entire transaction has completed and if not then roll back to the state before the transaction started.

The lessons to be learned from such considerations are that transactions should have the following properties:

*Atomicity*:   All the constituent operations of a transaction must either all have their effect on the database or make no change at all to the database.

*Independence*:   Each transaction should have an effect on the database independently of the interleaving of constituent operations of other transactions submitted concurrently.

These aims are assisted by DBMS operations of:

*Commit transaction*:   Signals a permanent change to the database when all constituent operations have been successfully completed.

**Table 2.1** Two non-atomic interleaving transactions that result in a *lost update*

| Transaction $T_1$ | Transaction $T_2$ | Balance in database | Program variable $X$ | Program variable $Y$ |
|---|---|---|---|---|
| | | £1000 | | |
| read balance from DB to $X$ | | £1000 | £1000 | |
| increment $X$ by £300 | | £1000 | £1300 | |
| | read balance from DB to $Y$ | £1000 | £1300 | £1000 |
| | decrement $Y$ by £400 | £1000 | £1300 | £600 |
| write $X$ to DB as new balance | | £1300 | £1300 | £600 |
| | write $Y$ to DB as new balance | £600 | £1300 | £600 |

*Rollback transaction*: Recovers the state of the database immediately prior to the transaction if there is any problem with the transaction.

Interleaving of database operations is clearly advantageous for performance, because the operations of a transaction may progress while more lengthy operations, such as disk input/output from other transactions, are in progress. If interleaving of transactions is not implemented, there will be long wait-times while not much is happening, and we would not be able to take advantage of a multiprogramming environment. However, if simultaneous execution of transaction operations is allowed, then precautions must be taken to prevent the occurrence of the kinds of problems discussed above. An important concept related to these considerations is that of a *serializable* schedule of interleaved transactions, where the schedule is equivalent to a schedule of the transactions applied independently and in strict sequence. Another useful notion is that of *locking*, where a transaction while in operation may lock out all other transactions to a part or all of the database. Widespread locking can slow the system down to unacceptable levels.

The aim of this section has been to give the reader an appreciation of the problems involved with the mechanics of data sharing in terms of transaction management and thus why the software to control this may be complex. Any of the general database texts given in the bibliography at the end of the chapter will have a detailed discussion of these issues.

### 2.1.6 Distributed databases

The architecture of a database system assumes a logically centralized data store. However, this store does not have to be physically centralized. With recent advances in communications, briefly discussed in Chapter 1, it is natural to have the data distributed at multiple sites. Such a technology results in a *distributed database* (DDB), supported by a distributed database management system (DDBMS). DDBs are appropriate in cases where the data are themselves naturally distributed (regional environmental data, for example). They allow the data to be shared globally while keeping some degree of control, for example on database modification, with the local sites. Local queries on local data will be optimized to have good performance, while global queries from local sites must expect to be slower. To improve global performance, it may be possible to replicate parts of the database at more than one site, although care must be taken when those data are modified to ensure that updates are propagated throughout the system. (It would lead to a loss of integrity if a modification to the DDB was committed at some sites but not at others. The mechanism for handling this is the *two-phase* commit.) Reliability of the system may be improved by distribution, because if a system goes down at one site, usage can still continue for the rest of the database. GIS applications are often thought of naturally in a distributed context.

The advantages of a DDB are clear, but they must be weighed against the increased complexity of the DDBMS. Examples of additional functionality required are:

- Remote access and transmission of queries and data using a communications network.

- Management of transactions that require parts of the database on multiple sites.
- Handling of system-down at individual sites, re-routing of queries.
- Handling of data replication at multiple sites.

The client–server architecture is natural and widespread in the context of architectures for DDBMS. The bibliography provides some references to further reading in this area.

## 2.2  RELATIONAL DATABASES

There are several models of the overall structure of a database system, including *hierarchical, network, relational, object-oriented,* and *deductive* database models. Hierarchical and network databases have been replaced by relational technology as the main paradigm. Object-oriented and deductive systems are emerging technologies and are the subject of later discussions. The focus of this section is the relational model, constructed by Ted Codd and introduced in his classic 1970 paper. Codd continued to work on relational databases in the 1970s, constructing several formalisms for interacting with them. Codd (1979) extended the model to incorporate more semantics, including null values, in what is often referred to as the 'Tasmanian model', RM/T (Relational Model/Tasmania).

### 2.2.1  The relational model

We have seen in the Nutty Nuggets example that a database holds not only primary data files but also connections between data in the files. In the traditional file-based approach, these connections would have been maintained by application programs, but it proves more efficient to hold the connections within the database. These connections are at the heart of the relational model. No single piece of data provides information on its own: it is the relationships with other data that add value.

The structure of a relational database is very simple (this is what makes it so powerful). A relational database is a collection of tabular *relations*, often just called *tables*. Table 2.2 shows part of a relation called FILM (from the CINEMA database given in full in Appendix 1) containing data of some old films, their names, directors, country of origin, year of release and length in minutes. A relation has associated

**Table 2.2**  Part of the relation FILM

| TITLE | DIRECTOR | COUNTRY | YEAR | LENGTH |
|---|---|---|---|---|
| Jaws | Spielberg | USA | 1977 | 125 |
| Star Wars | Lucas | USA | 1977 | 121 |
| American Graffiti | Lucas | USA | 1973 | 110 |
| Raiders of the Lost Ark | Spielberg | USA | 1981 | 115 |
| A Bridge Too Far | Attenborough | UK | 1977 | 175 |
| Manhattan | Allen | USA | 1979 | 96 |
| Kramer v Kramer | Benton | USA | 1979 | 105 |
| ... | ... | ... | ... | ... |

with it a set of *attributes*: in this case the attribute names are TITLE, DIRECTOR, CNTRY, YEAR and LENGTH, labelling the columns of FILM. The data in a relation are structured as a set of rows. A row, or *tuple*, consists of a list of values, one for each attribute. Each cell contains a single attribute occurrence, or *value*. The tuples of the relation are not assumed to have any particular order.

We make a distinction between *relation* scheme, which does not include the data but gives the structure of the relation, and *relation*, which includes the data. Data items in a relation are taken from *domains* (like data types in programming). Each attribute of the relation scheme is associated with a particular domain and in most current proprietary systems the possible domains are quite limited, comprising character strings, integers, floats, dates, etc. In our example, the attribute DIRECTOR might be associated with character string of length up to 20, say. We may now give some definitions.

A *relation scheme* is a set of attribute names and a mapping from each attribute name to a domain.

A *relation* is a finite set of tuples associated with a relation scheme in a relational database such that:

Each tuple is a labelled list containing as many data items as there are attribute names in the relation scheme.

Each data item is drawn from the domain with which its attribute type is associated.

A *database scheme* is a set of relation schemes and a **relational** *database* is a set of relations.

Relations have the following properties:

- The ordering of tuples in the relation is not significant.
- Tuples in a relation are all distinct from one another.
- Columns are ordered so that data items correspond to the attribute in the relation scheme with which they are labelled.

Most current systems also require that the data items are themselves *atomic*, i.e. they cannot be decomposed as lists of further data items. Thus, a single cell cannot contain a set, list or array of directors. Such a relation is said to be in *first normal form*. (So-called *non first normal form* (NFNF) databases are the subject of research. See, for example, Roth *et al.* (1988)). The *degree* of the table is the number of its columns. The *cardinality* of the table is the number of its tuples. As tuples come, go and are modified the relation will change, but the relation scheme is relatively stable. The relation scheme is usually declared when the database is set up and then left unchanged through the life-time of the system, although there are operations that will allow the addition, subtraction and modification of attributes.

The theory of a relational database so far described has concerned the structuring of the data into relations. The other aspects of the relational model are the operations that can be performed on the relations (database manipulation) and the integrity constraints that the relations must satisfy. The manipulative aspects will be considered next, after we have described our working example.

*Relational database example: CINEMA*

We will work with a hypothetical relational database called the CINEMA database. The idea is that data on the cinemas in the Potteries region are to be kept in a single relational database. The database holds data on cinemas and the films they are showing at a given time. The database is not *historic* for it only stores the current showings of films and does not keep records of past showings nor projections of future showings. The database scheme is as follows:

### CINEMA Database Scheme

CINEMA (<u>CIN ID</u>, NAME, MANAGER, TELNO, TOWN, GRID_REF)
SCREEN (<u>CINEMA ID, SCREEN NO</u>, CAPACITY)
FILM (<u>TITLE</u>, DIRECTOR, CNTRY, YEAR, LENGTH)
SHOW (<u>CINEMA ID, SCREEN NO, FILM NAME</u>, STANDARD, LUXURY)
STAR (<u>NAME</u>, BIRTH_YEAR, GENDER, NTY)
CAST (<u>FILM STAR, FILM TITLE</u>, ROLE)

Each relation scheme is given as its name followed by the list of its attributes. Thus, each cinema in the database is to have an identifier, name, manager name, telephone number, town location and Ordnance Survey Grid reference. Cinemas have screens, given by number, each with an audience capacity. Films have titles, directors, country of origin, year of release and length in minutes. A film may be showing on a screen of a cinema, and STANDARD and LUXURY refer to the two-tier ticket prices. Each film star has a name, year of birth, gender and nationality, and has roles in particular films. For brevity, the corresponding domains have been omitted. Note that the same attribute may have different names in different relations (e.g. CIN_ID and CINEMA_ID). There is a convention that the set of attributes constituting the primary key of the relation is underlined. A *key* is a minimal set of attributes whose values will serve to uniquely identify each tuple of the relation. Thus, the screen number is not sufficient to identify a particular screen uniquely as many cinemas may have 'Screen One', but the cinema identity and the screen number is sufficient and so the two attributes CINEMA_ID and SCREEN_NO form a key for relation SCREEN. (There may be several keys for a relation. Often, the system will accept only one, so we must choose, and this choice is called the *primary key*.) A sample database, using some old films, for a hypothetical week in the Potteries has been given in Appendix 1.

*Operations on relations*

We have seen that a relation is nothing more than a structured table of data items, each column of which is named by an attribute and has all its data items taken from the same domain. The operations that can be supported by a relational database are therefore fundamentally simple. They are the traditional set operations of **union**, **intersection** and **difference**, along with the characteristically relational operations of **project**, **restrict**, **join** and **divide**. The structure of these operations and the way that they can be combined is called *relational algebra* and was essentially defined by Codd (1970).

The relational model is *closed* with respect to all the above relational operations, because they each take one or more relations as input and return a relation as result. The set operations **union**, **intersection** and **difference** work on the relations as sets of tuples. Thus, if we have relations holding female and male film stars, then their union will hold stars of both genders and their intersection will be empty. Note that for the set operations, the relations must be *compatible*, in that they must have the same attributes, otherwise the new relation will not be well-formed.

The **project** operation is unary, applying to a single relation. It returns a new relation that has a subset of attributes of the original. The relation is unchanged, except that any duplicate tuples formed are coalesced. We use a simple syntax (there is no completely standard syntax for the relational operations) to describe this operation of the form

**project**(*relation*: *attribute list*).

For example, **project**(CINEMA: NAME, TOWN) returns the relation shown in Table 2.3(a) and **project**(CINEMA: NAME) returns the relation shown in Table 2.3(b). Note that in the second case the two identical tuples containing the value 'Regal' have been coalesced into a single tuple.

The **restrict** operation is also unary, this time working on the tuples of the table rather than the columns, and returns a new relation which has a subset of tuples of the original. A condition specifies those tuples required. The syntax used here is:

**restrict**(*relation*: *condition*)

For example, the films released after 1978 can be retrieved from the database using the expression **restrict**(FILM: YEAR > 1978). This will return the relation shown in Table 2.3(c). Operations can be combined, but the notation becomes cumbersome.

**Table 2.3** Results of relational projections and restrictions

(a)

| NAME | TOWN |
|---|---|
| Majestic | Stoke |
| Regal | Hanley |
| Regal | Newcastle |

(b)

| NAME |
|---|
| Majestic |
| Regal |

(c)

| TITLE | DIRECTOR | CNTRY | YEAR | LENGTH |
|---|---|---|---|---|
| Raiders of the Lost Ark | Spielberg | USA | 1981 | 115 |
| Manhattan | Allen | USA | 1979 | 96 |
| Kramer v Kramer | Benton | USA | 1979 | 105 |
| French Lieut's Woman | Reisz | UK | 1981 | 123 |

(d)

| DIRECTOR |
|---|
| Spielberg |
| Allen |
| Benton |
| Reisz |

For example,

**project(restrict(**FILM: YEAR > 1978): DIRECTOR)

returns the directors of films released after 1978, as shown in Table 2.3(d).

With the join operation, the relational database begins to merit the term 'relational'. The binary operation **join** takes two relations as input and returns a single relation. This operation allows connections to be made between relations. There are several different kinds of relational join but we describe only the *natural join* of two relations, defined as the relation formed from all combinations of their tuples that agree on a specified common attribute or attributes. We use the syntax

**join(**$rel_1$, $rel_2$: $att_1$, $att_2$)

to indicate that relations $rel_1$ and $rel_2$ are joined on attribute combinations $att_1$ of $rel_1$ and $att_2$ of $rel_2$. For example, to relate details of films with the screens on which they are showing, relations SHOW and FILM are joined on the film title attribute in each relation. The expression is:

**join(**SHOW, FILM: FILM_NAME, TITLE)

The resulting relation, shown in Table 2.4(a), where some of the attribute names have been abbreviated for print clarity, combines tuples of SHOW with tuples of FILM, provided that the tuples have the same film name. Notice that the join has not repeated the duplicate attribute. If we only require the directors of films showing at cinema 1, we may project the relation to have only attributes CINEMA_ID, SCREEN_NO, FILM_NAME and DIRECTOR; then restrict the relation to tuples whose CINEMA_ID has value '1'. The result is shown in Table 2.4(b).

The last example may be used to demonstrate an important property of relation operations: it matters in which order they are done, both in terms of result and performance. The join operation is the most time-consuming of all relational operations, because it needs to compare every tuple of one relation with every tuple of another. To extract data for Table 2.4(b) we performed operations join, restrict and

**Table 2.4** Results of relational joins, projections, and restrictions

(a)

| CIN | SCR | FILM_NAME | STD | LUX | DIRECT | CNTRY | YEAR | LNGTH |
|-----|-----|-----------|-----|-----|--------|-------|------|-------|
| 1 | 1 | The Godfather | £3.50 | £6.00 | Coppola | USA | 1972 | 175 |
| 1 | 2 | Superman | £3.50 | £6.00 | Donner | GB | 1978 | 143 |
| 1 | 3 | Last Tango in Paris | £3.00 | | Bertolucci | IT/FR | 1972 | 129 |
| 2 | 1 | The Godfather | £4.00 | £5.00 | Coppola | USA | 1972 | 175 |
| 2 | 2 | Jaws | £3.00 | £5.00 | Spielberg | USA | 1977 | 125 |
| 3 | 1 | Last Tango in Paris | £4.00 | | Bertolucci | IT/FR | 1972 | 129 |

(b)

| CINEMA_ID | SCREEN_NO | FILM_NAME | DIRECTOR |
|-----------|-----------|-----------|----------|
| 1 | 1 | The Godfather | Coppola |
| 1 | 2 | Superman | Donner |
| 1 | 3 | Last Tango in Paris | Bertolucci |

project. In fact it would have been more efficient to have first done a restrict operation on the SHOW table, then joined the resulting smaller table to FILM, and then projected. The result would be the same but the retrieval would perform better, because the join involves smaller tables. In technical terms, the query should be *optimized* to perform as efficiently as possible. Query optimization is a critical study for high-performance databases and has a very large literature. The general database texts in the bibliography will provide starting points.

### 2.2.2 SQL

The Structured or Standard Query Language (SQL) provides users of relational databases with the facility to define the database scheme (data definition), and then insert, modify and retrieve data from the database (data manipulation). The language may be used either on its own as a means of direct interaction with the database, or embedded in a general purpose programming language. SQL arose out of the relational database language SEQUEL (Chamberlin and Boyce 1974; Chamberlin *et al.* 1976), which itself was based on SQUARE. The most recent SQL standard is SQL/92 (also called SQL2) (ISO, 1992) but currently no system supports all of the very large range of features in the SQL2 standard. There is a large effort to move forward to SQL3 later in the decade: see the final chapter for some remarks on SQL3 and its application to geo-databases. This section gives the reader an introduction and feel for SQL2, without making any pretension to completeness. Some notes on SQL standards are provided as part of Appendix 2.

*Data definition using SQL*

The data definition language (DDL) component of SQL allows the creation, alteration and deletion of relation schemes. It is usual that a relation scheme is altered only rarely once the database is operational. A relation scheme provides a set of attributes, each with an associated data domain. SQL allows the definition of a domain by means of the expression (square brackets indicate an optional part of the expression)

> **CREATE DOMAIN** *domain-name data-type*
> [*default definition*]
> [*domain-constraint-definition-list*]

The user specifies the name of the domain and associates the name with one of a supplied set of domains that include character string, integer, float, date, time and interval types. The default definition allows the user to specify the value to be assigned to any data item not explicitly entered during the construction of a tuple: a usual default value is NULL. The domain-constraint-definition-list acts as an integrity constraint by restricting the domain to a set of specified values. An example of the definition of a domain for the attribute GENDER is as follows:

> **CREATE DOMAIN GENDER CHARACTER(1)**
> **CHECK VALUE IN {'M', 'F'};**

There are further commands to alter and drop domains that we will not describe here.

A relation scheme is created as a set of attributes, each associated with a domain, with additional properties relating to keys and integrity constraints. For example, the relation scheme CAST may be created by the command:

```
CREATE TABLE CAST
    (FILM_STAR          STAR,
    FILM_TITLE          FILM_TITLE,
    ROLE                ROLE,
    PRIMARY KEY         (FILM_STAR, FILM_TITLE),
    FOREIGN KEY         (FILM_STAR)
    REFERENCES STAR(NAME)
    ON DELETE CASCADE
    ON UPDATE CASCADE,
    FOREIGN KEY         (FILM_TITLE)
    REFERENCES FILM(TITLE)
    ON DELETE CASCADE
    ON UPDATE CASCADE,
    CHECK (FILM_STAR IS NOT_NULL),
    CHECK (FILM_TITLE IS NOT_NULL);
```

This statement begins by naming the relation scheme (called a table in SQL) as CAST. The attributes are then defined by giving their name and associated domain (assume that we have already created domains STAR, FILM_TITLE and ROLE). The primary key is next given as the attribute combination FILM_STAR, FILM_TITLE. A *foreign key* is a primary key of another relation contained in the given relation. In our example, there are two foreign keys: FILM_STAR and FILM_TITLE. Thus, for example, FILM_TITLE occurs as the primary key (TITLE) of the FILM relation. Referential integrity is maintained by ensuring that if a film is deleted (or updated) from the FILM relation, then any reference to it must also be deleted (or updated) in CAST. Finally, two further integrity checks are added to limit the insertion of NULL values on data entry to the ROLE attribute only: any attempt to insert a row with no entries for FILM_STAR and FILM_TITLE will be disallowed.

There are also SQL commands to alter a relation scheme by changing attributes or integrity constraints and to delete a relation scheme.

### Data manipulation using SQL

Having defined the schemes, the next step is to insert data into the relations. These SQL commands are quite straighforward, allowing insertion of single or multiple tuples, update of tuples in tables (e.g. change values of the CNTRY attribute in FILM from 'UK' to 'GB', or update a salary in a personnel database), and deletion of tuples.

Data retrieval forms the most complex aspect of SQL: a large book could be written on this topic alone. Our treatment is highly selective, giving the reader a feel for SQL in this respect. The general form of the retrieval command is:

SELECT *select-item-list*
FROM *table-reference-list*
[WHERE *condition*]

[GROUP BY *attribute-list*]
[HAVING *condition*]

A simple example of data retrieval, already considered in the relational algebra section, is to find the names of all directors of films that were released after 1978. The SQL expression is:

SELECT DIRECTOR
FROM FILM
WHERE YEAR > 1978;

The SELECT clause serves to project (how confusing!) on the required DIRECTOR attribute. The FROM clause tells us from which table the data are coming, in this case FILM. The WHERE clause provides the restrict condition.

Relational joins are effected by allowing more than one relation (or even the same relation called twice with different names) in the FROM clause. For example, to find details of films and where they are showing, we would give the following:

SELECT CINEMA_ID, SCREEN_NO, FILM_NAME, DIRECTOR
FROM SHOW, FILM
WHERE SHOW.FILM_NAME = FILM.TITLE;

In this case, the WHERE clause provides the *join condition* by specifying that tuples from the two tables are to be combined only when the values of the attributes FILM_NAME in SHOW and TITLE in FILM are equal. A more complex case, using all the clauses of the SELECT expression is the following expression that retrieves the average lengths of USA films made by directors that have at least two films in the database.

SELECT DIRECTOR, AVG(LENGTH)
FROM FILM
WHERE CNTRY = 'USA'
GROUP BY DIRECTOR
HAVING COUNT(*) > 1;

AVG and COUNT are built-in SQL functions. Apart from AVG(LENGTH), the first three lines of code act to retrieve the directors of USA films. The GROUP BY clause serves to logically construct a table where the tuples are in groups, one for each director. This table, shown in Table 2.5(a), is not a legal first normal form relation because values in some cells are not atomic. But we are concerned with the average film length of each director and will eventually project out titles. An intermediate stage with average lengths calculated is shown in Table 2.5(b). Finally, the HAVING clause comes into play to operate as a condition on the groups in the grouped relation. It selects only groups that have a tuple (indicated by *) count of at least two. The final stage of the retrieval, after projecting out the TITLE attribute is shown in Table 2.5(c).

Many of the features of SQL have been omitted in this whistle-stop tour. The standarded for SQL2 is about 600 pages in length. We have only given examples of the interactive form of SQL that could be used to directly retrieve data. There is also an embedded form that is used within conventional programs, written in languages such as C. It is also possible to define external views and provide different access for different classes of user, as discussed earlier in the chapter.

**Table 2.5**  Stages in the evaluation of an SQL query to the
CINEMA database

(a)

| TITLE | DIRECTOR | LENGTH |
|---|---|---|
| Jaws | } Spielberg | 125 |
| Raiders of the Lost Ark | | 115 |
| Star Wars | } Lucas | 121 |
| American Graffiti | | 110 |
| Manhattan | Allen | 96 |
| Kramer v Kramer | Benton | 105 |
| The Godfather | Coppola | 175 |
| All the President's Men | Pakula | 138 |
| Marathon Man | Schlessinger | 126 |

(b)

| TITLE | DIRECTOR | AVG(LENGTH) |
|---|---|---|
| Jaws | } Spielberg | 120 |
| Raiders of the Lost Ark | | |
| Star Wars | } Lucas | 115 |
| American Graffiti | | |
| Manhattan | Allen | 96 |
| Kramer v Kramer | Benton | 105 |
| The Godfather | Coppola | 175 |
| All the President's Men | Pakula | 138 |
| Marathon Man | Schlessinger | 126 |

(c)

| DIRECTOR | AVG (LENGTH) |
|---|---|
| Spielberg | 120 |
| Lucas | 115 |

### 2.2.3  Distributed relational databases

We have already briefly discussed distributed databases in general. This section
looks at some of the issues that arise when the database technology is relational.
The distributed relational database ideal is to have a system where each node may
act independently as a relational database or the nodes may act together in such a
way that users view the system as a single integrated relational database. Thus, a
distributed relational database is sometimes called a *virtual* system, because *it should
behave as if the data are stored in one physical repository* even though in reality the
data are scattered around the nodes. The software that manages this is known as a
distributed relational database management system. Date (1995) lists 'twelve objec-
tives' of a distributed relational database.

*Local autonomy*:  Each site should have the capacity to act independently as a
relational database, using its own share of the data.

*No central site*: No site is to be distinguished as the master on which other sites are dependent for critical functionality (distinguished from data).

*Continuous operation*: If one node goes down, others may act as temporary replacements, thus increasing reliability and availability of the database.

*Location independence*: A user at a node ideally will have no concept of where data are physically stored. This is desirable because programs and data then become completely portable between sites.

*Fragmentation independence*: *Relation fragmentation* is a key idea in distributed relational systems. If data are to be stored at several sites, then the relations must be broken up. Data can then be stored at nodes that access them most frequently. Fragmentation of relations should be hidden from users. Two kinds of fragmentation are possible:

*horizontal fragmentation*: distributes tuples of a relation between nodes.

*vertical fragmentation*: distributes attributes of a relation between nodes.

*Replication independence*: It may be desirable for efficiency reasons to store the same data at different nodes (*data replication*). This has to be handled very carefully: for example, an update at one node but not at another may lead to an inconsistency. The users should have no feel that data are replicated.

*Distributed query processing*: Optimization is an important issue, because there may be several ways to execute a query especially if there is replication.

*Distributed transaction management*: Support for recovery and concurrency is more complex in a distributed setting.

*Hardware, operating system and network independence*: These apply to any distributed system.

*DBMS independence*: Ideally, the distributed system should support a heterogeneity of proprietary DMBS at its nodes. Indeed, this is one of the most useful aspects of a distributed system, pulling in different types of users and making the whole more than the sum of the parts. Spatial database software could be part of the distributed community.

To illustrate some of these ideas, imagine that we want to extend our Potteries CINEMA database to a nationally distributed system (Figure 2.8). (Some specific aspects of this example are not realistic. The reader is asked to be charitable and suspend disbelief temporarily, because the principles are still valid.) Thus, both Newcastle upon Tyne and London join Stoke in the enterprise. The idea is to keep the film details (FILM, CAST, STAR) replicated at each node, but horizontally fragment the CINEMA, SCREEN and SHOW relations so that nodes only hold tuples relating to their regions. Thus, the London site would hold all tuples of the FILM, CAST and STAR relations as in Appendix 1. In addition, it would hold a subset of the CINEMA, SCREEN and SHOW as in Table 2.6(a), (b), (c).

A user of the CINEMA database would have an independent database of information about her cinemas, screens, shows and films. In addition, she could have a view of other information nationwide, for example to check which are the currently popular shows.

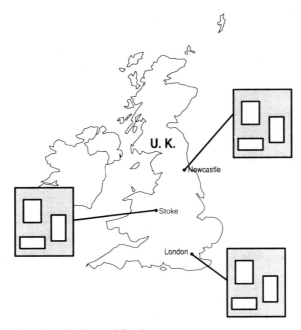

**Figure 2.8**   CINEMA distributed relational database.

**Table 2.6**   Extra data at the London node of the CINEMA distributed relational database

(a): CINEMA (London fragment)

| CINEMA_ID | NAME | MANAGER | TELNO | TOWN | GRID_REF |
|---|---|---|---|---|---|
| 3 | Imperial | Sue Hurd | 0171364892 | Chelsea | AB 123456 |
| 7 | Regal | Fred Jones | 0171593654 | Victoria | AB 876543 |
| 9 | Majestic | Dianne Coe | 0171296528 | Euston | AB 283746 |

(b): SCREEN (London fragment)

| CINEMA_ID | SCREEN_NO | CAPACITY |
|---|---|---|
| 3 | 1 | 900 |
| 3 | 2 | 300 |
| 7 | 1 | 600 |
| 9 | 1 | 600 |

(c): SHOW (London fragment)

| CINEMA_ID | SCREEN_NO | FILM_NAME | STANDARD | LUXURY |
|---|---|---|---|---|
| 3 | 1 | The Godfather | £3.00 | £5.00 |
| 3 | 2 | Jaws | £3.50 | £7.00 |
| 7 | 1 | Last Tango in Paris | £4.00 | |
| 9 | 1 | Jaws | £4.00 | £5.00 |

### 2.2.4   Relational databases for spatial data handling and extensible relational systems

In an unmodified state relational databases are unsuitable for spatial data management. Although relational databases have all the characteristics required for general
corporate data management, there are difficulties when the technology is applied to
spatial data. Some of the problems are listed:

*Structure of spatial data*:   Spatial data have a structure that does not naturally
fit with tabular structures. Vector areal data are typically structured as boundaries composed of sequences of line segments, each of which is a sequence of
points. Models of spatial data will be constructed in later chapters and the problems become apparent. In general, to reconstruct a spatial object requires the
joining of several tables, with consequent performance overheads. Typically,
many such spatial objects are required quickly, for example to show a map on
the screen, and it is difficult to get the desired performance with standard relational database technology.

*Indexes*:   Indexing questions will be considered in Chapters 5 and 6. An index
makes possible the fast access to data. Relational databases provide indexes that
perform well with standard tabular data (e.g. a file of stock items), but specialized
indexes are required for spatial data and these are not directly supported by
proprietary relational DBMS.

*SQL*:   The appropriateness of SQL as a spatial database interaction language is
considered in Chapter 7.

Spatial data handling is one of several application types that are not well handled
by relational databases. Other examples are computer-aided design and manufacture (CAD/CAM), engineering databases and office information systems (OIS).
What such applications have in common is a requirement for more complex structures than pure relations, along with application specific operations to be applied to
these structures and appropriate indexing techniques. An *extensible RDBMS* is
designed to provide specialist users with the facilities to add extra functionality,
specific to their domains. Such facilities include:

user-defined data types

user-defined operations on the data

user-defined indexes and access methods

active database functions (e.g. triggers)

Many of these aims are shared by object-oriented databases, to be discussed later.
The characteristic of extensible RDBMS is that the extra functionality is built upon
the relational model with as little change as possible.

Extensible RDBMS are still in the research domain. Two important projects
have been POSTGRES (Stonebraker and Rowe, 1986) and IBM Starburst (Haas *et
al.*, 1989). The POSTGRES project from the University of California at Berkeley
has resulted in the MIRO system (Stonebraker, 1993), which combines extensible
RDBMS functions with some object-oriented and active database ideas. All these
projects have been applied to spatial data management, for example Sequoia 2000
(Anderson and Stonebraker, 1994), built on the Illustra object-relational DBMS.

## 2.3   DATABASE CONSTRUCTION

In the last chapter (section 1.3.8), we saw a typical staged approach to constructing an application for an informtion system. Analysis of requirements results in a conceptual model of the system. This model is turned into a design (logical model), which may then be implemented using a particular proprietary system. This section looks at this process in general, but with an eye to implementation in a relational database.

A database management system is general purpose software that must be customized to meet the needs of particular applications. In order to do this, we have to have a precise idea of the way that information is structured in the system and the kinds of algorithms that will act upon the data. Here, we are not concerned with the actual data in the database, but with the kinds of data that we expect. For example, in a cartographic application, we are not so much concerned with individual data items, 'London', 'New York', 'France', 'Ben Nevis', as with data types, **city**, **country**, and **mountain**. We abstract from information system content to information structure.

Following requirements analysis, there are three main stages in the process of customizing a database system for a particular application.

1.   Construction of the conceptual model (*conceptual data modelling*).
2.   Translation of the conceptual model into the logical model (*database design*).
3.   Translation of the design into a system that works on a particular DBMS and platform (*implementation*).

The first stage, conceptual data modelling, takes the requirements (already elicited or elicited as the customization progresses) and transforms then into a general model of the information structure. The model of an information system at this level of abstraction is called a *conceptual data model* and is independent both of the actual items of information and of the manner in which the information system is implemented. Database design tailors the conceptual model to the particular kind of DBMS on which the system will be implemented, called a *logical data model*. For example, if the DBMS is relational, then part of the second stage will be the creation of relation schemes. Thus, the logical model may be mapped directly to an implementation, while itself being independent of the details of physical implementation. Implementation constructs the *physical data model*, which comprises all the implementation details for the system, for example details of how the data files are stored on particular storage media.

### 2.3.1   Conceptual data modelling

A conceptual model is a model of the projected system that is independent of implementation details. Methods that are to be useful for the construction of a conceptual model must enable the modeller to express the structure of the information in the system, that is, the types of data and their interrelationships. Such structural properties are often termed *static*, but the system will also have a *dynamic* component related to its behaviour in operation. For example, the allowable transactions with the system shape its dynamic properties.

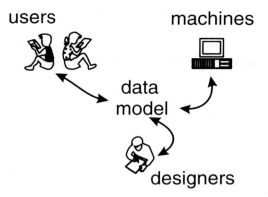

**Figure 2.9** The conceptual data model as mediator between users, designers and machines.

The correctness (integrity) of the information in a system is often a critical factor in its success. Correctness is maintained as much as possible by the specification of integrity constraints that impose conditions on the static and dynamic structures of the system. A data model should allow the specification of a range of integrity constraints.

A good conceptual data model will act as a mediator between users, systems specialists and the machine level (see Figure 2.9). An analyst or designer is able to use a good modelling tool as an efficient means of communication between him- or herself and potential users. This will aid the design of the system. Also, when the system is eventually implemented, the conceptual data model may act as a guide to the structure of the data in the system for potential users. This is the role of conceptual data model as source of system *metadata*. Spatial information systems are becoming ever more complex in their internal structure. Comprehensible data models as a source of metadata for the system provide a possible solution to this problem. In summary, conceptual models:

- Provide a framework that allows an expression of the structure of the system which is clear and easy to communicate to non-specialists.

- Contain sufficient modelling constructs so that the complexity of the system may be captured as completely as possible.

- Have the capability for translation into implementation-dependent models (i.e. logical and physical data models), so that the system may be designed and built.

*Entity-relationship model*

One of the most compelling and widely used approaches to forming a conceptual model of an information system was originally proposed by Chen (1976) and is known as the *entity-relationship model* (*entity-attribute-relationship model*, or *E-R model*). The entity-relationship model has been enlarged to the *extended entity relationship (EER) model*. The *object-oriented* model is a further extension of the entity-relationship model in which are treated not only structural aspects of the information in the system, but also dynamic behavioural components. The EER and object models are considered in later sections.

Imagine that we are designing a database that contains spatially referenced information about a specific location (the Potteries region, for example), and that a requirement is for the system to contain data on administrative units (towns, districts, wards, etc.), transportation networks (road, rail, canals, etc.), physical features (lakes, rivers, etc.) and spatially referenced attributes (areas and heights of physical features, populations of towns and cities, traffic loads on stretches of roads, railways and canals, etc.). How would we make a start? Well, we have already started, in that we have elicited some requirements of the system, and expressed these requirements in the form of collections of entities about which our system will have information and the relevant properties of such entity collections. This is the simple and powerful idea behind the entity-relationship model.

An *entity type* is an abstraction that represents a class of similar objects, about which the system is going to contain information. In our example, some of the entity types might be **town**, **district**, **road**, **canal** and **river**. (We have used the typographic convention that for conceptual modelling, types are rendered in bold, values in quotes. Constituents of a relational database (relation and attribute names) are rendered in upper-case). We make a distinction between the *type* of an entity and an *occurrence* or *instance* of an entity type. For example, we have entity type **town** and occurrences such as 'Newcastle-under-Lyme', 'Hanley' and 'Stoke'.

When we write about the town of Newcastle-under-Lyme, we are not being absolutely precise, since of course 'Newcastle-under-Lyme' is not a town but a data item which serves to name the town. The data item 'Newcastle-under-Lyme' is associated with a particular occurrence of the entity **town** as its name. Thus, entity types have properties, called *attribute types* that serve to describe them. Entity type **town** has attribute types **name**, **population**, **centroid**, etc., and there might exist a particular occurrence of **town** with attribute occurrences 'Newcastle-under-Lyme', etc. The attachment of attribute types to an entity type may be represented diagrammatically, as in Figure 2.10. Entity types are shown in rectangular boxes. (The E-R diagrammatic notation presented here is common but not standard. There are many variations in use.)

An important characteristic of an entity is that any occurrence of it should be capable of unique identification. The means of unique identification is through the value of a subset of its attribute types. In the preceding paragraph, we made a distinction between an occurrence of an entity and its attribute value. In everyday use of language, this distinction is not explicitly made. For example, if we were lost, we would probably ask a question like 'In which direction is London?', rather than 'In which direction is the city with name London?' In this context, there is no need to distinguish the name 'London' from the place that it names. However, if I asked

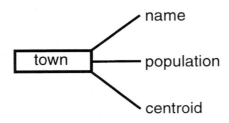

**Figure 2.10**   An entity type and its attribute types.

someone 'What is the population of Cambridge?', he or she would be entitled to ask 'Do you mean Cambridge, UK, or Cambridge, Massachusetts?' The problem is that a city's name is not always sufficient to identify it uniquely. An attribute type, or combination of attribute types, which serves to identify an entity type uniquely is termed an *identifier*. In our example, attribute type **centroid**, which gives the grid reference of the centroid of a town, is an identifier of **town**.

There could be several identifiers, in which case it may be useful to distinguish a particular identifier from among the candidates. The requirement to choose is often determined by implementation questions at lower levels in the modelling process, such as the necessity for a primary index for a file. Strictly, these are not questions for consideration at this stage. All that should be borne in mind is that an entity type, by definition must have a least one identifier. As in the notation for a relation scheme, the attributes comprising the chosen identifier are often underlined.

So far, the model comprises a number of independent entity types, each with an associated set of attribute types, one or more of which serve to identify uniquely each occurrence of the type. The real power in this model comes with the next stage, which provides a means of describing connections between entity types, and develops a diagrammatic language called *entity-relationship diagrams*, or *E-R diagrams*, for expressing properties of the model. (The type-occurrence distinction is important, but becomes cumbersome if made explicit on every occasion. If there is no danger of confusion we might omit the words 'type' and 'occurrence'.) For example, suppose that we have defined the two entity types:

- **town** with attributes **<u>centroid</u>, name, population**
- **road** with attributes **<u>road id</u>, class, start_point, end_point**

A question that we might want to ask of our finished system is 'Which towns lie on which roads?' This question can only be answered by forming a link between towns and roads upon which the towns lie. This connection is called a *relationship*. (It is unfortunate that the terms 'relationship' (connection between entities) and 'relation' (table in a relational database) are so similar. We have tried to be precise in usage of each of these terms.) A *relationship type* connects one or more entity types. In the example, the relationship type will be called **lies_on**. A *relationship occurrence* is a particular instance of a relationship type. Thus, the incidence of the town Hinckley with the A5 road from London to Anglesey is an occurrence of the relationship **lies_on**. The relationship **lies_on** between entities towns and roads may be shown in an E-R diagram as shown in Figure 2.11. Relationships are shown in diamond-shaped boxes.

Relationships may have their own attributes, which are independent of any of the attributes of the participant entities. In the above example, the relationship **lies_on** might have the attribute **length**, which gives the length of the road within the town boundary. A further attribute might indicate whether the road goes through the town centre or by-passes the town.

The entity-relationship model allows the expression of a limited range of integrity constraints. Relationship types are subdivided into many-to-many, many-to-one and one-to-one relationships. The relationship **lies_on** is an example of a *many-to-many* relationship, because each city may have several (potentially more than one) roads passing through it and each road may pass through several (potentially more than one) cities. This constraint on the relationship is shown diagrammatically by the $M$ and $N$ on each side of the relationship.

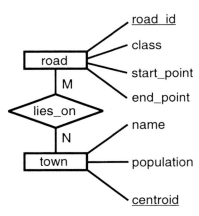

**Figure 2.11** Two entities and a many-to-many relationship.

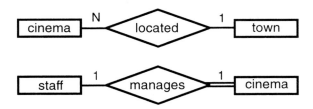

**Figure 2.12** Many-to-one and one-to-one relationships.

Not all relationships are many-to-many. For example, consider the relationship **located** between types **cinema** and **town** shown as an E-R diagram in Figure 2.12 with attributes omitted. Each cinema is located in at most one town, but a town may contain several (potentially more than one) cinemas. Such a relationship is called a *many-to-one* relationship. The positioning of the symbols *N* and 1 indicates the nature and direction of the relationship.

The third and final relationship category is exemplified by the relationship **manages** between **staff** and **cinema**. In this relationship, a member of staff may be the manager of at most one cinema and a cinema may have at most one manager. This is a *one-to-one* relationship. The form of the E-R diagram is shown in Figure 2.12. The positioning of the two 1's indicates the nature of the relationship.

In allowing the modeller to express information about the category of a relationship, the E-R modelling tool is providing facilities for the expression of *cardinality conditions* upon entity occurrences entering into relationships. The entity-relationship model also allows the expression of another class of integrity constraint, namely *participation conditions*. Entity types are involved in relationships with other (or the same) entity types. If, for a moment, we consider a single entity type *E* and a single relationship *R* in which it is involved, there is an important distinction to be made regarding the way that *E* participates in *R*. It may be that every occurrence of *E* participates in the relationship. In this case, we say that the *membership class* or *participation* of *E* in *R* is *mandatory*. On the other hand, if not every occurrence of *E* need participate in the relationship *R*, we say that the participation is *optional*.

For example, the participation of entity **staff** in **manages** is optional, because not every member of staff in an imagined database system will be a manager (there may also be office staff, projectionists, etc., in the full system). However, the participation of entity **cinema** in **manages** is mandatory, because we will assume that every cinema must have a manager. These conditions are displayed on the E-R diagram by using double lines for mandatory and single lines for optional participation.

All the relationships that we have given as examples so far have been binary, in that they have connected together precisely two entity types. It sometimes happens that these two types are the same. In this case, the relationship is one that relates an entity type to itself. Such a relationship is called *involutory*. For example, the information that roads intersect other roads may be represented by the relationship **intersect** between entity type **road** and itself. A relationship that connects three entity types is called a *ternary* relationship. Imagine that data are to be stored in our Potteries system showing the usage that different bus companies make of major roads in the counties of the region. We have data on the daily totals of vehicle-miles along major roads, for each bus company, recorded for each district in the region. We can model this using a ternary relationship **usage** connecting entities **bus_company**, **road** and **district**, with attributes **date** and **vehicle_miles**. This is shown in Figure 2.13. Participation conditions apply as with the binary case but with extra complexity. In this example, bus companies may have routes along roads in several districts; districts may have many roads through them used by several bus companies; and roads may pass through several districts and be used by several bus companies. This relationship is many-to-many-to-many.

A *dependent* or *weak entity* is an entity type that cannot be identified uniquely by the values of its own attributes, but relies on a *parent entity* type for its identification. The complete identification occurs through an *identifying relationship* between the dependent and weak entities. A good example of this phenomenon is the entity type **screen** in the CINEMA database. The attribute **screen_no** is insufficient to identify a particular screen; after all, 'Screen One' tells us very little. **screen_no** is a *partial identifier* for the dependent entity. We need the **cinema_id** to complete the identification. The notation for this is shown in Figure 2.14. Weak entities and identifying relationships are distinguished by double-lined boxes and diamonds. A partial identifier is underlined with a dotted line. A dependent entity always has mandatory participation in the identifying relationship, otherwise some occurrences may not be identified. For the same reason, the relationship is one-to-one or many-to-one from dependant to parent.

The E-R model is now applied to the example of the CINEMA database. Imagine that we are at the analysis stage of system development and that a set of requirements has been captured. There is a need to hold information about cinemas

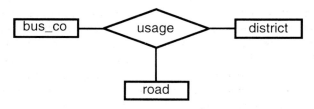

**Figure 2.13** A ternary relationship.

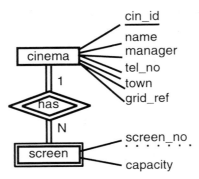

**Figure 2.14**   **cinema** and **screen** as parent and dependent entity types.

and the films that they show. Choosing entities and attributes for a model is a matter of judgement, often with more than one acceptable solution. Sometimes it is difficult to decide whether to characterize something as an attribute or an entity. In the modelling of the CINEMA database, there are several such choices. For example, should we make the town in which a cinema is located an entity or attribute? Some guidelines include:

- If the data type is relatively independent and identifiable, with its own attributes, then it is probably an entity: if it is just a property of an entity, then it is an attribute.

- If the data type enters into relationships with other entities (apart from being a property of one entity), then it is probably an entity.

In the case of **town**, because, in our system, it is just a name and a property of the cinema, we choose to make it an attribute of **cinema**. If **town** had possessed its own attributes, such as **population**, then we would probably have made it an entity and constructed a relationship between entities **town** and **cinema**. The same reasoning applies to **manager**, which we choose to be an attribute of **cinema** since only the manager's name is involved (it may be less confusing to call this attribute **manager_name**). Initial investigation reveals that the following entity types and their attributes are needed:

**cinema**   **cinema_id**, **name**, **manager**, **tel_no**, **town**, **grid_ref**

**screen**   **screen_no**, **capacity**

**film**   **title**, **director**, **cntry**, **year**, **length**

**star**   **name**, **birth_year**, **gender**, **nty**

where most of the attribute names are clear, except that attribute **year** of **film** denotes the year that the film was released and attribute **nty** of **star** denotes the nationality of the star. It is taken that **cinema_id** is by definition an identifier of the entity **cinema**. Identifiers for the other entities are less clear. After discussion with users, it becomes clear that for the purposes of this system a film is to be identified by its title (the assumption is that there cannot then be more than one film with the same title in the database) and a star by his or her name. Screens are not identified by their number alone and must also take the identifier of the cinema for complete identification.

Further analysis of the requirements reveals that each screen may be showing a film and that the cinemas have a two-tier ticket pricing scheme of standard and luxury seats: the prices depend upon the film and on which screen it is shown. Also, film stars are cast in particular roles in films. We can therefore construct some relationships between entities:

**show** between **screen** and **film**, with attributes **standard, luxury**

**cast** between **star** and **film**, with attribute **role**

There are also some system rules that we can glean from the requirements and common knowledge about the film industry.

1. Each cinema may have one or more screens.
2. Each screen is associated with just one cinema.
3. Each screen (not cinema) may be showing zero or one films.
   (The database is not to hold past or future information.)
4. A film may be showing at zero, one or several screens.
5. A film star may be cast in one or more films.
6. A film may contain zero, one or more stars.

All this information may be modelled and represented by the E-R diagram shown in Figure 2.15. The dependent entity and identifying relationship between **screen** and **cinema** have been discussed earlier. The relationship **shows** between **screen** and **film** is constrained by system rules 3 and 4 to be many-to-one with optional participation from **screen** and **film**). The relationship **cast** between **star** and **film** is constrained by system rules 5 and 6 to be many-to-many with mandatory participation from **star** and optional from **film**.

The E-R model is still the most widely used modelling tool for database design. Measured against the three criteria of a modelling approach given at the beginning of section 2.3.1, it scores highly. As a method based upon the intuitive notions of entity, attribute and relationship, it is easily grasped by non-specialists and thus

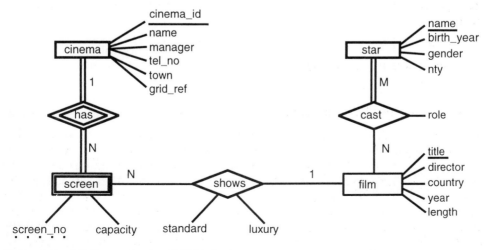

**Figure 2.15**   E-R diagram for the CINEMA database.

provides an excellent means of communication between systems analysts and users during the specification of the system. It also provides a simple means of communicating metadata to potential users. On the third criterion, we shall see later in the chapter how readily a conceptual data model framed using E-R may be translated into the relational logical model.

It is the second criterion, regarding the existence of sufficiently powerful modelling constructs, where the shortcomings of the basic E-R model are sometimes said to lie. This is particularly the case where the information system to be modelled does not fit into the standard corporate database pattern. As it is precisely such systems that are the concern of this book, we move on to look at extensions of the E-R model that set out to remedy this deficiency. The price to be paid for any increase in modelling power that such extensions can provide is the slight loss of the natural feel and simple diagrams that the basic E-R model possesses.

### The extended entity-relationship (EER) model

The *extended* or *enhanced* entity-relationship (EER) model has features additional to those provided by the standard E-R model. These include the constructs of *subclass*, *superclass* and *category*, which are closely related to *generalization* and *specialization* and the mechanism of *attribute inheritance*. This allows the expression of more meaning in the description of the database and leads towards object-oriented modelling; however, the object-oriented approach models not only the structure but also the dynamic behaviour of the information system.

An entity type $E_1$ is a *subtype* of an entity type $E_2$ if every occurrence of type $E_1$ is also an occurrence of type $E_2$. We also say that $E_2$ is a *supertype* of $E_1$. The operation of forming subtypes from types is called *specialization*, while the converse operation of forming supertypes from types is called *generalization*. Specialization and generalization are inverse to each other.

Specialization is useful if we wish to distinguish some occurrences of a type by allowing them to have their own specialized attributes or relationships with other entities. For example, the entity **travel_mode** may have generic attributes and relationships, such as a relationship with type **traveller**, and entities **canal** and **railway** may both be modelled as subtypes of **travel_mode** with their own specific properties. Entity **canal**, as well as having the attributes of **travel_mode**, might have attributes **tunnel_count**, **lock_count** and **min_width**, which for each canal give the totals of tunnels and locks, and the minimum width. Entity **railway** might have attributes additional to those of **travel_mode** such as **gauge**, which gives the gauge of each **railway** occurrence. Subtypes might also have additional relationships with other entity types. For example, **railway** might have a relationship with **rolling_stock**, indicating that particular pieces of rolling stock operate on particular stretches of railway.

The E-R diagram can be extended to show specialization. Figure 2.16 shows the above example in diagrammatic form. The subtype-supertype relationship is indicated by the subset symbol $\subset$. The figure also shows that **canal** and **railway** are *disjoint* subtypes of **travel_mode**, indicated by the letter 'd' in the circle. Types are *disjoint* if no occurrence of one is an occurrence of the other and vice versa. In our example, no **travel_mode** can be both a canal and a railway. On the other hand, Figure 2.16 also shows two overlapping subtypes. Entities **road_user** and **rail_user** are overlapping subtypes of entity **traveller**, indicated by the letter 'o' in the circle,

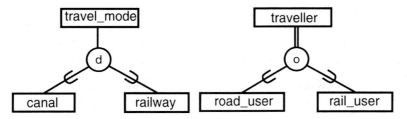

**Figure 2.16**   Specialization in the EER model.

and in our example allowing for the possibility that a traveller may use both road and rail. We also see from the double lines (mandatory participation) that the system model assumes that every traveller must use at least one of road or rail.

Specialization may be summarized as the creation of new types such that:

- Each occurrence of the subtype is an occurrence of the supertype.
- The subtype has the same identifying attribute(s) as the supertype.
- The subtype has all the attributes of the supertype, and possibly some more.
- The subtype enters into all the relationships in which the supertype is involved, and possibly some more.

The entire collection of subtype-supertype relationships in a particular EER model is called its *inheritance hierarchy*.

Generalization is the reverse modelling process to specialization. Given a set of types, we sometimes want to identify their common properties by generalizing them into a supertype. It might be that our initial modelling led us to entities **road_user** and **rail_user**. We then realized that these types had many common features, for example having similar identifiers and other attributes, such as **name**, **age**, **gender**, **address**, also a common relationship to entity **town** in which they lived. The commonalities between **road_user** and **rail_user** may be 'pushed up' into a new generalized entity **traveller**. The resulting model is the same in case of either specialization or generalization, but the process of arriving at the model is different.

A technical problem with generalization occurs when we wish to generalize from heterogeneous entity types. The difficulty arises in finding a set of identifying attributes for the generalized type, because the subtypes may all have different identifiers. Consideration of this leads to the concept of *categorization*, where several heterogeneous entities (with their own attributes and relationships) combine into a common *category*. This is discussed in Elmasri and Navathe (1994). Categorization actually adds considerable modelling power, because it provides the ability to represent the inner complexity of the entities. Entities need no longer be treated as atomic and indecomposable. This construction is sometimes treated as an enhancement of the E-R model, but it takes us so far from our original conception of entities that we prefer to consider this extension as a feature of the object model, which we will shortly introduce.

*E-R for spatial information*

The E-R model or its extension may be used to model configurations of spatial entities. The structure that we use as an example is a simplification of the model underlying most vector-based GIS (see the NAA representation in Chapter 5). Figure 2.17 shows a finite region of the plane partitioned into sub-regions. Each sub-region, which we shall call an *area*, is bounded by a collection of *directed arcs*, which have *nodes* at their ends. Figure 2.18 shows an E-R model for these types of configurations. The entities, relationships and subtype relationships in the model are:

*Entities and attributes:*
        **area:**                 **area_id**
        **right_area**
        **left_area**
        **directed_arc:**      **arc_id**
        **node:**               **node_id**
        **begin_node**
        **end_node**

*Relationships:*
        **left_bounded**
        **right_bounded**
        **begins**
        **ends**

*Subtypes:*
        **left_area, right_area** are subtypes of **area**
        **begin_node, end_node** are subtypes of **node**

In order to relate a node to a directed arc, we specify the nodes that begin and end the directed arc. To relate an area to a directed arc, we specify the areas to the left and right of it. Subtypes of **area** and **node** are created so that these relationships can be easily defined. Note that both specializations are overlapping: a node may begin one directed arc and end another; an area may be the left-area of one directed arc and the right-area of another. Also the mandatory nature of the participation of nodes and areas in the specialization relationships indicates that each node must either begin or end at least one arc (this disallows isolated nodes) and each area must be to the left or right of at least one arc (disallows areas with no bounding arc). All relationships are many-to-one between **directed arc** and the related entities. Thus, a directed arc must have exactly one area on its left and one area on its right:

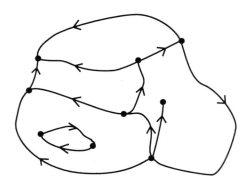

**Figure 2.17**    Planar configuration of nodes, directed arcs and areas.

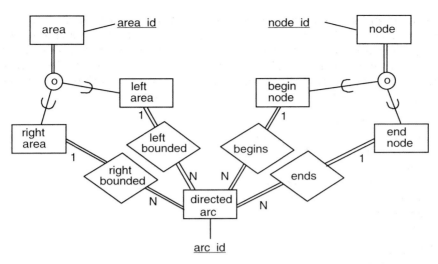

**Figure 2.18**   EER model of planar configurations of nodes, directed arcs and areas.

it must have exactly one begin and one end node. These constraints imply, for example, that a directed arc cannot exist without begin and end nodes and that there must be an exterior area defined for the finite planar partition. Further details of these kinds of spatial data models are discussed in Chapter 5.

### 2.3.2   Relational database design

Database design, in the case of relational databases, concerns the construction of a relational database scheme. This section will consider some of the principles upon which the design is based and show how an E-R model can be transformed into a relational database scheme. The central question is 'What characterizes a good set of relations for the applications required?'. Two advantageous features are:

- Lack of redundant data in relations that wastes space in the database and causes integrity problems.
- Fast access to data from relations.

These two features essentially trade-off space (and integrity) against time. Fast access to data results from few joins (the relational join is the most expensive of relational database operations), which implies few relations. At one extreme, it is possible to have many small relations that must be joined on retrievals that link them together. On the other hand, we could have just a few relations and a minimal number of joins. The problem with over-large relation schemes is shown by the following example from the CINEMA database.

Suppose that we decide to have all the information about stars and the films in which they are cast in a single relation in the database. The scheme would look something like:

STAR_FILM (FILM_TITLE, DIRECTOR, FILM_CNTRY, RELEASE_YEAR, LENGTH, STAR_NAME, BIRTH_YEAR, STAR_GENDER, STAR_NTY, ROLE)

**Table 2.7** Some rows and columns of the STAR_FILM relation

| FILM_TITLE | DIRECTOR | | | STAR_NAME | BIRTH_YEAR | | | ROLE |
|---|---|---|---|---|---|---|---|---|
| All the President's Men | **Pakula** | | | Dustin Hoffman | 1937 | | | Carl Bernstein |
| Marathon Man | Schlessinger | | | Dustin Hoffman | 1937 | | | Babe Levy |
| Kramer v Kramer | Benton | | | Dustin Hoffman | 1937 | | | Ted Kramer |
| All the President's Men | **Pakula** | | | Robert Redford | 1937 | | | Bob Woodward |
| A Bridge Too Far | Attenborough | | | Robert Redford | 1937 | | | |
| The Great Gatsby | Clayton | | | Robert Redford | 1937 | | | Gatsby |

This is a relation scheme of large degree, but if there were lots of retrievals requiring star and film details together, then this solution might suggest itself. The problem is that this scheme results in redundant duplication of data. Table 2.7 shows an example of the problem, with just a few attributes of the relation given. The items shown in bold involve redundant multiplicity of data. Thus, Pakula is shown as the director of *All the President's Men* in two tuples, while Dustin Hoffman's and Robert Redford's birth years are both given in triplicate. Such redundancy of information causes two problems for the database:

- Redundant repetition of data implies wasted space.

- Loss of integrity can result from certain transactions. For example, if Dustin Hoffman's year of birth is changed in one cell but not in the other two, database inconsistency results.

The solution to the problem of unnecessary repetition of data lies in the concept of *normalization*, which concerns appropriate structuring of relation schemes in a database. The specific problems in the STAR_FILM relation above can be solved by splitting the scheme, so that the film data are held in one relation and the star data are held in another. There is also a need for a third relation to hold the castings of stars in films, that is to make the connections between the two tables of information. The modified relation schemes are:

FILM (<u>TITLE</u>, DIRECTOR, CNTRY, YEAR, LENGTH)

STAR (<u>NAME</u>, BIRTH_YEAR, GENDER, NTY)

ROLE (<u>FILM STAR</u>, <u>FILM TITLE</u>, ROLE)

and the relations with the tuples from Table 2.7 appropriately distributed are shown in Table 2.8(a), (b) and (c). There is now no redundant repetition of data.

Apart from avoiding redundant duplication of information, decomposition of relation schemes has the advantage that smaller relations are conceptually more manageable and allow separate components of information to be stored in separate relations. This can also be helpful in a distributed database. Of course, relations cannot be split arbitrarily. Relations form connections between data in the database and inappropriate decomposition can destroy these connections. Database designers have constructed a hierarchy of *normal forms* for relational databases, where higher normal forms require a higher level of decomposition of the constituent database relations. Normal forms are guidelines for database design. In any logical database design, the level of normal form (i.e. degree of relation decomposition) must be balanced against the decrease in performance resulting from the need to reconstruct relationships by join operations.

An E-R model of a system may be transformed into a set of relation schemes that are a good first pass at a suitable database design. The general principle, with modifications in some cases, is that each entity and relationship in the E-R model results in a relation in the database scheme. An example will show the concepts involved in the transformation from E-R model to logical model. Consider the E-R model of the CINEMA database given in Figure 2.15. The independent entities and their attributes provide the following partial set of relation schemes:

CINEMA (<u>CIN ID</u>, NAME, MANAGER, TEL_NO, TOWN, GRID_REF)

**Table 2.8**   The STAR_FILM relation split into the three relations FILM, STAR, and CAST

(a): FILM

| TITLE | DIRECTOR | CNTRY | YEAR | LENGTH |
|-------|----------|-------|------|--------|
| A Bridge Too Far | Attenborough | UK | 1977 | 175 |
| Kramer v Kramer | Benton | USA | 1979 | 105 |
| The Great Gatsby | Clayton | UK | 1974 | 140 |
| All the President's Men | Pakula | USA | 1976 | 138 |
| Marathon Man | Schlessinger | USA | 1976 | 126 |

(b) STAR

| NAME | BIRTH_YEAR | GENDER | NTY |
|------|-----------|--------|-----|
| Dustin Hoffman | 1937 | M | USA |
| Robert Redford | 1937 | M | USA |

(c) CAST

| FILM_STAR | FILM_TITLE | ROLE |
|-----------|-----------|------|
| Dustin Hoffman | All the President's Men | Carl Bernstein |
| Dustin Hoffman | Marathon Man | Babe Levy |
| Dustin Hoffman | Kramer v Kramer | Ted Kramer |
| Robert Redford | All the President's Men | Bob Woodward |
| Robert Redford | A Bridge Too Far | |
| Robert Redford | The Great Gatsby | Gatsby |

FILM (<u>TITLE</u>, DIRECTOR, CNTRY, YEAR, LENGTH)

STAR (<u>NAME</u>, BIRTH_YEAR, GENDER, NTY)

Special care is required for the dependent entity Screen. We must add the identifier CINEMA_ID from Cinema to properly identify each tuple in the relation. This relation scheme is then:

SCREEN (<u>CINEMA ID</u>, <u>SCREEN NO</u>, CAPACITY)

It only remains to consider the relationships in the E-R model. Relationships result in relations in two different ways. In our example, each relationship results in a further relation. The relationship relation takes for its identifier the identifiers of the participating entities. Thus, the relationship **cast** leads to the relation scheme CAST with identifier (FILM_STAR, FILM_TITLE), where FILM_STAR is a renaming of NAME from relation STAR and FILM_TITLE renames TITLE from relation film. (The renaming is just to make the relation schemes more readable and less ambiguous.) The relation CAST also picks up the attribute ROLE from the **cast** relationship.

CAST (<u>FILM STAR</u>, <u>FILM TITLE</u>, ROLE)

The other relationship leads in a similar manner to the relation:

SHOW (<u>CINEMA ID</u>, <u>SCREEN NO</u>, <u>FILM NAME</u>, STANDARD, LUXURY)

The full database scheme for the CINEMA database is shown in section 2.2.2. It is given with a sample set of relations in Appendix 1.

It may not be necessary to provide a new relation for each relationship in the E-R model. For example, Figure 2.19 shows an E-R diagram of the vector spatial data model, given in EER form earlier in Figure 2.18. If we adopt the principle that each entity and relationship defines a relation, we have the following database scheme:

AREA (AREA_ID)

NODE (NODE_ID)

ARC (ARC_ID)

LEFT_BOUNDS (ARC_ID, AREA_ID)

RIGHT_BOUNDS (ARC_ID, AREA_ID)

BEGINS (ARC_ID, NODE_ID)

ENDS (ARC_ID, NODE_ID)

However, this database scheme can be simplified with no loss of connectivity (in the database structure sense). The **begins** relationship between **directed arc** and **node** has the property that each arc has exactly one begin node. Thus, rather than have the extra relation BEGINS in the database, it is simpler to add the identifier NODE_ID, in its capacity as a begin node, to the ARC relation. This construction is called *posting the foreign key*. In technical language we have posted the identifier of NODE as a foreign key into ARC. In fact, because of the nature of the four relationships, each leads to a posting of a foreign key into ARC. The resulting simplified database scheme is:

AREA (<u>AREA ID</u>)

NODE (<u>NODE ID</u>)

ARC (<u>ARC ID</u>, BEGIN_NODE, END_NODE, LEFT_AREA, RIGHT_AREA)

Posting a foreign key, rather than adding another relation, is always an option if the relationship between entities is as shown in Figure 2.20.

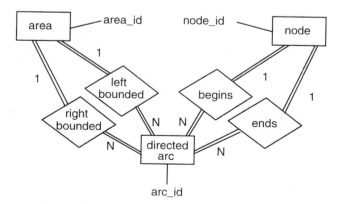

**Figure 2.19**　E-R model of planar configurations of nodes, directed arcs and areas.

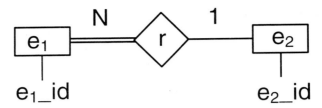

**Figure 2.20** A relationship where posting the foreign key is an option.

As the relationship $r$ is many-to-one and mandatory on the side of $e_1$, it is the case that each occurrence of $e_1$ relates to exactly one occurrence of entity $e_2$. Therefore, if we allow the posting of the foreign key of $e_2$ into $e_1$ to make the connection, then for each tuple in the relation for $e_1$, there will be one and only one value for the foreign key attributes, and therefore, there will be no possibilities of either null or repeating values.

### 2.3.3  Summary

Relational databases have a simple and formally convincing underlying model and are able to implement many aspects of the database philosophy discussed in the first part of the chapter. The relational model with associated interaction language SQL is currently by far the global leader among database paradigms. However, there are problems with more complex forms of data, which relational technology in its current standard form finds difficult to handle. Database construction is an art supported by many tools and methodologies. We have spent some time describing the data modelling approach provided by the entity-relationship model. This modelling approach has the advantage of simplicity and a simple transformation to an appropriate database scheme. However, there are other approaches, referenced in the bibliographic notes at the end of the chapter.

### 2.4  THE OBJECT-ORIENTED APPROACH TO SYSTEMS

This section introduces the object-oriented approach as an alternative to the relational model as a paradigm for information systems. The object-oriented approach to systems may be applied to any computer-based system at several levels. Thus, for example, there are object-oriented programming languages (OOPLA); object-oriented analysis and design methodologies (OODM) that further extend E-R and EER models; and object-oriented databases and database management systems (OODBMS). Many of the basic ideas of object-orientation were introduced over 20 years ago in the Simula programming language. Since then, many other languages with object-oriented features have been developed and object-oriented software development is today dominated by the use of the C++ programming language. As GIS has a basis in information systems technology, object-oriented ideas are introduced in this chapter and in the context of databases. Later sections of the book will contain some specific examples of the application of the object-oriented (OO) approach to GIS.

The staged representation of the information system construction process, from the model of the application domain, through analysis, design and implementation to the final system (see section 1.3.8) leaves us with an important difficulty that motivates the object-oriented approach at each of the stages. Computer science has developed the concept of *impedance mismatch* to explain certain problems with system development. The gap between the constructs that are available at different stages of the development process makes the transition from one stage to the next inefficient: information may be lost or a simple concept may get hidden in a complex modelling paradigm. An important instance of this problem occurs if an application domain model, expressed using high-level and domain-specific cons-tructs, must be translated into a low-level computational model in order that it can be implemented. Thus, the heartfelt cry of many users of information systems that computers force them to make square holes for their round and fuzzy-edged pegs. However, it is not the computers that force users into unnatural modes of expres-sion, but the primitive computational models in current use. In a nutshell, object-orientation is one of several attempts to raise the level of the computational modelling environment so that the impedance mismatch problem is lessened.

### 2.4.1  Foundations of the object-oriented approach

As might be expected, the concept of *object* is central to the object-oriented approach. It arises out of a desire to treat not just the static data-oriented aspect of information, as with the relational model, but also the dynamic behaviour of the system. As in the E-R model, the static aspect of an object is expressed by a collec-tion of named *attributes*, each of which may take a value from a pre-specified domain. A city object might have **name, centre, population** among its attributes. A particular city might take the value 'London' for the **name** attribute. The totality of attribute values for a given object at any time constitutes its *state.*

The dynamic, behavioural side of an object is expressed as a set of operations that the object will perform under appropriate conditions. For example, the idea of a region is captured not just by specifying a set of points or curves giving the boundary of its extent (the data), but also the operations that we can expect a region to support. Such operations might include the calculation of the region's area and perimeter, plotting the region at different scales and levels of generalization, the creation and deletion of regions from the system, operations that return the lineage of the region (when it was created, to what accuracy, etc.). For the object-oriented approach, the key notion is that

*object = state + functionality*

This is the minimum, without which we do not have the object-oriented approach.

As is stated in the ANSI Object Oriented Database Task Group Final Technical Report (1991), an object is something

> which plays a role with respect to a request for an operation. The request invokes the operation that defines some service to be performed.

A request is a communication from one object to another and, at the system level, it is implemented as a message between objects (Figure 2.21). Exactly how the object will respond to a message in any particular case is dependent on its state. For

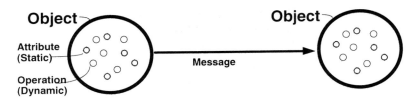

**Figure 2.21**    Objects encapsulate state and behaviour and communicate via messages.

example, a message to print a region may result in a picture that depends on the scale and the boundary data values.

An object is thus characterized by its *behaviour*, which is the totality of its responses to messages. Objects with similar behaviours are organized into *types*. Thus, we might have the type **region**, as above. At the design and implementation stages of the system life-cycle, such semantic notions must be represented by specific computational objects. Each such object will have data structures holding its attribute values and methods that implement specific operations. At this level, we often talk about *object classes*, which are groupings of objects with corresponding *data structures* and *methods*. Thus, an object class is an implementation construct while an object type is a semantic notion (see Table 2.9).

As has already been noted, different object-oriented models emphasize different constructs. The list below itemizes some of the more common ones. We have tried to distinguish between the roles of the construct at different stages of the modelling process.

*Object identity*

Object identity is primarily a system construct, although it does have resonances at all stages in the modelling process. At the application domain level, the issue is whether objects have unique identities, independent of their attribute values. This corresponds to a natural idea, which can be simply expressed by means of an example.

> My name is Mike, but I am not my name. My name may change, indeed every cell in my body may change, but my identity remains the same.

Object identity may be implemented at the system levels by insisting that each system object has a unique identifier, created when the object is created, never altered and only dropped when the object is destroyed. This identifier is additional to other attributes of the object. Almost all self-respecting object-oriented systems support object identity.

When modelling entities using the E-R model, it was stipulated that each entity should be capable of unique identification. Attributes which identified entities were therefore always present, and included in the list of attributes for an entity. Each occurrence of that entity was pinpointed uniquely by giving the value(s) for the identifying attributes. This is sometimes summed up by saying that the E-R model and the relational model are *value-based*. However, the object model is not value-based. An object occurrence retains its identity even if all its attributes change their values. One consequence of object identity is that each object has an internal identity, independent of any of its attribute values (and if the model is implemented, this

**Table 2.9** Terms for object specification and implementation

| Specification | Implementation |
| --- | --- |
| Object type | Object class |
| Operation | Method |
| Relationship | Message |

identify will be stored in the computer). Thus, it is unnecessary to provide instance variables solely for the purpose of identifying an object.

### Encapsulation

Encapsulation is a system construct (but also a programming philosophy). One of the primary original motivations behind object-oriented system design is *reuse*. The idea is that systems are more quickly developed if they can be constructed by bolting together a collection of well-understood and specified sub-components. The key to successful reuse is that sub-components should behave in a predictable way in any setting, and this can only be achieved if their internals are insulated from the external environment. This is the principle of *encapsulation*, where an object's state and the methods that it uses to implement its operations are not directly visible externally, but only by means of a clearly defined protocol prescribed in the definition of the object.

With encapsulation, the world external to an object is shielded from the internal nature of that object and only communicates with it via a set of predefined message types that the object can understand and handle. To take an example from the real world, I usually do not care about the state of my car under the bonnet (internal state of object type 'car') provided that when I put my foot on the accelerator (send message) the car's speed increases (change in the internal state leading to a change in the observable properties of the object). From the viewpoint external to the object, it is only its *observable* properties that are usually of interest. These properties are observed using the *object interface*. Each object in a single object class has the same object interface.

Encapsulation has important implications for object-oriented databases. Relational databases are based on the *call-by-value* principle, where tuples are accessed and connected to other tuples by means of their values. However, values contained within an object (part of the object's state) are encapsulated and only accessible indirectly by calling the appropriate methods. This level of indirection can make the performance of an OODBMS unacceptable if not handled properly: we shall return to this issue later.

### Inheritance

Inheritance is an important system and real-world construct. Reuse is also a motivating factor here. In the systems context, inheritance involves the creation of a new object (or object type) by modifying an existing object (or type). In real-world modelling, as discussed in the EER model, inheritance acts in two ways, one the inverse of the other:

**Generalization:**   abstracting the properties of several object types. For example, **village**, **town** and **city** generalize to **settlement**.

**Specialization:**   partitioning object types according to specific roles. For example, type **traveller** specializes to **road_user** and **rail_user**.

Object-oriented models will differ in the support they provide for these constructs. Attribute inheritance is a feature of the EER model: however, inheritance in an object-oriented setting also allows inheritance of methods. Thus, **triangle** and **rectangle** are subtypes of **polygon**. The subtypes inherit all the attributes *and operations* from the supertype as well as adding their own. In our example, **triangle** and **rectangle** may have specialized operations, **equilateral?** and **square?**, returning Boolean objects, that determine whether the shapes are equilateral or square, respectively. They may also have their own specialized methods for implementing operations (see operation polymorphism, below), for example the algorithm implementing the operation **area** may be different for **rectangle** and **triangle**, and possibly different again from the method for a general polygon. The entire inheritance structure is termed an *object type inheritance hierarchy* and is shown in diagrammatic form in Figure 2.22.

With *single inheritance*, a specialized object type inherits all the state and behaviour of a single generic type, possibly adding some state and behaviour of its own. For example, the type **displayed polyline** is a specialization of type **polyline**, having additional methods in order that the polyline may be displayed on a map sheet or on the screen. The notion of single inheritance may be extended to that of *multiple inheritance*, where a type is permitted to inherit state and behaviour from more than one supertype. Multiple inheritance requires more complex system support. Methods for an operation in a subtype may be inherited from more than one of its supertypes, and a decision has to be made about which method to use. Another kind of conflict arises when there are name clashes amongst inherited attributes or operations. There has to be some kind of protocol for resolving such conflicts. As we are more concerned with the object-oriented approach as a modelling tool rather than an implementation framework, we shall not discuss precise rules for such conflict resolution.

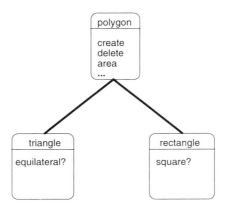

**Figure 2.22**   Object diagram showing inheritance of operations.

## Composition

Object composition is an important system and real-world construct, allowing the modelling of objects with complex internal structure. There are several ways in which a collection of objects might be composed into a new object.

**Aggregation**: An aggregate object is one that is made up of component objects. For example, an object of type **property** might be composed of objects of type **land parcel** and **dwelling**. To quote Rumbaugh *et al.* (1991), 'an aggregate object is semantically an extended object that is treated as a unit in many operations, although physically is made up of several lesser objects'. Sometimes, an extra constraint is imposed that the existence of the aggregate object is dependent upon the existence of its sub-objects, in that it cannot exist without them.

**Association**: An associated or grouped object is one that is formed from a set of other objects, *all of the same type*. For example, an object of type **districts** might be an association of individual **district** objects.

**Ordered Association**: When the ordering of component objects within the object set is important, it is more appropriate to use ordered association. For example **point_sequence** might be an object structured as a linear ordering of **point** objects.

## Polymorphism

Operation polymorphism is primarily a system construct, allowing the same operation to be implemented in different ways (by different methods) in different classes. For example, the operation **perimeter** may have different implementations for classes **square**, **triangle** and **circle**. Polymorphism becomes powerful in combination with inheritance. It provides flexibility for execution of processes in information systems, because operations need only be bound to implementations at run-time.

### 2.4.2 Object-oriented system modelling

The modelling tools associated with relational technology are sufficient for analysis of the kind of information structures that exist in many traditional applications. However, as the technology is applied to wider classes of applications, shortcomings in the E-R model approach become apparent. This is particularly true for applications where the entities themselves have a rich substructure or organization (for example, hierarchies). This is usually the case in spatial applications, where spatial types are themselves composite.

One of the great advantages of object modelling over entity modelling is that the objects in the modelling process can be the natural objects that we observe in the real world. Thus, for example, a partition of a finite region of the plane into areas can be modelled in a natural way as an aggregation of areas, arcs and nodes. These object types can then form the basis (as supertypes) for more application-dependent objects, such as land parcels, districts or roads. The E-R modelling approach is designed so that a system model may be easily translated into relational terms, but this forces some limitations into the modelling process. Another ingredient that object-oriented data models provide over and above the EER model is the modelling of behaviour in addition to the structural properties of information. A range of object-oriented modelling methodologies is being developed. Among the currently most popular are those by Booch (1994) and Rumbaugh *et al.* (1991). The notation and terminology below is general and not specific to a particular methodology.

An object type is declared by being given a name and an optional list of parameters. Then optionally follow details of inheritance, state, behaviour, and integrity constraints. The state of an object type consists of a collection of attributes and describes the static aspects or properties of the object type. It resembles the collection of attributes of an entity in the E-R model. However, *each occurrence of an attribute is itself an object*. The behaviour consists of a collection of operations that can be performed on objects in the object type. Operations define the behaviour of the objects in the type. Each operation will be implemented as a method, which acts on members of implemented object classes and returns a member of an object class. However, the details of the implementation of each operation are not the concern of the conceptual data model. Some general groups of operations are:

- *Constructors-destructors:*   Operations that add or delete objects from the type. For example, for type **road**, such an operation would insert new roads or delete old roads in the system as the road network changes.

- *Accessors:*   Operations that return objects whose properties are dependent upon the properties of the particular objects accessed. For example, the total length of an object in type **road** might be given by the accessor operation **length**.

- *Transformers.*   Operations that change objects. For example, for type **road**, a new survey might result in a reclassification of a road. This may be communicated to the appropriate object by means of the operation **update classification.**

As an example, we might begin to model the object type **road** as follows.

```
type road
   state
        name : string
        classification : string
        year built : date
        centre line : arc
        . . .
   behaviour
        create :  → road
        delete : road →
        length : road → real
        update classification : road → road
        display : road, scale →
        . . .
   end road
```

In this example, we have defined a type of object called **road**. It has attribute types **name, classification, year built, centre line**, amongst others. Each occurrence of each attribute type belongs itself to an object class, for example, object instance 'A1' is a road name, which belongs to the object class **string**. Type **road** has associated with it operations: for example, the operation **length** returns an object of type **real** when given an instance of class **road**. This integer value will give the length of a particular road, as calculated from its attributes using a method supplied for the object type. Operation **length** is an accessor operation. Operations **create** and **delete** are constructor-destructor operations. Operation **update classification** is a transformer operation. Operation **display** is interesting because it takes more than parameter: in this highly simplified example, we assume that a road can be displayed on the screen in a certain way depending on the attributes of the road and the scale of the display.

The modelling could continue by defining a subtype of **road** as **motorway**, with its own specific attributes and operations (as well as attributes and operations inherited from **road**). We assume some extra attributes, such as the collection of its service stations as a set of buildings. It may be that some of the methods will be different between **motorway** and **road**. For example, a motorway may be displayed differently upon a screen. The new type declaration may be written as:

```
type motorway
  inherits
    road
  state
    service stations: set (building)
    ...
end motorway
```

*Object modelling diagrams*

Entity-modelling methods have the ER diagram as the main modelling notation, but object-oriented approaches have yet to arrive at standardized modelling tools. Object-modelling diagrams must take account not only of attributes and relationships between objects, but also object behaviour under operations. As an earlier simple example showed (Figure 2.22), it is possible to extend inheritance diagrams, as in EER modelling, to diagrams that show the inheritance of attributes *and operations*.

Figure 2.23 shows a small example of an *Object Interconnection Network Diagram*. Such a diagram shows the object types and some of their internal

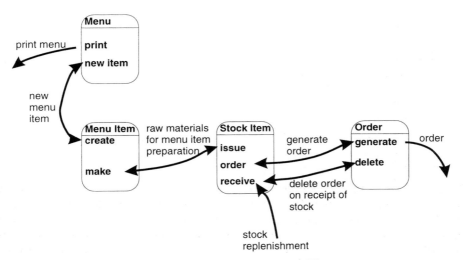

**Figure 2.23** An example of an Object Interconnection Network Diagram.

structure, but its primary purpose is to show connectivity between operations upon the object types. The example shows a small part of the vegetarian restaurant object-oriented system. Four object types are shown: **menu, menu item, stock item** and **order**. The menu may be printed, using the **print** operation. Adding new items to the menu by the operation **new item** connects with the **create** operation in **menu item**. Operation **make** prepares a menu item for consumption using raw materials generated by **issue** on **stock item**. Based upon the reorder level, operation **order** on **stock item** connects with operation **generate** on type **order** to raise an order for stock. Operation **receive** on **stock** takes in stock replenishment from suppliers and connects with operation **delete** on **order** to delete and archive the order.

Other object-oriented diagrams, not shown here, are the *Object Internal View*, showing the internal structures of objects and more of relevance to the system developer; the *Object Relationship Diagram (ORD)*, an extension of the ER-diagram discussed earlier; and the *Object Dependency Diagram (ODD)*, showing how objects are organized within an application. The Object Dependency Diagram shows the client-server relationship between objects.

### 2.4.3   Object-oriented DBMS

For GIS, it is especially pertinent to examine the role of object-orientation in information systems. Products labelled object-oriented databases have been available for purchase for some years. An object-oriented database management system (OODBMS), in addition to supporting the object constructions described above, must also provide a safe environment for the management of persistence for the object type inheritance hierarchy inhabiting it. Ideally, all the features available in a modern database should be provided by an OODBMS. These include:

- *scheme management*, including the ability to create and change class schemes;
- providing a usable *query environment*, including automatic query optimization and a declarative query language;
- *storage and access management*;
- *transaction management*, including control of concurrent access, data integrity and system security.

Unfortunately there are technological problems in matching some of these facilities with the system object model. For example, the provision of query optimization (and therefore the possibility of a non-procedural query language) is made difficult by the complexity of the object types in the system. No longer do we have a small number of global and well-defined operations, such as the relational operations of project, restrict and join, for which it is possible to estimate the cost of execution and so choose between different strategies for executing a query. In an object-oriented system, there may be a multitude of methods implementing operations, each without any measure of implementation cost. Another problem is the seeming incompatibility of indexing with the notion of encapsulation and object-identity. Here, the difficulty is that indexes rely on direct access to attribute values, but an object is only accessible via messages through its protocol and identified by its object-id.

Transactions in an object-oriented database may be of a much higher level of complexity than simple transactions with a relational database. For example, updating a portion of a map following a new survey may take hours or days. Also, due to the hierarchical nature of much object data, transactions may cascade downwards and affect many other objects. OODBMS usually have support for *long* and *nested* transactions. One proposed solution to the problems arising from long transactions is *version management*, which handles the currency of several maybe inconsistent versions of the system. *Scheme evolution*, the updating of the database scheme, is more complex with an object-oriented database, because usually an object-scheme is itself more complex than a relational database scheme. Performance has always been the sticking point for OODBMS. Many techniques, such as pointer swizzling and caching, have been developed and incorporated.

If performance problems can be solved, an OODBMS has great advantages. Some of these are technical and involve reductions in joins and possible speed-up using object-identifiers. However, for the user the advantage is that the conceptual model of the domain may be closer to the system model, thus minimizing *impedance mismatch*. To implement a rich conceptual model, using a limited system model is like playing a Beethoven piano sonata on the trumpet: the instrument does not match the conception and much of the intended meaning will be lost in the execution. In traditional systems, support for the conceptual model is low-level, close to the system model and far from the user's idea of the application domain. The object-oriented approach allows the much more expressive system model to be closer to the user's view of the application domain.

On a down-to-earth note, it must be said that even today, object-oriented databases are still rarely used for realistically large-scale applications. Many of the solutions found by research and development have yet to find their way to the market-place. However, there is enormous potential.

### 2.4.4   Objects are everywhere

The word 'object' is very common is computer-speak. Sometimes it is just used to sell a new system: even in less superficial cases it can have different meanings. This last section on the object-oriented approach briefly reviews some of these meanings.

- **Object-oriented programs**   allow the system construction process to be carried through using the object-oriented approach from analysis to implementation. The most common OOPLA is C + +. There is no guarantee for support of the persistence of the objects created.

- **Object-oriented user interfaces,**   such as the Macintosh graphics user interface (GUI), now form the basis of most systems. They are referred to as object-oriented mainly because they can deal with high-level representations and because, in some cases, they encapsulate methods with objects.

- **Object-oriented databases**   provide support for persistent objects as discussed above.

Object-oriented approaches have been promoted especially in areas for which the application of pure relational technology has been found to be problematic. Examples of such areas are computer-aided design (CAD), office information systems (OIS), software engineering support (CASE) and GIS. Each area is characterized by one or more non-classical features, such as structurally complex information, specialized graphical requirements, non-standard transactions or version control requirements. There is a small but growing number of proprietary GIS having object-oriented features: those that do exist have been highly successful for specialized markets and it is likely that more proprietary object-oriented GIS will follow.

## 2.5  NOTE ON OTHER PARADIGMS

Besides relational and object-oriented databases, there are other paradigms that offer some advantages and are actively being researched and developed. One of these is the deductive database. A *deductive database* provides persistence for computation using logic-based languages such as Prolog. The relational model can be viewed as allowing the expression of a set of facts, for example, 'Spielberg is the director of Jaws', 'Lucas is the director of Star Wars'. Deductive databases allow the storage of not just facts but more general propositions, such as 'Screen 1 in London and Screen 1 in Newcastle will always show the same films'. Storage of general propositions may avoid the storage of much data. Deductive databases also support deduction with their data. In order to add database functionality to a logic programming environment, there is a need for at least the following:

- Support for container data types, such as sets.
- Indexes that will handle data in non-relational form (e.g. retrieve quickly all the facts and rules with a particular characteristic).
- Support for recursive rules, and therefore for recursion in the query language.

Chapter 8 has a section on knowledge-based systems and their application to GIS.

### BIBLIOGRAPHIC NOTES

Overview books on general database technology are plentiful. Particularly recommended are Date (1995), the classic text, recently updated and very good on SQL; Elmasri and Navathe (1994), wide-ranging and full of information on relational, functional and object-oriented databases; and Howe (1989), detailed on entity-relationship analysis and relational database design. Ozsu and Valduriez (1991) and Bell and Grimson (1992) specialize in distributed databases. Object-oriented databases (OODBMS) are discussed in Khoshafian (1993) and Kim (1995). Those readers who wish to examine the deductive database paradigm will find material in the text by Ceri *et al.* (1990).

   The relational model for databases was constructed by Codd (1970). Codd worked at IBM San Jose and his colleagues Chamberlin and Boyce (1974) introduced the first version of SQL, then known as SEQUEL. Codd's original 1970 paper is quite readable and gives a good account of the first stages in the construction of the relational model. Codd's later work (1979) extended the relational model to incorporate null values and extra semantics. Readers interested in hierarchical and network databases can find information in the general database texts already referenced. Material on extensions to the relational model may be found in Rowe and Stonebraker (1987) for Postgres and later variants, and in Lohman *et al.* (1991) for IBM Starburst. Abiteboul *et al.* (1995) provides an advanced and theoretical database text, mainly focused on the relational model.

The entity-relationship model was introduced by Chen (1976). Constructs that enhanced Chen's original E-R model were developed by several authors: for example, generalization and aggregation (Smith and Smith, 1977) and the lattice of subtypes (Hammer and McCleod, 1981). An interesting and extensive discussion of the enhanced E-R model is given by Teorey, Yang and Fry (1986). The best general modern treatment of these ideas is the already recommended general database text of Elmasri and Navathe (1994). Other modelling approaches are described in Brodie *et al.* (1984) and in Nijssen (1976). A survey of modelling emphasizing the provision of constructs that allow the capture of more semantics from application domains is Hull and King (1987). Fernández and Rusinkiewicz (1993) apply the EER modelling approach to the development of a soil database for a geographical information system.

A general introduction to object-orientation is given in the textbook of Khoshafian and Abnous (1990). A more technical survey, well worth reading, is (Mattos *et al.* 1993). The object-oriented approach to systems analysis and design is fully treated in texts such as Booch (1994) and Rumbaugh *et al.* (1991). A general text written from the software engineering perspective is McGregor and Sykes (1992). Bekker (1992) considers most of the major data modelling paradigms. Among the increasing number of texts that treat object-oriented databases are Kim and Lochovsky (1989), Gray *et al.* (1992), Khoshafian (1993) and Bertino and Martino (1993).

# CHAPTER THREE

# Fundamental spatial concepts

What! Will the line stretch out to the crack of doom?

Shakespeare, *Macbeth*

## 3.1 INTRODUCTION: WHAT IS SPACE?

The notion of space is a difficult one. Gatrell (1991) defines it as 'a relation defined on a set of objects', which includes just about any structured collection. We all have an intuitive idea about the concrete space in which our bodies move. It is this idea that we wish to capture and extend, while avoiding the complete generalization implied by the definition above. A term used by many is 'physical space', perceived through our senses. Vision allows us to perceive spatial properties such as size, direction, colour and texture. All the senses provide us with a multi-modal understanding of physical space (Freksa, 1991). As noted by Nunes (1991), 'perhaps the most widely accepted convention of space is that of a container or framework in which things exist'. Chrisman recognized early on in the development of GIS (Chrisman, 1975, 1978) that space may be conceptualized in two distinct ways, either as a set of locations with properties (absolute space, existent in itself) or as a set of objects with spatial properties (relative space, dependent upon other objects). This dichotomy turns out to have far reaching implications for spatial modelling, where the philosophy of absolute space is modelled as a set of fields and relative space as collections of spatially referenced objects. It reaches right down to the implementation level, where the absolute and relative views are implemented using raster and vector technologies, respectively. Couclelis (1992) has observed that this dichotomous view of space is part of a more general ontology that has manifested as plenum versus atom and wave versus particle.

A phrase that is often used by workers in GIS is 'geographic space'. Again, definitions of this term are problematic. Almost tautologous is the definition that geographic space is absolute or relative space referenced to geographic scales. In the atomistic view, geographic space is the collection of geographic entities, referenced to a part of the Earth and taking part in geographic processes. Alternatively, geographic space is a collection of terrestrial locations over which geographic phenomena may take place.

Geometries provide formalisms that represent abstract properties of structures within space. Modern treatments of geometry, originating with Klein's Erlangen programme of 1872, are founded on the notion of invariance. Thus, geometries are distinguished by the group of transformations of space under which their propositions remain true. To illustrate, consider a space of three dimensions and our usual notion of distance between two points, then the set of all transformations that preserve distances, that is for which distance is an *invariant*, form a geometry. Into this set would fall translations and rotations, because the distance between two points is the same before and after a translation or rotation is effected. However, scalings (enlargements) would not be members because they usually change distance. Looking at this the other way round provides us with a definition of a geometry as the study of the invariants of a set of transformations. Thus, the invariants of the set of translations, rotations and scalings include angle and parallelism, but not distance.

In this chapter we will examine some of the geometries that are proving the most useful for GIS. As coordinatized Euclidean geometry provides a view of space that is intuitive, at least in Western culture, this is the starting point. We discuss next the most primitive space of all, just collections of objects with no other structure, and proceed to build up to richer geometries. Because of the nature of the topic, the treatment will of necessity be sometimes abstract and formal, but examples are provided along the way. Some of the texts in the bibliography provide further background.

### 3.2   EUCLIDEAN SPACE

Underlying the modelling of many geo-spatial phenomena is an embedding in a coordinatized space that enables measurements of distances and bearings between points according to the usual formulas (given below). This section describes this coordinatized model of space, called *Euclidean* space, which transforms spatial properties into properties of tuples of real numbers. Assume for simplicity a two-dimensional model (*Euclidean plane*). Set up a coordinate frame consisting of a fixed, distinguished point (*origin*) and a pair of orthogonal lines (*axes*), intersecting in the origin.

#### 3.2.1   Point objects

A *point* in the plane of the axes has associated with it a unique pair of real numbers $(x, y)$ measuring its distance from the origin in the direction of each axis, respectively. The collection of all such points is the *Cartesian plane*, often written as $\mathscr{R}^2$. It is often useful to view Cartesian points $(x, y)$ as *vectors*, measured from the origin to the point $(x, y)$, having direction and magnitude and denoted by a directed line segment (see Figure 3.1). Thus, they may be added, subtracted and multiplied by scalars according to the rules:

$$(x_1, y_1) + (x_2, y_2) = (x_1 + x_2, y_1 + y_2)$$

$$(x_1, y_1) - (x_2, y_2) = (x_1 - x_2, y_1 - y_2)$$

$$k(x, y) = (kx, ky)$$

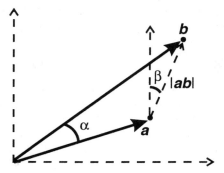

**Figure 3.1** Distance, angle and bearing between points (vectors) in the Euclidean plane.

Given a point vector, $x = (x, y)$, we may form its *norm*, defined as follows:

$$\|x\| = \sqrt{(x^2 + y^2)}$$

In a coordinatized system, measures of distance may be defined in a variety of ways (see section 3.5 on metric spaces). A *Euclidean plane* is a Cartesian plane with the particular measures of distance and angle given below. These measures form the foundation of most school geometry courses and refer to the 'as the crow flies' concept of distance. Given points (vectors) $a$, $b$ in $\mathcal{R}^2$, the distance from $a$ to $b$, $|ab|$ is given by (see Figure 3.1):

$$|ab| = \|a - b\|$$

Suppose that the points $a$, $b$ in $\mathcal{R}^2$ have coordinates $(a_1, a_2)$ and $(b_1, b_2)$, respectively. Then the distance $|ab|$ is precisely the Pythagorean distance familiar from school days, given by:

$$|ab| = \sqrt{((b_1 - a_1)^2 + (b_2 - a_2)^2)}$$

The *angle* $\alpha$ (see Figure 3.1) between vectors $a$ and $b$ is given as the solution of the trigonometric equation:

$$\cos \alpha = (a_1 b_1 = a_2 b_2)/(\|a\| \cdot \|b\|)$$

The *bearing* $\beta$ (see Figure 3.1) of $b$ from $a$ is given by the unique solution in the interval $[0,360[$ of the simultaneous trigonometric equations:

$$\sin \theta = (b_1 - a_1)/|ab|$$

$$\cos \theta = (b_2 - a_2)/|ab|$$

### 3.2.2 Line objects

Line objects are very common spatial components of a GIS, representing the spatial attributes of objects and their boundaries. The definitions of some commonly used terms related to straight lines follow.

*line*:   Given two distinct points (vectors) *a* and *b* in $\mathcal{R}^2$, the *line* incident with *a* and *b* is defined as the pointset $\{\lambda a + (1 - \lambda)b \,|\, \lambda \in \mathcal{R}\}$.

*line segment*:   Given two distinct points (vectors) *a* and *b* in $\mathcal{R}^2$, the *line segment* between *a* and *b* is defined as the pointset $\{\lambda a + (1 - \lambda)b \,|\, \lambda \in [0,1]\}$.

*half line*:   Given two distinct points (vectors) *a* and *b* in $\mathcal{R}^2$, the *half line* radiating from *b* and passing through *a* is defined as the pointset $\{\lambda a + (1 - \lambda)b \,|\, \lambda \geq 0\}$.

These definitions represent different types of lines in a parametrized form. The parameter $\lambda$ is constrained to vary over a given range (the range depending upon the object type). As $\lambda$ varies, so the set of points constituting the linear object is defined. Figure 3.2 shows some examples.

Not only straight lines are of interest for GIS. Straight lines may be specified by a single bivariate polynominal equation of degree one ($ax + by = k$). Higher degree bivariate polynomials specify further classes of one-dimensional objects. Thus, polynomials of degree two (quadratics of the form $ax^2 + bxy + cy^2 = k$) specify conic sections, which could be circles, ellipses, hyperbolas or parabolas. Cubic polynomials are used extensively in graphics to specify smooth curves (see Chapter 4).

### 3.2.3   Polygonal objects

A *polyline* in $\mathcal{R}^2$ is defined to be a finite set of line segments (called *edges*) such that each edge end-point is shared by exactly two edges, except possibly for two points, called the *extremes* of the polyline. If, further, no two edges intersect at any place other than possibly at their end-points, the polyline is called a *simple polyline*. A polyline is said to be *closed* if it has no extreme points. A (*simple*) *polygon* in $\mathcal{R}^2$ is defined to be in the area enclosed by a simple closed polyline. The polyline forms the *boundary* of the polygon. Each end-point of an edge of the polyline is called a *vertex* of the polygon. Some possibilities are shown in Figure 3.3. An extension of the definition would allow a general polygon to contain holes, islands within holes, etc.

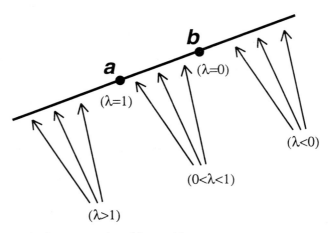

**Figure 3.2**  Parametrized representation of linear objects.

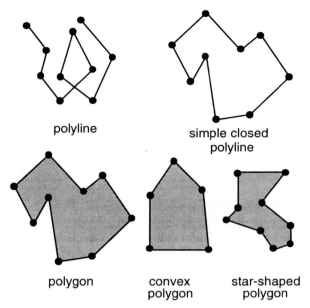

polyline

simple closed
polyline

polygon

convex
polygon

star-shaped
polygon

**Figure 3.3** Polylines and types of polygons.

Many different types of polygon have been defined in the computational geometry literature. A useful category is the set of *convex polygons*, each of which has an interior that is a convex set and thus has all its internal angles not greater than 180°. We note that the convex hull of a finite set of points is a convex polygon. For a convex polygon, every interior point is visible from every other interior point in the sense that the line of sight lies entirely within the polygon. A *star-shaped* polygon has the weaker property that there need exist only at least one point which is visible from every point of the polygon (see Figure 3.3). Convexity is discussed further in section 3.3.

The definition of *monotone* polygons depends upon the concept of *monotone chain*. Let chain $C = [p_1, p_2, \ldots, p_n]$ be an ordered list of $n$ points in the Euclidean plane. Then $C$ is *monotone* if and only if there is some line in the Euclidean plane such that the projection of the vertices onto the line preserves the ordering of the list. Figure 3.4 shows monotone and non-monotone chains. A polygon is a *monotone polygon* if its boundary may be split into two polylines, such that the chain of vertices of each polyline is a monotone chain. Clearly, every convex polygon is monotone. However, the converse of this statement is not true. It is not even true that every monotone polygon is star-shaped, as may be seen from the example in Figure 3.5.

Polygons may be triangulated, as may any region of the Cartesian plane. A *triangulation* of a polygon is a partition of the polygon into triangles that only intersect at their mutual boundaries. It is not too hard to show that a triangulation of a simple polygon with $n$ vertices that introduces an extra $m$ internal vertices (due to the triangulation) will result in exactly $n + 2m - 2$ triangles. If no internal vertices (sometimes called *Steiner points*) are introduced, the triangulation is called a *diagonal triangulation* and results in $n - 2$ triangles. We shall show that this simpler

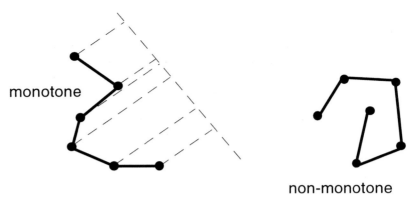

**Figure 3.4**   Monotone and non-monotone chains.

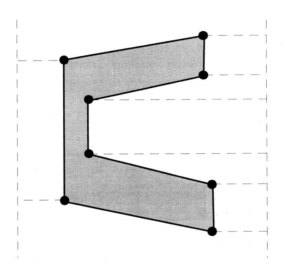

**Figure 3.5**   A monotone but not a star-shaped polygon.

formula is true by using an inductive argument on $n$, the number of vertices. The more general result is left as an exercise for the reader. For $n = 3$, there is but a single triangle and the result is clearly true. For the inductive step, take a general polygon $P$ with an arbitrary number $k > 3$ vertices. Suppose that the result holds for all polygons with less than $k$ vertices. Split $P$ down one of its triangulation edges into two smaller polygons $P'$ and $P''$ (this can be done since $k > 3$). Suppose that $P'$ and $P''$ have $k'$ and $k''$ vertices, respectively. Then:

1. $k' + k'' - 2 = k$ (as two vertices will be counted twice in adding up the vertices of $P'$ and $P''$).

2. The number of triangles in $P'$ is $k' - 2$ (as we are assuming the result holds for $P'$).

3. Number of triangles in $P''$ is $k'' - 2$ (as we are assuming the result holds for $P''$).

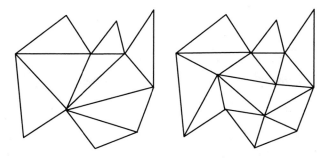

**Figure 3.6** A diagonal and non-diagonal triangulation of a polygon.

Adding (2) and (3), the number of triangles in $P$ is $k' - 2 + k'' - 2$, which simplifies to $k' + k'' - 2 - 2$, and this is $k - 2$ (from (1)). This is what was required to show.

Figure 3.6 shows a polygon with 11 vertices. As predicted by the formula, the left-hand triangulation results in 9 triangles and the right-hand triangulation that introduces four internal points results in 15 triangles.

### 3.2.4 Transformations of the Euclidean plane

*Translations, reflections and rotations*

This section describes some common transformations of the Euclidean plane. A transformation of $\mathscr{R}^2$ is a function from $\mathscr{R}^2$ to itself. Thus, every point of the plane is transformed into another (maybe the same) point. Some transformations preserve particular properties of embedded objects:

*Euclidean transformations* (*congruences*): preserve the shape and size of embedded objects. An example of a Euclidean transformation is a translation.

*Similarity transformations*: preserve the shape but not necessarily the size of embedded objects. An example of a similarity transformation is a scaling: all Euclidean transformations are also similarities.

*Affine transformations*: preserve the affine properties of embedded objects such as parallelism. Examples of affine transformations are rotations, reflections and shears: all similarity transformations are also affine.

*Projective transformations*: preserve the projective properties of embedded objects. An intuitive idea of a central projection is the action of a point light source, sending a figure to its projection upon a screen. For example, the projection of a circle may result in an ellipse. All affine transformations are also projective.

*Topological transformations* (*homeomorphisms, bicontinuous maps*): preserve topological properties of embedded objects. We shall study this class in detail later.

For some classes of transformations, formulas may be provided.

*Translation*:

    $(x, y) \rightarrow (x + a, y + b)$                     *(a, b* are real constants)

*Scaling*:

    $(x, y) \rightarrow (ax, by)$                         *(a, b* are real constants)

*Rotation through angle θ about origin*:

    $(x, y) \rightarrow (x \cos \theta - y \sin \theta, x \sin \theta + y \cos \theta)$,

*Reflection in line through origin at angle θ to x-axis*

    $(x, y) \rightarrow (x \cos 2\theta + y \sin 2\theta, x \sin 2\theta - y \cos 2\theta)$

*Shear parallel to the x-axis*

    $(x, y) \rightarrow (x + ay, y)$                     *(a* is a real constant)

### 3.3    SET-BASED GEOMETRY OF SPACE

The Euclidean plane is a highly organized kind of space, with many well-defined operations and relationships that can act upon objects within it. In this section, we retreat to the much more rarefied realm of set-based space, and then gradually build up more structure with topological and metric spaces.

### 3.3.1    Sets

The set-based model of space does not have the rich set of constructs of the Euclidean plane. It provides only *elements, collections* (or *sets*) of elements and the *membership* relation between an element and a collection in which it occurs. Even though this model is abstract and provides little in the way of constructions for modelling spatial properties and relationships, it is nevertheless fundamental to the modelling of any spatial information system. For example, relationships between different base units of spatial reference may be modelled using set theory. Counties may be contained within countries, which may themselves be contained within continents. Cities may be elements of countries. Such hierarchical relationships are adequately modelled using set theory. Sometimes, areal units are not so easily handled; for example, in the UK there is no simple set-based relationship between the postcoded areas, used for the distribution of mail, and the administrative units such as ward, district and county. This mismatch causes considerable problems when data referenced to one set of units are compared or combined with data referenced to a different set. The set-based model has as its basic constructs:

*Elements or members*:   The constituent objects to be modelled.

*Sets*:   Collections of elements to be modelled. Such collections, for computer-based models, are usually finite, or, at least countable.

*Membership*:   The relationship between the elements and the sets to which they belong.

It is to be noted that in classical set theory, an object is either an element of a particular set or it is not. There is no half-way house or degree of membership. If the binary on-off nature of the membership condition is lifted, and this is often more appropriate to geographic applications, then we are using the fuzzy set-based model. We shall return to this model in the final chapter.

From the basic constructs of element, set and membership, a large number of modelling tools may be constructed. We shall consider just a few here.

*Equality*:  A relationship between two sets that holds when the sets contain precisely the same members.

*Subset*:  A relationship between two sets where every member of one set is a member of the second. The relationship that set $S$ is a subset of set $T$ is denoted $S \subseteq T$.

*Power set*:  The set of all subsets of a set. The power set of set $S$ is denoted $\mathscr{P}(S)$.

*Empty set*:  The set containing no members, denoted $\varnothing$.

*Cardinality*:  The number of members in a set. The cardinality of set $S$ is denoted $\# S$.

*Intersection*:  A binary operation that takes two sets and returns the set of elements that are members of both the original sets. The intersection of sets $S$ and $T$ is denoted $S \cap T$.

*Union*:  A binary operation that takes two sets and returns the set of elements that are members of at least one of the original sets. The union of sets $S$ and $T$ is denoted $S \cup T$.

*Difference*:  A binary operation that takes two sets and returns the set of elements that are members of the first set but not the second set. The difference of sets $S$ and $T$ is denoted $S\backslash T$.

*Complement*:  A unary operation that when applied to a set returns the set of elements that are not in the set. The complement is always taken with reference to an (implicit) universal set. The complement of set $S$ is denoted $S'$.

Figure 3.7 shows by shading on set diagrams some of the Boolean set operations described above.

Some sets, particularly sets of numbers, are used so often that they have a special name and symbol. Some of these are listed in Table 3.1. The Boolean set **B** is used whenever there is a two-way choice to be made. This choice may be viewed as

$$A \cap B \qquad A \cup B \qquad A \setminus B$$

**Figure 3.7**  Set intersection, union and difference.

**Table 3.1**  Some distinguished sets

| Booleans | Two-valued set of true/false, 1/0 or on/off | B |
|---|---|---|
| Integers | Positive and negative whole numbers, including zero | Z |
| Reals | Measurements on the number line | $\mathscr{R}$ |
| Real plane | Ordered pairs of reals (often the Euclidean plane when viewed as points in space) | $\mathscr{R}^2$ |
| Closed interval | All reals between reals $a$ and $b$ (including $a$ and $b$) | $[a, b]$ |
| Open interval | All reals between reals $a$ and $b$ (excluding $a$ and $b$) | $]a, b[$ |
| Semi-open interval | All reals between reals $a$ and $b$ (including $a$ and excluding $b$) | $[a, b[$ |

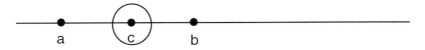

**Figure 3.8**  The rationals are dense.

between on and off, true and false, or one and zero, depending upon the context. The set of integers **Z** is used in discrete models. Sometimes, the interest may be in only the positive integers, $\mathbf{Z}^+$. For continuous models, the real numbers are required. In fact, it is provably impossible to capture the reals completely using a computer, therefore rational numbers are used in practice. Rational numbers are real numbers that can be expressed as a fraction, e.g. 123/46. Rational numbers have the property that they are *dense*, that is between any two rationals $a$ and $b$, where $a$ is less than $b$, no matter how close $a$ and $b$ are, it is always possible to find a third rational $c$ such that $a < c < b$. (See Figure 3.8.) In this way, the rationals are able to approximate to the modelling of continuous processes. Any particular computer implementation will place a restriction upon the precision of the rationals.

Some subsets of the reals are particularly useful. The *intervals* are connected sets of real numbers. They may or may not contain their end-points and are then called *closed* or *open*, respectively. It is also possible for intervals to be closed at one end and open at the other. Such intervals are called *semi-open* (or *semi-closed*). Thus, the closed interval [2, 5] denotes the set of all real numbers not less than two and not greater than five. The semi-open interval ]2, 5] denotes the set of all real numbers greater than two but not greater than five.

Returning to pure set theory, even though the set concept expresses the most rudimentary relationships and structuring of space, it is surprisingly difficult to capture the essence of a set in a few words. A first attempt might be 'a set is any collection of objects', but this lays itself open to Russell's Paradox, as follows.

Since a set is any collection of objects, consider the set $S$ of all sets that are not members of themselves. Most sets are not members of themselves, for example the set of all people is not a person, and so most sets will belong to $S$. The issue arises whether $S$ is a member of itself or not. Well, if $S$ is a member of itself, then by its definition it is not a member of itself: on the other hand, if $S$ is not a member of itself, then it must be a member of itself. Either way, we arrive at an *impasse*!

This paradox has spurred considerable efforts this century to find a more adequate definition of a set, but no definition has proved completely satisfactory.

### 3.3.2   Relations

Sets on their own are limited in their application to modelling. Life becomes more interesting when relationships between two or more sets are modelled. In order to provide these tools, a further set-based operation is defined.

*Product*:   A binary operation that takes two sets and returns the set of ordered pairs whose first elements are members of the first set and second elements are members of the second set. The product of sets $S$ and $T$ is denoted $S \times T$.

An example of a product set that has already been seen is the set of points in the Cartesian plane. Each point is represented as an ordered pair of real numbers, measuring its distance from a given origin in the direction of two axes. In set-theoretic terms, the collection of all such points is a product set, being the product of the set of real numbers with itself. This set is denoted by $\mathscr{R} \times \mathscr{R}$ or $\mathscr{R}^2$. This notion may be generalized to Cartesian 3-space $\mathscr{R}^3$ or indeed Cartesian $n$-space $\mathscr{R}^n$. A second example would be the set of points in the unit square in the Cartesian plane, with vertices $(0,0)$, $(0,1)$, $(1,0)$ and $(1,1)$. This set is the product of two intervals, $[0,1] \times [0,1]$. It is the case that $[0,1] \times [0,1] \subseteq \mathscr{R}^2$.

Product spaces provide a means of defining relationships between objects. Hence the following construction:

*Binary relation*:   A subset of the product of two sets whose ordered pairs show the relationships between members of the first set and members of the second set.

Suppose that $S = \{\text{Fred, Mary}\}$ and $T = \{\text{apples, oranges, bananas}\}$, then the formal way of expressing the relationship 'likes' between people and fruit is by means of the relation, constructed as a set of ordered pairs $\{(\text{Fred, apples}), (\text{Fred, bananas}), (\text{Mary, apples})\}$. This set of ordered pairs is a subset of the entire product space $S \times T$, containing six pairs.

Binary relations have many modelling applications, and there follow two examples demonstrating the concept. (Relations may in general apply to the product of more than two sets (as in the relations in the relational database model).) Suppose that we are given two sets of point references of Potteries places of interest and town centres, respectively. For the first relation, suppose that a place of interest $p$ be related to a town centre $t$ if $t$ is the nearest town centre to $p$. For the second example we consider a relation acting upon a single set, the places of interest. Suppose that a place of interest $p$ is related to $p'$ if $p'$ is the nearest place of interest to $p$. These two relations can be shown diagrammatically, as in Figure 3.9(a) and (b). An arrow is used to show the direction of the relation if there is any ambiguity.

Some binary relations between objects of the same set have special properties, as shown below:

*Reflexive relation*:   Where every element of the set is related to itself.

*Symmetric relation*:   Where if $x$ is related to $y$ then $y$ is related to $x$.

*Transitive relation*:   Where if $x$ is related to $y$ and $y$ is related to $z$, then $x$ is related to $z$.

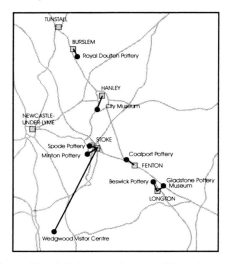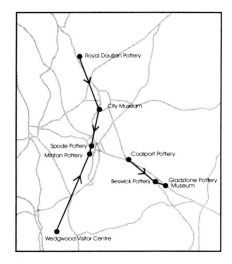

**Figure 3.9**    Relations on the sets of Potteries places of interest and town centres.

In the second example above, there is an implicit assumption that a place cannot be related to itself, therefore the relation is not reflexive. It is less obvious whether or not the relation is symmetric. However, an example shows that the relation 'has nearest place of interest' cannot be symmetric. There is a relation from Coalport to Beswick Pottery, since Beswick is the nearest place of interest to Coalport, but there is not a relation from Beswick to Coalport because Gladstone Pottery Museum is the nearest place of interest to Beswick. Also, the relation is not transitive, because if *p* has nearest place of interest *q* and *q* has nearest place of interest *r*, there is no guarantee, indeed it is most unlikely that *p* has nearest place of interest *r*.

A binary relation that is reflexive, symmetric and transitive is termed an *equivalence relation*. Another useful class of relations are *order* relations satisfying the transitive property and which are also irreflexive and antisymmetric. These further properties and partial orders are discussed in section 3.5.1.

### 3.3.3    Functions

A very useful modelling tool is the *function*, which is a special type of relation that has the property that each member of the first set relates to exactly one member of the second set. Thus a function provides a rule that transforms each member of the first set, called the *domain*, into a member of the second set, called the *codomain*. We use the notation:

$$f : S \rightarrow T$$

to mean that *f* is a function, *S* is the domain and *T* is the codomain. If the result of applying function *f* to element *x* of *S* is *y*, we write $y = f(x)$ or $f : x \mapsto y$. Figure 3.10 shows schematically the relationship between function, domain, codomain and image (defined shortly).

For example, suppose that *S* is the set of points on a spheroid and *T* the set of points in the plane. The field of map projections is essentially about functions with

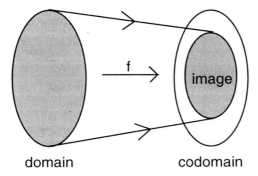

**Figure 3.10**   An abstract function *f*.

domain *S* and codomain *T*. Usually, the functions are constructed to leave some properties of the sets invariant, for example, lengths, angles or areas. The familiar Universal Transverse Mercator (UTM) is a function that preserves angles. The UTM projection has a further property that any two different points in the domain are transformed to two distinct points in the codomain. Such a function is said to be an *injection*. Not all functions are injections. Consider, for example, the function which computes the square of any integer. Then, as both 2 and −2 square to the same number 4, the square function is not an injection.

The set of outputs from the application of a function to points in its domain will form a subset of the codomain. Thus, the UTM will project a spheroid onto a finite subset of the plane. The set of all possible outputs is often called the *image* of the function. If the image actually equals the codomain, then the function is termed a *surjection*. The UTM projection is not a surjection, because the spheroid projects onto only a finite portion of the plane and not the whole plane. A function that is both a surjection and an injection is termed a *bijection*.

Injective functions have the special property that they have inverse functions. Consider again the UTM projection from the spheroid to the plane. Given a point in the plane that is part of the image of the transformation, it is possible to reconstruct the point on the spheroid from which it came. This reversal of the process allows us to form a new function whose domain is the image of the UTM and which maps the image back to the spheroid. This is called the *inverse* function.

### 3.3.4   Convexity

Convexity has already been discussed for polygons and is now generalized to the same property for arbitrary pointsets in the Euclidean plane. The same notion is also meaningful in Euclidean 3-space and the definition is very easily extendible to this case. The essential idea is that a set is convex if every point is visible from every other point within the set. To make this idea precise, we define visibility and then convexity.

Let *S* be a set of points in the Euclidean plane. Then point *x* in *S* is *visible* from point *y* in *S* if either *x* = *y* or it is possible to draw a straight line segment between *x* and *y* that consists entirely of points of *S*.

Let $S$ be a set of points in the Euclidean plane. The point $x$ in $S$ is an *observation point* for $S$ if every point of $S$ is visible from $x$.

Let $S$ be a set of points in the Euclidean plane. The set $S$ is *semi-convex (star-shaped* if $S$ is a polygonal region) if there is some observation point for $S$.

Let $S$ be a set of points in the Euclidean plane. The set $S$ is *convex* if every point of $S$ is an observation point for $S$.

Figure 3.11 shows the visibility relation within a set between three points $x$, $y$ and $z$. Points $x$ and $y$ are visible from each other, as are points $y$ and $z$. But, points $x$ and $z$ are not visible from each other. The visibility relation is reflexive and symmetric, but not transitive. Also observe that any convex set must be semi-convex (but not conversely). Figure 3.12 gives some examples of sets that are not semi-convex, semi-convex but not convex, and convex, respectively. It is noteworthy that the intersection of a collection of convex sets is also convex, and therefore any collection of convex sets closed under intersection has a minimum member. This leads to the definition of a *convex hull* of a set of points $S$ in $\mathscr{R}^2$ as the intersection of all convex sets containing $S$. From above, the convex hull must be the unique smallest convex set that contains $S$ (Figure 3.13). A convex hull of a finite set of points is always a polygonal region.

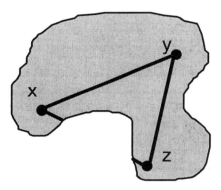

**Figure 3.11**   Visibility between points $x$, $y$ and $z$.

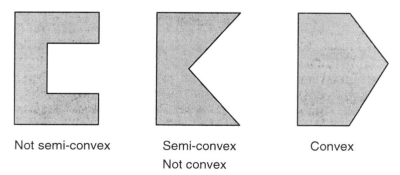

Not semi-convex          Semi-convex          Convex

Not convex

**Figure 3.12**   Degrees of convexity in pointsets.

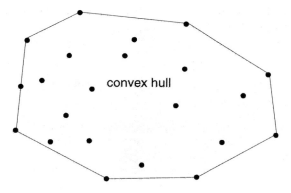

**Figure 3.13**    Convex hull of a pointset.

### 3.4    TOPOLOGY OF SPACE

The word 'topology' derives from Greek and literally translates to 'study of form'. This may be contrasted with the word 'geometry', which translates to 'measurement of the earth'. Topology is a branch of geometry, concerned with a particular set of geometrical properties – those that remain invariant under topological transformations. So, what is a topological transformation? In fact, it is a very general notion that can be applied to many different kinds of space.

#### 3.4.1    Topological spaces

To gain some topological intuition, imagine the Euclidean plane to be an unbounded sheet of fine quality rubber that has the ability to stretch and contract to any desired degree. Imagine a figure drawn upon this rubber sheet. Allow this sheet to be stretched but not torn or folded. Certain properties of the original figure will remain while others will be lost. For example, if a polygon were drawn upon the sheet and a point was drawn inside the polygon, then after any amount of stretching the point would still be inside the polygon; on the other hand, the area of the polygon may well have changed. We say that the property of 'insideness' is a topological property (because it is invariant under rubber sheet transformation) while 'area' is not a topological property. The transformation induced by stretching on a rubber sheet is called a *topological transformation* or *homeomorphism*. Thus, we have the following definitions:

A *topological property* is one which is preserved by topological transformations of the space.

*Topology* is the study of topological transformations and the properties that are left invariant by them.

Table 3.2 lists topological and non-topological properties of objects embedded in the Euclidean plane. For object types we take arcs, loops and areas. Later, these types will be defined more carefully: for now, assume the obvious meanings to arc

**Table 3.2** Topological and non-topological properties of objects in the Euclidean plane with the usual topology

| Topological | A point is at an end-point of an arc. |
|---|---|
| | An arc is a simple arc (the arc does not cross over itself). |
| | A point is on the boundary of a region. |
| | A point is in the interior of a region. |
| | A point is in the exterior of a region. |
| | An area is open (excludes all its boundary). |
| | An area is closed (includes all its boundary). |
| | An area is simple (has no holes). |
| | An area is connected (given any two points in the area, it is possible to follow a path from one point to the other such that the path is entirely within the area). |
| | A point is within a loop. |
| Non-topological | Distance between two points. |
| | Bearing of one point from another point. |
| | Length of an arc. |
| | Perimeter of an area. |
| | Area of an area. |

(possibly curved linear object), loop (closed arc) and region (two-dimensional piece of plane, possibly with holes and islands).

The discussion will cover two branches of topology: *pointset* (or *analytic*) topology and *algebraic* (combinatorial or *geometric* topology). In pointset topology, the focus, as one would expect, is on sets of points and in particular on the concepts of neighbourhood, nearness and open set. We shall see that several important spatial relationships, such as connectedness and boundary may be expressed in pointset topological terms. The other important branch of topology that has been applied to spatial data modelling is algebraic topology, in particular the theory of simplicial complexes. Even though these ideas may at times seem rarefied and far removed from spatial databases, in fact they do form the basis of several prominent conceptual models for spatial systems currently in use or under development. It is certainly true that the construction of sound and lasting generic spatial models relies on knowledge of the material that is introduced here. The reader is encouraged to gain further understanding of this area by sampling material from the bibliography provided at the end of the chapter.

### 3.4.2 General pointset topology

It is possible to define a topological space in many different ways. The definition below is based upon a single primary notion, that of neighbourhood. A set upon which a well-defined notion of neighbourhood is provided is then a topological space. It turns out that all the familiar topological properties are definable in terms of this single concept. Given any set, the approach is to define a collection of its subsets, constituting the neighbourhoods, and thus provide a neighbourhood topology on the set. The formal definition is now given.

Let $S$ be a given set of points. A *topological space* is a collection of subsets of $S$, called *neighbourhoods*, that satisfy the following two conditions.

*T1.*   Every point in $S$ is in some neighbourhood.

*T2.*   The intersection of any two neighbourhoods of any point $x$ in $S$ contains a neighbourhood of $x$.

Figure 3.14 shows the two conditions of a topological space in action. Neighbourhoods are shown surrounding each point in the set and two neighbourhoods are shown overlapping and containing in their intersection another neighbourhood.

By far the most important example of a topological space, for our purposes, is the so-called *usual topology* for the Euclidean plane, so called because it is the topology that naturally comes to mind with the Euclidean plane and corresponds to the 'rubber-sheet' topology introduced earlier. It is possible to define other (unusual?) topologies on the Euclidean plane, and examples of these follow the usual topology.

*Example 1 (The usual topology of the Euclidean plane)*

Define an *open disc* to be a set of points bounded by a circle in the Euclidean plane, but not including the boundary. An example is given in Figure 3.15. The convention is that a hatched line at the boundary indicates that the boundary points are excluded whereas a continuous line indicates that boundary points are included.

Define a neighbourhood of a point $x$ in $\mathscr{R}^2$ to be any open disc that has $x$ within it (see Figure 3.16). We now show that, under this definition of neighbourhood, $\mathscr{R}^2$ is a topological space. To check that condition *T1* for a topological space holds, it is

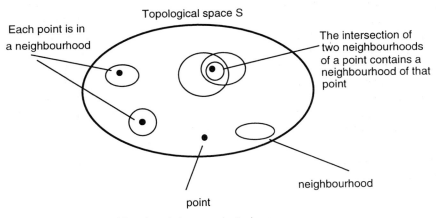

**Figure 3.14**   Points and neighbourhoods in a topological space.

**Figure 3.15**   An open disc in the Euclidean plane with the usual topology.

**Figure 3.16**   A neighbourhood of x in $\mathscr{R}^2$ with the usual topology.

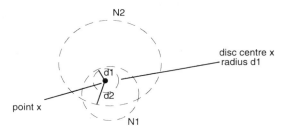

**Figure 3.17**   Condition *T2* is satisfied for $\mathscr{R}^2$ with the usual topology.

sufficient to observe that every point in $\mathscr{R}^2$ can certainly be surrounded by an open disc. For *T2*, take any point $x$ in $\mathscr{R}^2$ and surround it by two of its neighbourhoods (open discs with $x$ inside), $N_1$ and $N_2$. Now, $x$ will lie in the intersection of these two neighbourhoods and it is always possible to surround $x$ with an open disc entirely within this intersection. To see this, let $d_1$ be the minimum distance of $x$ from the boundary of $N_1$ and $d_2$ be the minimum distance of $x$ from the boundary of $N_2$. Then, the open disc with centre $x$ and radius the minimum of $d_1$ and $d_2$ will contain $x$ and lie entirely in the intersection of $N_1$ and $N_2$ (see Figure 3.17) Thus *T1* and *T2* are satisfied, and under this definition of neighbourhood, $\mathscr{R}^2$ is a topological space, called the *usual topology* for $\mathscr{R}^2$.

*Example 2 (The discrete topology)*

Let $S$ be any set, and define the neighbourhoods to be *all* the subsets of $S$. Conditions *T1* and *T2* may easily be verified to confirm that this neighbourhood structure defines a topological space. The space is called the *discrete topology*, as the smallest neighbourhood of each point $x$ of $S$ is $\{x\}$ and so each point of $S$ is separated by a neighbourhood from every other point.

*Example 3 (The indiscrete topology)*

This is the other extreme to example 2. Let $S$ be any set, and define the only neighbourhood to be the set $S$ itself. Again, this may easily be verified to be a topological space, called the *indiscrete topology*.

*Example 4 (The usual topology of the real number line)*

The usual topology of the Euclidean plane may be scaled up or down to Euclidean space of any dimension. This is illustrated with the usual topology on the Euclidean line. Let $\mathscr{R}$ be the set of real numbers, as usual. For any real number $x$, define a neighbourhood of $x$ to be any open interval containing $x$. This is the one-

dimensional equivalent of the usual topology on the Euclidean plane, and may be
seen to satisfy properties $T1$ and $T2$ in a similar way.

*Example 5 (Travel time topology)*

Here is an example more obviously related to geographic information. Let $S$ be the
set of points in a region of the plane. Suppose that the region contains a transportation network and that we know the average travel time between any two points in
the region using the network, following the optimal route. (For the purposes of this
example, we need to assume that the travel time relation is symmetric: that is, it
must always be the case that the travel time from $x$ to $y$ is equal to the travel time
from $y$ to $x$.) For each time $t$ greater than zero, define a $t$-zone around point $x$ to be
the set of all points reachable from $x$ in less than time $t$. As an illustration, Figure
3.18 shows a 5-zone, 10-zone and 15-zone around the Spode Pottery. Let the neighbourhoods be all $t$-zones (for all times $t$) around all points. Then, clearly $T1$ is
satisfied, since each point will have some $t$-zones surrounding it. The argument that
$T2$ is satisfied is similar to that used for the usual topology of the Euclidean plane
and is omitted. It is noteworthy that the symmetry of the travel time relation is
required at a critical stage in the argument. The travel time measure between two
points is an example of a metric and the travel time topology is a special case of the
topology that can be induced by any metric on a space.

It is surprising that out of the single primitive notion of neighbourhood it is
possible to construct all the features and properties of a topological space. This

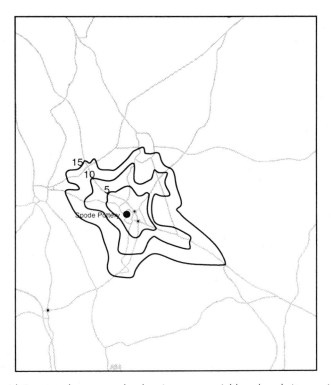

**Figure 3.18**  Travel time topology example, showing some neighbourhoods ($t$-zones).

section describes some of these constructions, beginning with the definition of 'near-ness'. Many topologists use the phrase 'limit point' to replace our use of 'near point'. This treatment is very similar to that of Henle (1979).

Let $S$ be a topological space. Then $S$ has a set of neighbourhoods associated with it. Let $X$ be a subset of points in $S$ and $x$ an individual point in $S$. Define $x$ to be *near X* if every neighbourhood of $x$ contains some point of $X$.

For example, in the Euclidean plane with the usual topology, let $C$ be the unit open disc, centre the origin, $C = \{(x, y) \,|\, x^2 + y^2 < 1\}$. Then the point (1,0), although not a member of set $C$, is near to $C$, since any open disc (no matter how small) that surrounds (1,0) must impinge into $C$. In fact, any point on the circumference of $C$ is near to $C$ as indeed is any point inside $C$. However, any point exterior to $C$ and not on the circumference is not near $C$, because it will always be possible to surround it with a neighbourhood that separates it from $C$ (see Figure 3.19).

A fundamental topological invariant of any set is its boundary, and this notion is constructed out of our primitives. First, open and closed sets are introduced using the neighbourhood idea. It will emerge that an open set is one that does not contain its boundary, whereas a closed set is one that contains all its boundary. Related notions of interior and closure are presented, along with the topological definition of boundary itself. It may seem far removed from reality to be discussing such concepts, but the notion of ownership of part or all of a boundary is key in land ownership issues and land information systems.

Let $S$ be a topological space and $X$ be a subset of points of $S$. Then $X$ is *open* if every point of $X$ can be surrounded by a neighbourhood that is entirely within $X$.

Let $S$ be a topological space and $X$ be a subset of points of $S$. Then $X$ is *closed* if it contains all its near points.

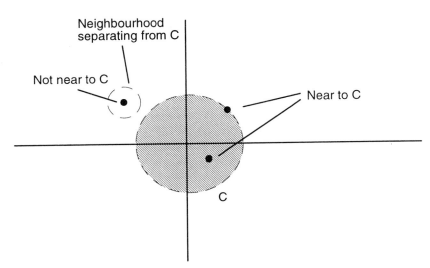

**Figure 3.19**  Points near and not near to the open unit circle $X$ in the Euclidean plane with the usual topology.

The open unit disc $C$ above is open, since any point, no matter how close to the circumference, may be surrounded by a neighbourhood made sufficiently small that it is entirely within $C$. However, $C$ is not closed, because points on the circumference are near points to $C$ but not contained in $C$. For $C$ to be closed, it would have to include its circumference. This leads us to the next definition.

> Let $S$ be a topological space and $X$ be a subset of points of $S$. Then the *closure* of $X$ is the union of $S$ with the set of all its near points. The closure of $X$ is denoted $X^-$.

Clearly the closure of a set is itself closed, in fact the closure of set $X$ is the smallest closed set containing $X$. In our example, the closure of the open unit disc $C$ is formed by annexing its circumference. Thus $C^- = \{(x, y)\,|\,x^2 + y^2 \leq 1\}$. This set is called the *closed unit circle*. We may also force a set to be open by stripping away its unwanted near points, as follows:

> Let $S$ be a topological space and $X$ be a subset of points of $S$. Then the *interior* of $X$ consists of all points that belong to $X$ and are not near points of $X'$, the complement of $X$. The interior of set $X$ is denoted $X^0$.

Notice that for a point $x$ to be near to the complement of set $X$, it must be the case that each neighbourhood of $x$ impinges upon $X'$. Therefore, a point in $X$ that is not a near point of $X'$ has at least one neighbourhood of it that is entirely within $X$. Thus, the interior of a set is open, in fact, the interior of a set $X$ is the largest open set contained in $X$. As an example, if $D$ is the closed unit disc, then $D^0$ is the open unit disc. Further examples of open sets in the Euclidean plane with the usual topology are given in Figure 3.20.

We are now sufficiently prepared to define the boundary of a set in purely topological terms.

> Let $S$ be a topological space and $X$ be a subset of points of $S$. Then the *boundary* of $X$ consists of all points that are near to both $X$ and $X'$. The boundary of set $X$ is denoted $\partial X$.

Let point $x$ be a member of $\partial X$. Since $x$ is near to $X$, then $x$ must be in $X^-$. Since $x$ is near to $X'$, then $x$ cannot be in $X^0$. Thus $\partial X$ is the set difference of $X^-$ and $X^0$. In the case of any unit disc in the Euclidean plane with the usual topology, its boundary is its circumference, as we would expect. As a further example using the

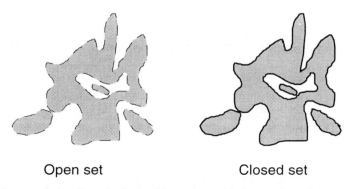

Open set                              Closed set

**Figure 3.20**   Open and closed sets in the Euclidean plane with the usual topology.

Euclidean plane with the usual topology, suppose that $S$ is the connected region of the plane containing a single hole shown in Figure 3.21. We see that the outer boundary of $S$ is excluded from $S$, but the inner boundary of $S$ (i.e. the boundary of the hole) is contained in $S$. Then its interior, closure and its boundary are shown in Figure 3.21. $S^0$ contains all the points of $S$ excluding its inner boundary. $S^-$ includes both inner and outer boundaries. $\partial S$ is the union of the inner and outer boundaries of $S$.

It is important to realize that it is not possible to consider the topological properties of sets in exclusion from the larger spaces in which they are embedded. To illustrate this point, consider a finite length of straight line. If we take the line to be embedded in a two- (or more) dimensional Euclidean space, as shown in Figure 3.22(a), then its interior is the empty set because it contains no open sets, and its boundary and closure are both the line itself. On the other hand, if the same line is embedded in a one-dimensional Euclidean space, that is, in the real number line, as shown in Figure 3.22(b), then its interior is the line excluding its end-points, its closure is the line itself, and its boundary consists of the end-points of the line.

We conclude this brief excursion into pointset topology by considering the notion of connectedness. In fact, pointset topology recognizes several different kinds of connectedness. This section defines a simple form based directly upon the neighbourhood properties of a topological space. The next section includes a description of further forms, including weak, strong and path-connectedness.

Let $S$ be a topological space and $X$ be a subset of points of $S$. Then $X$ is *connected* if whenever it is partitioned into two non-empty disjoint subsets, $A$ and $B$, then either $A$ contains a point near $B$ or $B$ contains a point near $A$, or both.

Consider the three sets shown in Figure 3.23(a), (b) and (c). In the case of (a) and (b), no matter how we choose to partition either of them into two, the partition will always satisfy the condition of the definition. The worst case is to partition the set in

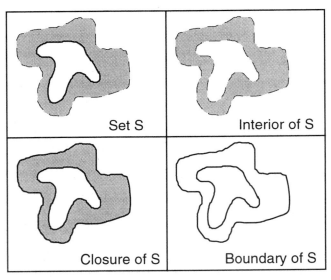

Set S    Interior of S    Closure of S    Boundary of S

**Figure 3.21**   Interior, closure and boundary of a region in the Euclidean plane with the usual topology.

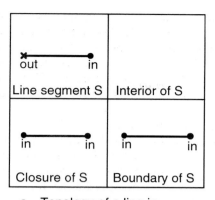 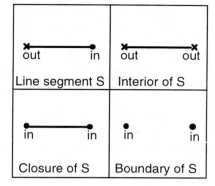

a. Topology of a line in      b. Topology of a line in
Euclidean 2-space           Euclidean 1-space

**Figure 3.22** Interior, closure and boundary of a line segment in Euclidean 2-space and 1-space with the usual topologies.

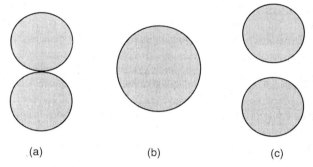

   (a)                 (b)                 (c)

**Figure 3.23** Sets connected (a, b) and disconnected (c) in the Euclidean plane with the usual topology.

(a) into its upper and lower discs. Suppose that we allow the upper disc to include the point of intersection of the two discs. Then this point is certainly a near point of the lower disc. The sets in (a) and (b) are each connected, however (c) shows a set that is not connected. To see this, partition the set in (c) into its upper and lower component discs. Then no point of the upper disc is a near point of the lower disc and no point of the lower disc is a near point of the upper disc. This example shows that the topological definition of connectedness accords with our intuition.

This section has concentrated its examples on the Euclidean plane with the usual topology, and this will continue in section 3.4.3. However, the concepts are completely general and may be applied to any topological space. We give some examples of the application of the concepts of open, closed, boundary and connectedness with respect to other topologies.

Recall that the discrete topology on space $S$ defined the neighbourhoods to be all the subsets of $S$. Let $X$ be any subset of $S$, then the only near points of $X$ are the points of $X$ itself, since any point not in $X$ may be surrounded by the neighbourhood containing just that point, and this neighbourhood does not impinge upon $X$.

Thus $X$ is closed. $X$ is also open, because we can surround any point of $X$ with the neighbourhood containing just that point, and this neighbourhood is entirely within $X$. The discrete topology is therefore odd in that every set within it is both open and closed.

This is not the case for the indiscrete topology. Let $S$ be a set upon which the indiscrete topology is defined and let $X$ be any subset of $S$. Then, the only neighbourhood in $S$ is $S$ itself, therefore every point in $S$ is a near point of $X$. Thus, unless $X$ is either empty or equal to $S$, it is neither open nor closed.

The travel time topology provides a further example. A sample open set is the set of all points less than one hour's travelling time from a specified point, say Spode Pottery. This set has boundary the set of all points that are exactly one hour from Spode Pottery and closure the set of all points having travel time from Spode Pottery not greater than one hour.

### 3.4.3    Pointset topology of the Euclidean plane

The Euclidean plane with the usual topology provides by far the most important example of a topological space, for the purposes of GIS. The Euclidean plane can be generalized in a straightforward manner to Euclidean 3-space or indeed to Euclidean $n$-space, and the natural topologies would apply in the same way. Thus, for 3-space, the neighbourhoods are the open spheres, that is, spheres that do not include their surfaces. Three-dimensional topological spaces are important for 3D GIS, which is discussed in Chapter 8. Returning to $\mathcal{R}^2$, where the neighbourhoods are the open discs, we now explore further the notion of homeomorphism, introduced in section 3.4.1, and look at some more open and closed sets. Homeomorphisms are defined here solely with reference to $\mathcal{R}^2$, but the concept is completely general and can be applied to any topological space. Throughout this section, we assume the usual topology.

A *homeomorphism* (or *topological transformation*) of $\mathcal{R}^2$ is a bijection of the plane that transforms each neighbourhood in the domain to a neighbourhood in the range. Furthermore, any neighbourhood in the range must be the result of the application of the transformation to a neighbourhood in the domain. Put more simply and intuitively, a homeomorphism corresponds to the notion of a rubber sheet transformation, which stretches and distorts the plane without folding or tearing.

If we apply a homeomorphism to a pointset $X$ and pointset $Y$ results, we say that $X$ and $Y$ are *topologically equivalent*. Thus, in Figure 3.24, the disc $S$ and the area $T$ are topologically equivalent, but neither is topologically equivalent to the area $U$. From an intuitive point of view, it is clearly possible (at least, in the mind's eye, with a lot of stretching) to transform $S$ to $T$ (and back again) by stretching and contracting a rubber sheet. However, the only option to arrive at the area $U$ from $S$ is to tear the sheet, so as to form the hole in $U$. Tearing is not allowed and therefore $S$ and $U$ are not topologically equivalent. Set $V$, which is formed by gluing the disc to itself at a single point, is homeomorphic to none of the sets $S$, and $T$ and $U$.

Many mapping ideas are based upon the idea of homeomorphism. For example, the map of the Potteries' bus routes is shown on the left side of Figure 3.25 as a similarity transformation (neglecting the Earth's curvature) of the actual routings and on the right as a topological transformation of the actual routings. The two bus route maps are topologically equivalent.

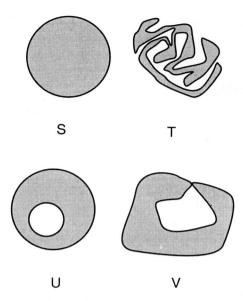

S                 T

U                 V

**Figure 3.24**   Topologically equivalent and inequivalent planar objects.

Properties that are preserved by homeomorphisms are called *topological invari-ants*. From the definition of a homeomorphism, it is clear that the configuration of neighbourhoods of a space is a topological invariant. Other constructs, such as open set, closed set and boundary, because they are defined purely in terms of neighbour-hoods, are also topological invariants.

The paradigm for all open sets in the Euclidean plane is the unit open disc, with centre the origin of coordinates. Therefore, all areas topologically equivalent to this are open as well. Any set homeomorphic to the unit open disc we term an *open cell*. Any set homeomorphic to the closed unit disc is a *closed cell*. An open cell is clearly an open set, and similarly a closed cell is clearly a closed set. In general, if a set of points owns its entire boundary then it is closed, and if it owns none of its boundary then it is open.

Connectedness has already been defined in the context of a general topological space. As connectedness is defined purely in terms of the topology, it is a topological invariant. Essentially, a connected set is 'all of a piece'. Table 3.3 below shows a list of connected sets in the Euclidean plane under the usual topology. Even though all the sets in the table are connected, they are not all homeomorphic to each other. A cell is not homeomorphic to an annulus; the rubber sheet would need to be torn to make the hole. Connected sets that do not have holes are called *simply connected*, and being simply connected is a topological invariant. Cells are simply connected; annuli are not simply connected.

Two basic one-dimensional object types are the straight-line segment and the circle. (Throughout this text, we make a distinction between a circle and a disc: a circle is the boundary of a disc.) Using the notion of homeomorphism, we may generalize these two basic object types as follows. A *simple arc* is topologically equivalent to a straight-line segment: it is clearly connected. As the homeomorphism from a straight-line segment to a simple arc is a bijection, it cannot be possible for a

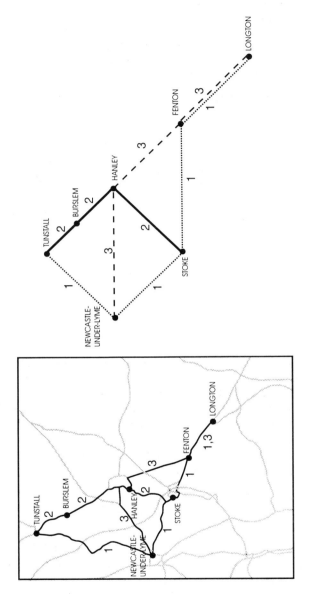

**Figure 3.25**   A topological map of the Potteries bus routes.

**Table 3.3**  Some connected sets in $\mathcal{R}^2$

Euclidean plane
Half-plane
Straight-line segment
Two intersecting line segments
Infinite straight line
Circle
Cell
A cell glued to itself at a single boundary point
Two cells glued at a single boundary point
Two cells connected by a line segment
Annulus

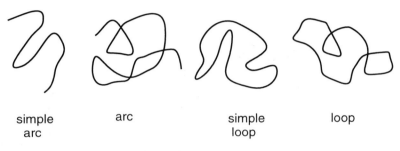

simple          arc          simple          loop
arc                          loop

**Figure 3.26**  Some one-dimensional planar objects.

simple arc to cross over itself or for its end-points to be coincident. If the condition on no self-crossings is relaxed, then the resulting one-dimensional object is termed an *arc*. If the end-points are coincident and self-crossings are not allowed, then the object is termed a *simple loop*. If the end-points are coincident and self-crossings are allowed, then the object is termed a *loop*. A simple loop is topologically equivalent to a circle. Examples are shown in Figure 3.26.

In 1887, the mathematician Jordan stated a famous theorem about simple loops, now known as the Jordan Curve Theorem.

> Given any loop, then the complement of the loop is not connected, but is partitioned into two connected components, one of which is bounded (called the *inside* of the loop) and one not bounded (called the *outside* of the loop).

This proposition may seem so blindingly obvious as to be the sort of thing that gives mathematicians a bad name. In fact, it is quite difficult to prove convincingly. The problem lies in the wide variety of shapes that qualify as loops. For example, the snowflake of von Koch is a loop but does not have a defined slope at any point on it. (Technically, it is *nowhere differentiable*.) The snowflake results from the infinite recursion of the process shown in Figure 3.27. We shall not attempt to prove the Jordan Curve Theorem in this text, but point the interested reader to the texts given in the bibliographic notes at the end of the chapter. The effort in proving this theorem resulted in the development of techniques that were instrumental in the

**Figure 3.27**    The first three stages in the construction of the von Koch snowflake curve.

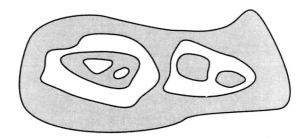

**Figure 3.28**    Holes within holes within holes...

birth of topology as a major branch of mathematics. It has a practical application in the field of GIS, being the foundation of the point-in-polygon operation.

Moving up a dimension, we have already defined the *cell* as the primary two-dimensional topological object, topologically equivalent to the disc and simply connected. Cells, though basic to topological work, are the subject of one of the most intriguing of results in introductory topology. Imagine a cell laid out as a rubber shape on the plane. Now deform the cell by twisting, stretching, even folding, but not tearing, and replace the cell on the plane in such a way that it is entirely within its original outline. Then, there will always be at least one point that has not moved from its original position! Such a point is called a *fixed point* and the result is known as Brouwer's Fixed Point Theorem. (Fixed point theorems have played an important role in computer science, in particular underpinning the theory of recursion.)

Another notable class of areal objects is the class of annuli (cells with a single hole). Of course, we can increase the number of holes, arriving at double annuli, triple annuli, etc. We can also allow the holes to be occupied by further objects (islands) (see Figure 3.28).

The following paragraphs explore further the concept of connectedness, particularly as it relates to areal objects. The topological property of connectedness was defined earlier, in section 3.4.2. We begin by providing a different definition.

A set in a topological space is *path-connected* if any two points in the set can be joined by a path that lies wholly in the set.

This definition relies on the definition in topological terms of the notion of 'path', and while this is not hard to do, it takes us some way from our main themes. Instead we suggest some intuitive ideas about a path as an unbroken and 'well-behaved' curve (in the sense of not doing odd things like getting infinitely tangled anywhere). In the Euclidean plane with the usual topology, for 'path' we may read 'simple arc'.

A natural first question is 'Given that there are two notions on connectedness, namely connectedness (as defined earlier) and path-connectedness, do these notions define the same property?' The answer to this is, in general, no. Although it is possible to show that every path-connected set is connected, there are sets that are connected but not path-connected. However, in the special case of the Euclidean plane with the usual topology, each example of a connected but not path-connected set is pathological, involving an infinite number of twists and turns. Therefore, for practical purposes, we may identify notions of connectedness and path-connectedness, certainly for the areal objects that we will shortly define. Path-connectedness is a more intuitive notion than pure connectedness, and therefore can be used as a test for connectedness in practical cases. To summarize, test for connectedness by asking the question: Given any two points in the set, is it possible to move from one point to the other along a path entirely within the set?

Many applications of spatial analysis require classes of planar objects that are purely areal, that is not mixtures of points, lines and areas. Also, they do not have isolated missing points (*punctures*) or arcs (*cuts*). Interestingly, it is possible to define the notion of a purely areal object using only topological notions.

Let $X$ be a set of points in the Euclidean plane under the usual topology. Then define the *regularization* of $X$ to be the closure of the interior of $X$, that is $\text{reg}(X) = X^{0-}$.

The regularization process has the effect of eliminating from a set any pathological and non-areal features. Consider the example shown in Figure 3.29(a), that is an amalgamation of a punctured and cut cell with some arcs and isolated points.

The regularization of the set in Figure 3.29(a) removes all cuts, punctures, extraneous arcs and isolated points. Regularization first finds the interior of the object, which will remove exterior arcs and points. Taking the closure will then remove cuts and punctures. What remains is always a closed purely areal object. In our example, shown in Figure 3.29(b), the result is a cell.

The regularization concept can now be used to characterize pure area. If an object is already purely areal, then regularizing it will have no effect. An object for

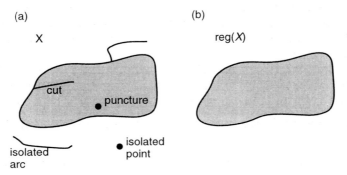

**Figure 3.29**  A spatial object comprising an area with cut and puncture, arcs and points, and its regularization.

which regularization has no effect is termed *regular closed*. The regular closed sets are exactly the purely areal objects that we require. The formal definition follows.

Let $X$ be a set of points in the Euclidean plane under the usual topology. Then $X$ is *regular closed* if and only if $X^{0-} = X$.

Having finally arrived at a topological characterization of objects which are purely areal, we now reconsider in more detail the notion of connectedness. Figure 3.30 shows three connected sets. There are some clear differences in the kind of connectedness here. In the first two cases, given any two points in each set, there are few constraints upon the path of connection from the first point to the second. All that is required is that the path starts at the first point, stays within the set and ends at the second point. However, in the case of set $Z$, if the two points are in the upper portion and lower portion respectively, then the path is constrained to pass through one of the two points on the horizontal diameter. This difference is expressed by saying that $X$ and $Y$ are *strongly connected*, but that $Z$ is *weakly connected*. To arrive at a formal definition, note that $Z$ may be made disconnected by removing a finite number of points (in fact, the two points on its horizontal diameter). However, no matter how large a finite number of points we remove from $X$ and $Y$, they will remain connected. The formal definitions now follow.

A set $X$ in the Euclidean plane with the usual topology is *weakly connected* if it is possible to transform $X$ into an unconnected set by the removal of a finite number of points.

A set $X$ in the Euclidean plane with the usual topology is *strongly connected* if it is not weakly connected.

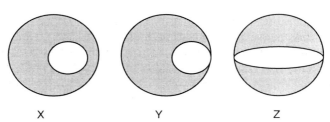

X          Y          Z

**Figure 3.30**   Three connected sets.

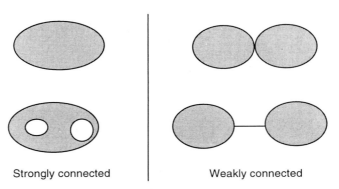

Strongly connected          Weakly connected

**Figure 3.31**   More strongly and weakly connected sets.

Figure 3.31 shows some more strongly and weakly connected sets. The notions of strong and weak connectedness play an important role later in the categorization of planar objects in the object-based approach to spatial modelling.

Thus concludes this excursion into pointset topology. The other topological area to be considered is algebraic, or combinatorial topology. In some ways, this is more pertinent to computer-based models, because the often finite and discrete structures that arise in combinatorial topology are highly suitable for representation in computer-based data structures.

### 3.4.4  Combinatorial topology of the Euclidean plane

We gave Brouwer's Theorem on fixed points as a paradigm for a result of analytic or pointset topology. There is another type of topology, called combinatorial topology, a typical result of which is Euler's famous formula:

Given a polyhedron with $f$ faces, $e$ edges and $v$ vertices, $f - e + v = 2$.

A very similar formula applies to an arrangement of cells in the plane. Remove a single face from the polyhedron above, and apply a 3-space homeomorphism to flatten the shape onto the plane. What results is a configuration of cells, with arcs forming common boundaries and nodes forming the intersection points of the arcs. Since we have removed a face from the polyhedron (it has actually become the exterior to the cellular configuration), we may simply modify the Euler's formula for the sphere to derive Euler's formula for the plane.

Given a cellular arrangement in the plane, with $f$ cells, $e$ edges and $v$ vertices, $f - e + v = 1$.

Figure 3.32 shows an example of a planar configuration of cells, with 7 faces, 17 edges and 11 nodes.

The clear topological content of these results becomes apparent when we observe that no matter how the surface of the sphere is divided into polyhedral arrangements, the result of $f - e + v$ is always two, and for planes the result is always one. Thus, the number two characterizes a sphere and distinguishes it from a plane. If we

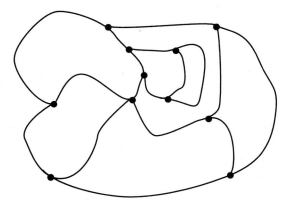

**Figure 3.32**  Planar cellular arrangement where $f = 7$, $e = 17$ and $n = 11$.

were to perform the same exercise on the surface of a torus (doughnut shape), the result is always zero. The result of $f - e + v$ is called the Euler characteristic of a surface.

### Simplexes and complexes

As the next chapter will discuss, much of the current work on generic models of space is undertaken using variations of the planar cellular arrangements described above. Theoretically, the most fundamental formal model uses the notion of a *simplicial complex*. In the two-dimensional case, simplicial complexes are simple triangular network structures in the Euclidean plane. The constructions performed in this section are all planar, but the ideas can be generalized to higher-dimensional structures.

- A *0-simplex* is a set consisting of a single point in the Euclidean plane.
- A *1-simplex* is a closed finite straight-line segment.
- A *2-simplex* is a set consisting of all the points on the boundary and in the interior of a triangle whose vertices are not collinear.

An *n*-simplex is said to have *dimension n*. The *vertices* of a simplex are defined as: for a 0-simplex the point itself, for a 1-simplex its end-points, and for a two-simplex the vertices of the triangle. A *face* of a simplex $S$ is a simplex whose vertices form a proper subset of the vertices of $S$. Figure 3.33 shows examples of 0-, 1- and 2-simplexes. Of course, the dimensionality can be extended beyond two. Thus, a 3-simplex would be a tetrahedron. The *boundary* of a simplex $S$, written $\partial S$ is the union of all its faces. For example, suppose that a 2-simplex $S$ has vertices $x$, $y$ and $z$. Then the faces of $S$ are the three 1-simplexes $xy$, $xz$, $yz$ and the three 0-simplexes $x$, $y$, $z$. The boundary of $S$ is the union of these faces, thus corresponding to the usual pointset topological definition of boundary.

Simplexes are to be the building blocks of larger structures, the simplicial complexes. Complexes are built out of simplexes in a way that is now made precise. A *simplicial complex C* is a finite set of simplexes satisfying the properties:

A face of a simplex in $C$ is also in $C$.

The intersection of two simplexes in $C$ is either empty or is also in $C$.

Figure 3.34 shows examples of configurations, of which two are simplicial complexes and two are not. In fact, the complexes on the right are formed by adding sufficient nodes and edges to the configurations on the left to make them satisfy the simplicial complex formation rules – a form of 'completing the topology'. In the case of the 2-simplexes *abc* and *def*, their intersection is not a face of either simplex. This is

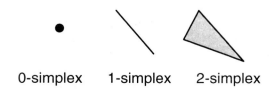

0-simplex     1-simplex     2-simplex

**Figure 3.33**  Examples of 0-, 1- and 2-simplexes.

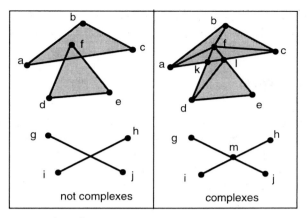

**Figure 3.34**   Aggregations of simplexes.

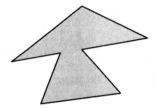

**Figure 3.35**   Planar embedding of a simplicial complex from Figure 3.34.

rectified by adding nodes *k* and *l* and decomposing the original simplexes *abd* and *def* into simplexes *akf*, *afb*, *bfc*, *cfl*, *fkl*, *dkl* and *del*. These simplexes, along with their faces *ab*, *af*, *ak*, *bc*, *bf*, *cf*, *cl*, *dk*, *dl*, *de*, *el*, *fl*, *fk*, *kl*, *a*, *b*, *c*, *d*, *e*, *f*, *k* and *l*, form a simplicial complex. The 1-simplexes in the lower part of the figure are enhanced in a similar way by adding node *m*.

The set of points that are contained in the constituent simplexes of a simplicial complex is a set in the Euclidean plane, called the *planar embedding* of the complex. Figure 3.35 shows the planar embedding of the upper complex in Figure 3.34

Even though we have given an abstract presentation of simplicial complexes, such structures are common in the spatial sciences. One-dimensional complexes are the graphs of Section 3.5. Two-dimensional complexes may be used to model the triangulated irregular networks (TINs) used in terrain modelling and indeed any areal objects.

The *dimension* of a simplicial complex is the maximum dimension of its constituent simplexes. For an *n*-complex, *C*, the boundary of *C*, $\partial C$, is a simplicial complex of dimension $n - 1$. We would like to be able to calculate the boundary of a simplicial complex. In order to do this, it is easier to work with 'oriented' complexes. An *n*-simplex, ($n > 0$), may always be given an orientation. Thus, the orientation of a 1-simplex is indicated by an arrow showing a direction along the length of the segment. The orientation of a 2-simplex is shown by a circular clockwise or anticlockwise arrow (Figure 3.36).

The orientation of a 2-simplex determines the orientation of all its constituent 1-simplexes. An *oriented simplicial complex* is a simplicial complex in which every

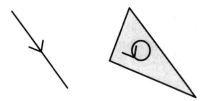

**Figure 3.36**   Examples of oriented 1- and 2-simplexes.

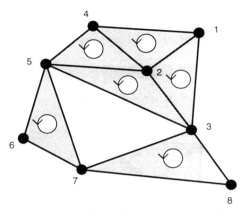

**Figure 3.37**   The oriented simplicial complex C.

simplex is consistently oriented. Note that an oriented one-dimensional simplicial complex is a directed graph.

It is now possible to introduce a purely symbolic method of calculating the boundary of a simplicial complex. Let the 0-simplex consisting of the point, $x$, be denoted $x$. Let the oriented 1-simplex, with start point $x$ and end point $y$, be denoted $xy$. Let the oriented 2-simplex, with vertices in order being $x$, $y$ and $z$, be denoted $xyz$. Introduce linear combinations of simplexes (called *chains*) of the form $a_1 S_1 + \ldots + a_m S_m$ where the $S_1, \ldots, S_m$ are simplexes and the $a_1, \ldots, a_m$ are integers, and define $-xy$ to be $yx$. Then define a boundary operator $\partial$ as follows:

$$\partial(x) = 0$$

$$\partial(xy) = y - x$$

$$\partial(xyz) = xy + yz + zx$$

Now extend the boundary operator $\partial$ to operate on any linear combinations of simplexes by means of the rule:

$$\partial(a_1 S_1 + \ldots + a_m S_m) = a_1 \partial(S_1) + \ldots + a_m \partial(S_m)$$

In other words, the boundary of a linear combination of simplexes is equal to the same linear combination of the boundaries of the component simplexes. This construction is illustrated using the simplicial complex $C$ in Figure 3.37. As a formal chain, we can express $C$ as $123 + 142 + 452 + 253 + 567 + 378$, and by the rules for calculating the boundary operator above:

$$\partial(C) = \partial(123) + \partial(142) + \partial(452) + \partial(253) + \partial(567) + \partial(378)$$

$$= 12 + 23 + 31 + 14 + 42 + 21 + 45 + 52 + 24 + 25 + 53 + 32 + 56$$

$$+ 67 + 75 + 37 + 78 + 83$$

$$= 31 + 14 + 45 + 53 + 56 + 67 + 75 + 37 + 78 + 83$$

since $12 + 21 = 12 - 12 = 0$, etc. This calculation agrees with the expected result, as seen in Figure 3.37. Computations such as these form the basis of the branch of algebraic topology called homological algebra.

Well-behaved objects inhabiting the Euclidean plane (under the usual topology) have been described as regular closed sets for two dimensions, arcs for one dimension and points for two dimensions. In terms of the topology developed here, all these sets may be grouped under the umbrella of triangulations. A *triangulation* in the Euclidean plane is a set that is topologically equivalent to the planar embedding of a simplicial complex. (The term 'triangulation' is used differently in this section from elsewhere in the book. Elsewhere, a triangulation is a planar tessellation of a surface into triangles. The definition in this section is more general.) Thus, a triangulation is an amalgamation of cells (topologically equivalent to 2-simplexes), simple arcs (topologically equivalent to 1-simplexes), and points (topologically equivalent to 0-simplexes). Figure 3.38 shows an example of a triangulation (on the right) and a complex from which it is derived (on the left), maybe modelling the shipping routes between three islands. Simplicial complexes provide a sound and complete framework for analysis of the topology of amalgamations of cells, arcs and points in the plane, and therefore part of the theoretical underpinning for GIS.

Triangulations have associated with them a famous result known as the Four Colour Theorem. This states that any triangulation on the plane or the surface of the sphere may be coloured with four colours in such a way that no two adjacent cells (i.e. cells that share a common boundary) have the same colour. The Four Colour Theorem has only a tenuous connection with the theme of this text, yet its 'colourful' history deserves comment. Its origins are unclear, but it was certainly known in the mid-nineteenth century. Despite the fact that it is so easily stated and grasped, it was not proved until the last decade, and then only with the aid of a massive amount of computational power and complex arguments. Like the Jordan Curve Theorem, some propositions are easy to make but extremely hard to justify!

Triangulations provide a means of describing the topological make up of planar spatial objects. These final few paragraphs on combinatorial topology discuss another approach, known as the *combinatorial map*. Most GIS practitioners

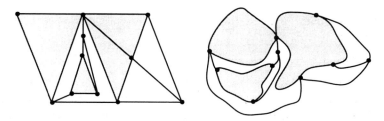

**Figure 3.38**   A triangulation and associated simplicial complex.

are familiar with the standard node-arc-polygon representation of planar configurations, where arcs have associated with them left and right polygons. This representation begins to give a topological description of a planar object, in particular in terms of its adjacency relationships. However, it falls short of a full topological representation in at least two ways.

- The more detailed connectivity of the object is not explicitly given. Thus, there is no explicit representation of weak, strong or simple connectedness.

- The representation is not *faithful*, in the sense that two different topological configurations may have the same representation.

The notion of a combinatorial map goes some way to meeting these two shortcomings. The problem is that the same shape may be viewed in different ways. Consider the weakly connected cellular arrangement in Figure 3.39. Is it a disc with an ellipse removed or the union of two crescents? The notion of a combinatorial map helps in this regard. Assume that the boundary of a cellular arrangement is decomposed into simple arcs and nodes that form a network (see the next section). Give a direction to each arc so that travelling along the arc, *the object bounded by*

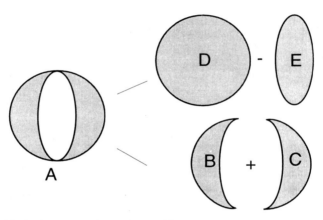

**Figure 3.39**    Ambiguity in representing a planar object.

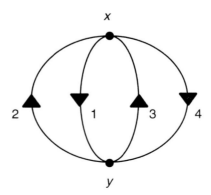

**Figure 3.40**    Directed graph associated with the object in Figure 3.39.

*the arc is to the right of the directed arc.* Thus, the object above has a boundary network as shown in Figure 3.40. The final step is to provide a rule for the order of following the arcs: *after following an arc into a node, move anticlockwise around the node and leave by the first outward arc encountered.* Thus, in our example, suppose the traversal begins at x and moves along arc 1, when arriving at y we rotate anticlockwise around y in our search for the first outbound arc. This is arc 2. Now arriving back at x, we choose arc 1. As arc 1 is already in our list (cycle), the process moves on. We start with an arc not already traversed and repeat the cycle generating process. When there are no more arcs to be traversed, the process halts. Thus, we have found two cycles [1, 2] and [3, 4].

The weakly connected figure in Figure 3.39 has been represented as a union of crescents, each homeomorphic to a cell. It is possible to show that for any cellular configuration, the map provides a set of cycles, each corresponding to a component cell of the configuration. Thus, the representation is unique and unambiguous. We shall return to combinatorial maps in section 5.3.4 when considering computer representations (DCELs and object-DCELs) of spatial objects.

### 3.5   NETWORK SPACES

The founder of the systematic study of topology is generally recognized to be the mathematician Leonard Euler, who in 1736 solved a famous problem of his time called the Königsberg Bridge Problem. Figure 3.41 shows a rough map of the Pregel River of Königsberg, in which two islands are linked to each other and the river banks by seven bridges. The problem is to trace a circuit which crosses each of the bridges exactly once and arrives back at the start. A few experiments will quickly convince you that the task is impossible – no such route exists without getting your feet wet. However, the step from believing that there was no route to demonstrating this conclusively was not so easy.

Euler succeeded in proving that this was an impossible task; or, in other words, that the problem is unsolvable. In order to do this, he built a spatial model of the Königsberg bridges that abstracted out all but the topological relationships between the bridges. Euler's model of the Königsberg bridges is represented in Figure 3.42.

The black circles represent *nodes* or *vertices*. They are labelled w, x, y and z, and are abstractions of the regions of dry land. The lines represent *arcs* or *edges*, which

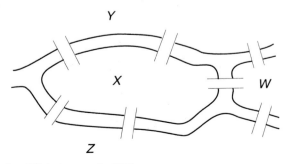

**Figure 3.41**   Bridges in a Königsberg park, 1736.

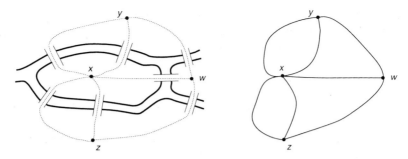

**Figure 3.42**   Graph-theoretic model of the bridges in a Königsberg park.

are abstractions of the direct routes between the regions of dry land and require the use of one bridge in each case. The complete model is called a *network* or *graph*. Euler proved that it was impossible to start from a node, trace a path along the edges of the graph, traversing each edge exactly once, and end up at the starting node. The argument he used is remarkably simple and relies on the odd/even number of edges issuing from each node. We can see that, apart from the start and end nodes (actually the same in this case), the path through a node must come in along one edge and out along another edge. This accounts for two edges incident with that node. Thus, the number of edges incident with each intermediate node must be even, if the problem is to be solvable. However, referring to the graph in Figure 3.42, *none* of the nodes is incident with an even number of edges. Thus, the graph-theoretic problem is unsolvable, and so the original problem relating to the Königsberg bridges is also unsolvable.

### 3.5.1   Abstract graphs

A *graph G* is defined as a finite non-empty set of *nodes* together with a set of unordered pairs of distinct nodes (called *edges*). If $x$ and $y$ are nodes of $G$ and $e = \{x, y\}$ is an edge of $G$, then $e$ is said to *join* $x$ to $y$, or to be *incident* with $x$ and $y$. Similarly $x$ and $y$ are *incident* with edge $e$.

A graph is a highly abstracted model of spatial relationships, and represents only connectedness between elements of the space. However, in many situations such a model can be very useful, particularly if we allow some extensions, as follows:

A *directed graph* is a graph in which each edge is assigned a direction. Directed edges are often indicated by arrowed lines on a diagram.

A *labelled graph* is a graph in which each edge is assigned a label (maybe a number or string). Such labels are usually indicated on a diagram near to the appropriate edges.

One can imagine the usefulness of a directed graph in modelling the road network in a city centre, where there are many one-way streets. A labelled graph can model a host of situations, for example, distances, travel times or traffic usage of roads in a network.

It is usual to show the linkages of a graph by a diagram. The graph $G$ in Figure 3.43 consists of the six nodes $a$, $b$, $c$, $d$, $e$ and $f$ and the nine edges $ab$, $bd$, $cd$, $ac$, $af$, $be$, $ce$, $df$ and $ef$. The *degree* of a node is the number of edges with which it is incident, and so the degree of all the nodes in $G$ is three. A *path* between two nodes is a connected sequence of edges between the nodes, and is usually denoted by the nodes that it passes through. Examples of paths between nodes $a$ and $d$ in $G$ are $afd$, $acd$, $abd$, $abecafd$ and $abefd$. A *connected* graph is such that there always exists a path between any two of its nodes: $G$ is connected.

Two graphs may show exactly the same connectivity relationships, and such graphs are said to be *isomorphic*. Sometimes isomorphism can be hard to detect, because a graph can disguise itself quite well. Thus, graph $H$ in Figure 3.44 is isomorphic to graph $G$ in Figure 3.43, because it has precisely the same configuration of nodes and edges. (In this case, we have made the problem easier by labelling the nodes to show the isomorphism.)

A path from a node to itself traversing at least one edge is called a *cycle*. Examples of cycles in $G$ (or $H$) are $abeca$ and $abdfeca$. A graph which has no cycles is called an *acyclic* graph. A very useful set of graphs are the trees. A *tree* is a connected acyclic graph. Figure 3.45 shows the three non-isomorphic trees with five nodes.

A *rooted tree* has one of its nodes, the *root*, distinguished from the others. Rooted trees (often we omit the word 'rooted') are often drawn with the root at the top and

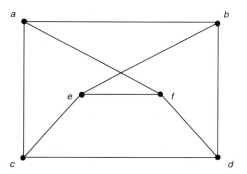

**Figure 3.43**   A graph $G$ with six nodes and nine edges.

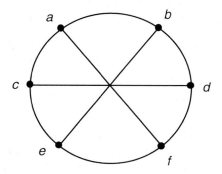

**Figure 3.44**   The graph $H$, isomorphic to $G$.

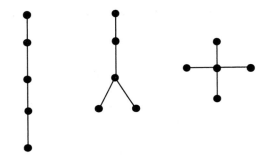

**Figure 3.45**   The three non-isomorphic trees with five nodes.

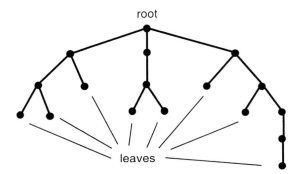

**Figure 3.46**   Rooted tree with six levels and eight leaves.

nodes occupying successive levels down, depending upon their distance (in terms of path) from the root. Nodes immediately below the root are termed *immediate descendants* of the root; they themselves have descendants, etc. A node with no descendants is termed a *leaf*. Figure 3.46 shows an example of a layered rooted tree. Trees provide some useful data structures for computational purposes.

It is possible to extend the notion of path and cycle to directed graphs. Thus, there is a *(directed) path* from node *a* to node *b* if there is a sequence of correctly oriented directed edges leading from *a* to *b*. A *(directed) cycle* is a (directed) path from a node to itself. A class of graphs useful for a wide range of applications is the class of directed graphs that have no cycles: such graphs are termed *directed acyclic graphs* or *dags*. One reason for the importance of the class of dags is that a dag defines a partial order on the graph nodes. A *(strict) partial order* is a special form of transitive relation on a set that satisfies the following additional properties:

*Irreflexive*:   Where every element of the set is not related to itself.

*Anti-symmetric*:   Where if *x* is related to *y* then *y* is not related to *x*.

Partial orderings are a generalization of the relation 'greater than', between two real numbers. No number can be greater than itself (irreflexive); if *x* is greater than *y*, then it cannot be that *y* is greater than *x* (anti-symmetric); and *x* greater than *y* greater than *z* implies *x* greater than *z* (transitive). In the case of dags, define node *a* to be related to *b* if there is a path from *a* to *b*: it can be checked that this relation is a partial order.

### 3.5.2   Planar graphs

A further level of information may be added to the graph-theoretic model by considering the embedding of the graph in the Euclidean plane. A *planar graph* is a graph that can be embedded in the plane in a way that preserves its structure: in particular, its edges are embedded as arcs and arcs may only intersect at nodes of the graph. Figure 3.47 shows a planar graph, and a non-planar graph for which no rearrangement of arcs in the plane will preserve the original connectivity and not lead to a crossing not at a node. In general, there are many topologically inequivalent planar embeddings of a planar graph in the plane. Figure 3.48 shows three embeddings of the planar graph of Figure 3.47. The upper two embeddings are

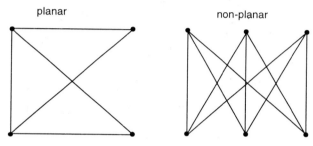

**Figure 3.47**   Planar and non-planar graphs.

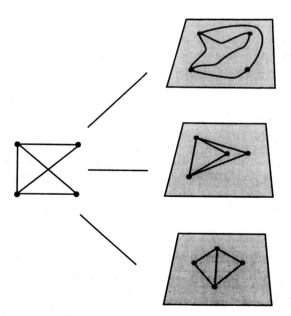

**Figure 3.48**   Three planar embeddings of a planar graph.

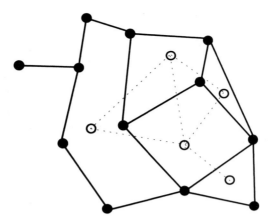

**Figure 3.49**   A graph and its dual.

homeomorphic, but not homeomorphic to the lowest, with respect to the usual
topology of the Euclidean plane. There is no way, using topological transformations
(rubber-sheet geometry) to move the node of degree 2 inside the triangle to a position outside the triangle.

   This last discussion raises an interesting consideration. Configurations that are
taken to be equivalent in one model may very well be inequivalent in another. The
second and third configurations are equivalent in a graph-theoretic sense, both
being identically connected. However, when viewed as embeddings in the plane, they
are inequivalent in the topological sense (and certainly in the metric sense). It all
depends upon the level of abstraction.

   A planar embedding of a planar graph determines a subdivision of the plane into
*faces* or *regions*. A simple integrity constraint upon a planar graph discussed in an
earlier section is given by the Euler formula for the plane, which provides a relationship between the number of nodes, $n$ arcs $a$, and faces $f$, as $f - a + n = 1$.

   A useful concept associated with planar embedded graphs is that of *duality*. The
dual $G^*$ of a planar graph $G$ is obtained by associating with each face in $G$ a node in
$G^*$. Two nodes in $G^*$ are connected by an edge if, and only if, their corresponding
faces in $G$ are adjacent. Given an edge $e$ in $G$, the dual edge $e^*$ to $e$ joins the nodes
in $G^*$ corresponding to the two faces in $G$ incident with $e$. Figure 3.49 shows a
planar graph (nodes filled in black, edges marked with continuous lines) and its dual
(nodes shown unfilled and edges marked with dotted lines).

   When the planar graph $G$ is a diagonal triangulation of a polygon (with no
Steiner points), then the dual graph $G^*$ has the properties that the degree of each
node is no more than three (since triangles have three sides). Moreover, $G^*$ is a tree.
We will show this by arguing that $G^*$ is acyclic and connected. That the degree of
each node of $G^*$ is no more three follows immediately from the fact that a triangle
has three sides. To see that the dual graph $G^*$ is acyclic, suppose that there were a
cycle of edges in $G^*$. This would imply that the triangulation $G$ contained a cycle of
faces, and this cycle must surround at least one node of $G$. But the only nodes of $G$
are vertices of the polygon and must therefore be on the boundary of the polygon.
This contradiction implies that there can be no cycles in $G^*$. $G^*$ is clearly connected
and therefore a tree.

## 3.6 METRIC SPACES

This section explores the kind of properties that a model of space should have if it is to include the concept of distance between objects in the space. Such a space is called a *metric space*. The formal definition follows, but it should be noted that it does not always accord with our common-sense notion of distance.

A pointset $S$ is said to be a *metric space* if there exists a function, **distance**, that takes ordered pairs $(s, t)$ of elements of $S$ and returns a real number **distance** $(s, t)$ that satisfies the following four conditions:

1. For each pair $s$, $t$ in $S$, **distance** $(s, t) > 0$ if $s$ and $t$ are distinct points and **distance** $(s, t) = 0$ if $s$ and $t$ are identical.

2. For each pair $s$, $t$ in $S$, the distance from $s$ to $t$ is equal to the distance from $t$ to $s$, **distance** $(s, t) =$ **distance** $(t, s)$.

3. For each triple $s$, $t$, $u$ in $S$, the sum of the distances from $s$ to $t$ and from $t$ to $u$ is always at least as large as the distance from $s$ to $u$, that is: **distance** $(s, t) +$ **distance** $(t, u) \geq$ **distance** $(s, u)$.

Put into more informal language, the first condition stipulates that the distance between points must be a positive number unless the points are the same, in which case the distance will be zero. The second condition ensures that the distance between two points is independent of which way round it is measured. The third condition states that it must always be at least as far to travel between two points via a third point rather than to travel directly.

In order to motivate this definition, we give below some possible distance functions and consider them with respect to properties (1) − (3) above. Let $S$ be a set of cities on the globe and distance between two cities in $S$ defined as follows (see Figure 3.50).

*Geodesic distance*: The distance 'as the crow files', i.e. the distance along the great circle of the Earth passing through the two city centres.

*Spherical Manhattan distance*: The difference in their latitudes plus the difference in their longitudes. (The name 'Manhattan distance' arises because, in Manhattan, the street configuration may be modelled as a collection of straight lines in two perpendicular directions.)

*Travelling time*: The minimum time that it is possible to travel from one city to the other using a sequence of scheduled airline flights (assuming that each city has at least one airport).

*Lexicographic*: The absolute value of the difference between their positions in a list of cities in a fixed gazetteer.

The first property of a metric space is not controversial, and satisfied by any self-respecting distance function. The distance function assigns to an ordered pair of spatial elements a real number, called the *distance* between the two elements, and sensible distances cannot be allowed to be negative. Also, the distance between an element and itself is always zero, whereas the distance between two distinct elements in always greater than zero. All the distance functions in our city example possess these properties.

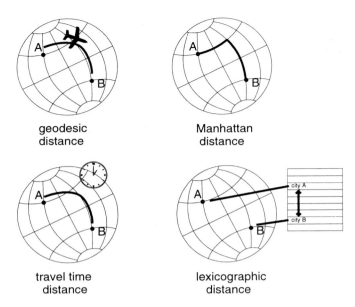

Figure 3.50   Distances defined on the globe.

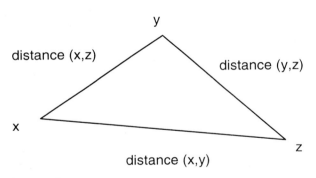

Figure 3.51   Triangle inequality.

The second property specifies that in a metric space, distance is symmetric, that is, the distance from $a$ to $b$ is always the same as the distance from $b$ to $a$. While the geodesic, Manhattan and lexicographic distances satisfy this condition, the travelling time distance is not symmetric. It is perfectly possible, and indeed (due to prevailing winds) usual for the flight times between two cities to be different in each direction. A distance function which obeys properties (1) and (3) is called a *quasi-metric*.

The third property is called the triangle inequality due to the configuration shown in Figure 3.51. In plain English, the triangle inequality implies that it is never any further to go by a direct route rather than an indirect route. All the above examples obey that triangle inequality.

So, the collection of cities together with the geodesic function, the Manhattan distance function and the lexicographic distance function are all metric spaces; while the collection of cities with the travel time function is not a metric space. The archetypal example of a metric space is Euclidean space, where the distance between two points is defined by the Pythagorean formula given early in this chapter. This distance function can easily be extended to higher dimensions.

For a further example of a space that is not a metric space, consider a plane terrain with no prevailing wind in which some areas are easy walking (maybe covered with well-cropped grass) and others more difficult (maybe rocky). The distance function is defined as the average time it takes to walk in a straight line from one point to another. We may assume that conditions (1) and (2) hold, but it is unlikely that (3) holds. It might well be the case that it is faster to walk round a nasty stretch than to walk straight across it (see Figure 3.52). Thus, the triangle inequality does not hold and this space is not a metric space.

### 3.6.1   Topology of metric spaces

It turns out that a metric space has a natural topology. Let $S$ be a metric space with distance function $d$. For each point, $x$ in $S$ and each real number $r$, define the *open ball* $B(x, r)$ to be the set of points whose distance from the point $x$ is less than the number $r$. Expressed formally:

$$B(x, r) = \{y \mid d(y, x) < r\}$$

Define the set of *neighbourhoods* to be the set of open balls. It is not hard to verify that this defines a topology for $S$. In the case where we have the Euclidean metric, this example reduces to the usual topology of the Euclidean plane. In the case where we have the Manhattan metric, then the open ball $B(x, r)$ will contain all points for which the sum of the horizontal and vertical distances from $x$ is less than $r$. In the case of the Manhattan metric applied to a flat planar space, the open balls $B(x, r)$ will be squares of diagonal $2r$, as shown in Figure 3.53.

Travel time measures have some interesting properties. We have seen that in general, they do not lead to metric spaces because they are not necessarily symmetric (for example in a one-way traffic system). Let us for the moment make the simplifying assumption that distances are symmetric, so a topology may be defined as above. This is the travel-time topology introduced in section 3.4.2. Computing topological neighbourhoods can be illuminating in this case. Figure 3.54 shows a

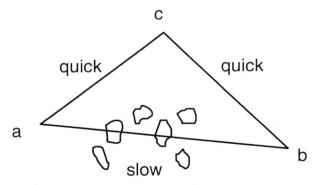

**Figure 3.52**   More haste, less speed in a non-metric space.

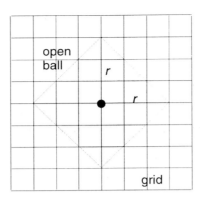

**Figure 3.53**   An open ball induced by the Manhattan metric.

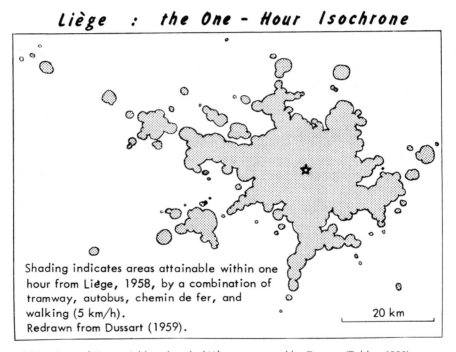

**Figure 3.54**   A travel time neighbourhood of Liège, computed by Dussart (Tobler, 1993).

travel time neighbourhood (*t*-zone) of Liège, computed by Dussart and redrawn and considered by Tobler (1993).

The shaded region shows the area within one hour's travel time of the centre of Liège in 1958 by the common means of travel available at that time. This 'neighbourhood' is not even connected, being the disjoint union of a finite number of cells. This type of pattern arises because of the discontinuous nature of travel on public transport, boarding and alighting at fixed and discrete points. As Tobler shows us, such travel time configurations are not amenable to modelling within the framework of Euclidean space, not even by continuous transformations resulting from

Gaussian metrics. Such problems force us to use network models. Later chapters further develop this theme.

### 3.7   END-NOTE ON SPHERES

Many of the models in this chapter have been founded upon the plane. Such models, when applied to geographic reality, will be useful only in limited localities where the region may be approximated to a plane surface. The trend is towards global information systems, and for such systems to be well-founded it is essential that they have available the geometry of other surfaces, in particular the surface of the sphere. (The Earth is of course not exactly spherical, but a sphere at least provides a first approximation. The Earth is homeomorphic to the sphere in the usual Euclidean 3-space topology.)

The first observation is that the surface of the sphere, although embedded in Euclidean 3-space, is itself two-dimensional. Thus, any point on it may be uniquely specified by two numbers. A common system of coordinates for points on the surface of the sphere are the familiar latitude and longitude. Assume that the sphere is embedded in Euclidean 3-space with its centre at the origin of coordinates $O$, as shown in Figure 3.55. Suppose that $P$ is an arbitrary point on the surface of the sphere and $OP$ is the radial line from $O$ to $P$. Project line $OP$ onto the line $OQ$ in the $xy$-plane, as shown in the figure. Let $\theta$ by the angle that $OP$ makes with the $x$-axis and $\phi$ be the angle that $OP$ makes with $OQ$. Then $\theta$ is the longitude of the point $P$ and $\phi$ is its latitude. The great circle of the sphere in the $xy$-plane is the *equator* and the great circle in the $xz$-plane is the *meridian*.

The usual topology of the surface of the sphere is similar to the usual topology of the Euclidean plane. Neighbourhoods are sets of points with distance from a fixed point, measured along a geodesic (great circle), less than a constant. One interesting

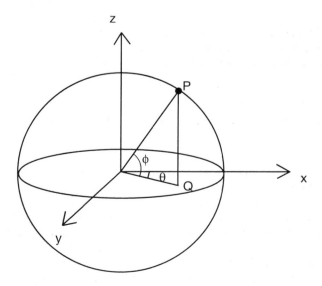

**Figure 3.55**   Coordinatization of the sphere (latitude and longitude).

difference is that on the sphere, of course, there is no notion of an unbounded exterior. Thus, the Jordan Curve Theorem is modified so that both the interior and exterior are bounded.

## BIBLIOGRAPHIC NOTES

In the author's view, still the best general text on geometry is by Coxeter (1961), which gives a fascinating overview of the variety that geometry offers, from basic work with triangles and polygons, through tessellations and two-dimensional crystallography to the platonic solids and golden section. Later chapters take us into more advanced realms, with chapters on projective, hyperbolic, differential, geodesic and four-dimensional geometries.

Mathematically inclined readers can find much detailed information about metric and pointset topological spaces in Sutherland (1975) and Armstrong (1979). There are now many readable elementary texts on the theory of graphs (for example, Harary, 1969). Algebraic, or combinatorial, topology can be quite inaccessible to all but the mathematically gifted. A book that has withstood the test of time as a readable elementary introduction is by Giblin (1977). A further book that introduces topology in an algebraic and combinatorial framework is by Henle (1979). Simplicial complexes are also discussed in chapters in Hoffmann (1989), the focus of which is computational geometry and solid modelling, but it contains in its early chapters much interesting material concerning embeddings of simplicial complexes in three dimensions. Tobler (1993) has some thought-provoking observations about travel time structures.

# CHAPTER FOUR

# Models of spatial information

... the first law of geography: everything is related to everything else, but near things are more related than distant things

Waldo Tobler, 1970

This chapter considers the high-level modelling from Chapter 2 in the context of spatial information, taking advantage of some of the general spatial concepts developed in Chapter 3. Section 4.1 begins by setting spatial models in the context of the general modelling process. A characteristic of GIS is the dichotomy of spatial models (field *versus* object), the effect of which cascades down into spatial data structures and implementations. This chapter looks in detail at both the field model (section 4.2) and the object model (section 4.3) of spatial information. A brief epilogue provides some cases where the field and object division is not so clear-cut.

## 4.1 THE MODELLING PROCESS AND SPATIAL DATA MODELS

This section sets the scene by examining the modelling process in a general setting. The field *versus* object dichotomy is described and some data quality issues are set out.

### 4.1.1 Models, domains and morphisms

The word 'model' has been used freely in varying contexts in the preceding chapters. A model is an artificial construction in which parts of one domain (*source domain*) are represented in another domain (*target domain*). The constituents of the source domain may, for example, be entities, relationships, processes, or any other phenomena of interest. The purpose of the model is to simplify and abstract from the source domain. Constituents of the source domain are translated by the model into the target domain and viewed and analysed in this new context. Insights, results, computations, or whatever has taken place in the target domain, may then be interpreted in the source domain. A simple example of a model is a flight simulator. Objects in the real world such as an aeroplane, its instrument panel, sounds, movements, views from the cockpit and the navigation space, are simulated in an artificial environment. The pilot may manipulate the model environment, for example by

145

simulating a landing to Martha's Vineyard in bad weather; this experience within the target domain may then be transferred back to experience with flying real planes.

The key notions that make particular models useful are how closely they can simulate the source domain and how easy it is to move between the two domains. The mathematical concept behind this is *morphism*, being a function from one domain to another *that preserves some of the structure in the translation*. Those readers old enough to remember school work with logarithms have an excellent example of a morphism. The logarithm function translates the positive real number multiplicative structure to the real number additive structure (remember: to multiply two numbers, take the logs and add). It is then possible to return to the original domain with the result (take the anti-log).

Cartography and way-finding provide a second example. Suppose that the geographic world is the source domain, modelled by a map (target domain). A user needing to travel from Edinburgh to London by road, consults and analyses the map, then translates the results of these analyses back to navigate through the UK road network. If the map is a good model of the real road network then the user's journey may be smooth.

The modelling process can be shown schematically as in Figure 4.1. The left-hand oval represents the source domain to be modelled. In this example, suppose that the source domain is part of the electrical supply network. Suppose further that we wish to perform some network analyses, such as predict current flows in the case of a

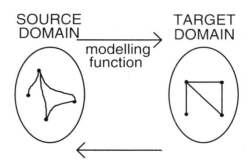

**Figure 4.1**   The modelling process as source domain, modelling function and target domain.

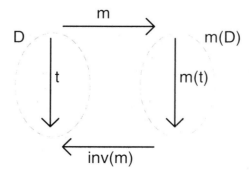

**Figure 4.2**   Modelling as a morphism.

break at some point. The appropriate model in this case may be a mathematical network structure, so this forms the target domain. The modelling function associates elements of the source domain with elements of the target domain. Network transformations and analyses can be made in the target domain and results translated back and interpreted in the source domain.

Figure 4.2 shows the modelling process more abstractly in terms of morphisms. Source domain $D$ is modelled using the modelling function $m$. A transformation $t$ in the source domain is modelled by the transformation $m(t)$ in the target domain. The result of the transformation in the target domain is then reinterpreted in $D$, using the inverse $\text{inv}(m)$ of the modelling function. The whole process 'works' if the model accurately reflects the transformation $t$ in domain $D$. This is expressed by the equation:

$$\text{inv}(m) \circ m(t) \circ m = t$$

where $\circ$ indicates function composition. This relationship can be expressed more simply as:

$$m(t) \circ m = m \circ t$$

Such structural relationships are the subject of the mathematical theory of categories. The configuration in Figure 4.2 is called a *diagram*, and the satisfactory way that the functions (*arrows*) work together in the diagram allows us to say that the diagram *commutes*. Such flights of abstraction are helpful in general but take us rather too far from the main thread of this chapter. Herring (1991b) gives a useful discussion of the application of category theory to system modelling.

For GIS, models operate in a wide range of different situations, from models of particular application domains (e.g. transportation models) to specific computer-based models of the physical information in the system. Thus, Figure 4.1 can be elaborated to the more general scheme, shown in Figure 4.3. Moving from left-to-right, the world of the application domain is the subject of a *domain model*. Such models range from the simple to the highly complex and are constructed by domain experts. Examples are:

- *Administrative areas domain*: a region of the plane is divided and subdivided into regions representing areas in an administrative hierarchy (e.g. counties, districts, wards).

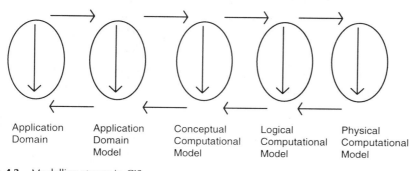

| Application Domain | Application Domain Model | Conceptual Computational Model | Logical Computational Model | Physical Computational Model |

**Figure 4.3**  Modelling stages in GIS.

- *Electrical supply networks domain*:  possibly a simple network model, but populated with complex entities modelling switching, substations, transformers, variable voltages, etc.
- *Global environmental domain*:  highly complex models of climate, terrain, etc.

The *conceptual computation model* takes into account the computational context. This is where E-R and object modelling can play a role – for example, electricity substations may be represented as object types with attributes, behaviour and relationships to other object types. The conceptual model is usually constructed by combinations of computer specialists (analysts) and domain experts.

The *logical computational model* takes into account not just the generalized role of computation but a specific paradigm or paradigms. If relational databases are to be used to hold the non-spatial data on electricity supply, then the objects/entities of the conceptual model will need to be represented as relation schemes. Constructing this model is the task of the information system designer.

The *physical computational model*, constructed by the system developer, represents the implementation of the above models in a specific system and platform.

### 4.1.2  Model quality: accuracy and precision

The modelling formalism introduced in section 4.1.1 above allows a clear distinction to be made between two components of data quality: accuracy and precision. *Accuracy* concerns level of match between source and target domains in the modelling approach, whereas *precision* is related to *resolution* and concerns the level of granularity in which measurements can be made in the target domain. The *resolution* of a domain is the smallest measurement that it is possible to register in the domain. All models have shortcomings and do not provide a complete representation of the domain of interest (otherwise, they would not be models but the domain itself). Lack of accuracy can occur in any model: it arises from a mismatch between domains (i.e. the domain is not appropriate or modelling function is not accurate). Such discrepancies may be qualitative, for example differences of classification in the two domains, or quantitative.

Spatial models represent domains in which there is an important spatial component. Quantitative accuracy has always been a problem for spatial data, because such errors propagate rapidly and unpredictably. Quantitative errors may arise because of differences of resolution in the two domains. Chrisman (1991) has provided a typography of error in spatial models. Chrisman's error components include:

- *Error in position*:  Are the locations of the spatial elements of the model compatible (to some level of accuracy) with their locations in the application domain?
- *Error in attribute value*:  Are the non-spatial attributes of the elements in the model compatible (to some level of accuracy) with their properties in the application domain?
- *Logical consistency* (e.g. holding to topological rules):  Are the elements (spatial and non-spatial) in the model consistent with each other?
- *Completeness*:  Does the modelling domain handle all the elements of the application domain that are required?

Apart from questions of accuracy and precision, it should always be kept in mind that the models are just models and no more than models – they are not the real world. Too great a reliance on any model can restrict our understanding of the world: geographic phenomena are not just occurrences of predefined spatial object types. Couclelis (1992) notes that 'The points, lines and polygons that do exist in the geographic world are practically all human artefacts, falling into two broad categories: (a) *engineering* works such as roads, dykes, runways, railway lines, and surveying landmarks, and (b) administrative and property boundaries.'

### 4.1.3  Field or object?

As discussed in Chapter 3, there are two broad and opposing classes of models of geographic information (Chrisman, 1975, 1978; Peuquet, 1984). The class of *field-based* models treats such information as collections of spatial distributions, where each distribution may be formalized as a mathematical function from a spatial framework (for example, a regular grid placed upon an idealized model of the Earth's surface) to an attribute domain. Patterns of topographic altitudes, rainfall and temperature fit into this view. The class of *object-based* models treats the information space as populated by discrete, identifiable entities, each with a geo-reference. These two model types result in two opposed GIS implementation approaches, namely vector and raster. Couclelis (1992) observed that 'the object-vs. field debate in GIS closely parallels a much more fundamental controversy in the philosophy of science, that between the atomic and the plenum ontologies'. This manifests itself, for example, in the particle (atomic, object-based) *versus* wave (field, field-based) physical theories of the earlier twentieth century.

From the viewpoint of the relational model (Chapter 2), measurable geographic phenomena may be recorded as collections of tuples: each tuple contains a list of values measured in each of the dimensions and each value conforms to the structure of some domain of data. Figure 4.4 shows a collection of tuples recording annual weather conditions at different locations. The tuple records the location (probably the identifier), mean temperature, temperature variation, mean wind speed, wind

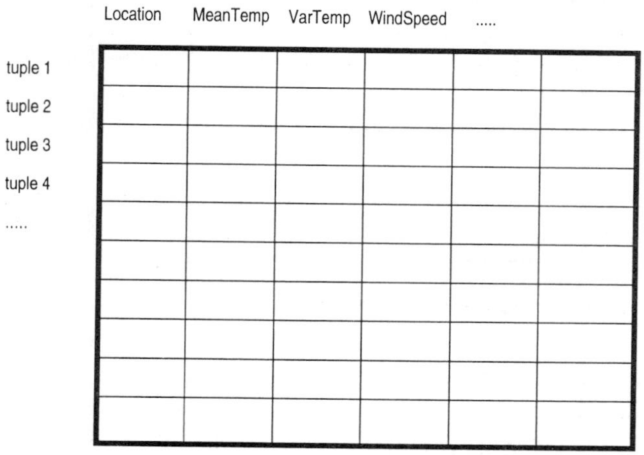

**Figure 4.4**  Relation containing annual climate data.

direction and mean rainfall. Some attributes may be calculated from others and therefore need not be explicitly stored in the tuple (e.g. graphical position may be calculated from real-world position at a given scale and level of generalization). This viewpoint, while being general is not necessarily the most helpful: a large collection of tuples does not provide any immediate evidence of pattern and may involve intractably large amounts of information. The field-based and object-based approach are attempts to impose structure and pattern on such data.

The field-based approach treats the information as a collection of fields. Each field defines the spatial variation of an attribute in the relation as a function from the set of locations (*spatial framework*) to an attribute domain. In the example of Figure 4.4, the fields (Figure 4.5) include mean temperature, rainfall and wind speed. Note that it is the *function* that is the field, not the set of values.

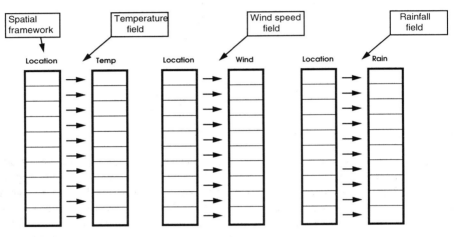

**Figure 4.5** Field-based approach to geographic phenomena.

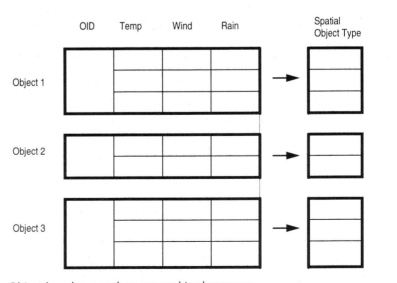

**Figure 4.6** Object-based approach to geographic phenomena.

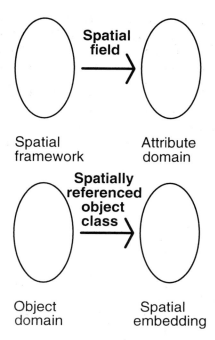

**Figure 4.7**   Spatial field and object are inverse constructs.

The field-based approach conceptualizes the relation as divided into variations of single or multiple attributes (columns). The alternative object-based approach clumps a relation as single or groups of tuples. In the example of Figure 4.4, we may characterize certain groups of measurements of climatic variables as falling into a finite set of types. Thus (and purely fictitiously), Southern Maritime may have a higher than average temperature, medium-to-high wind and rain. Southern Maritime is then an object that will have some spatial references (the locations that experience that climate type). This structuring of the relation is shown in Figure 4.6. We have given each set of clumped tuples a single object identifier to show that they all belong to the same object.

It is interesting that the field-based and object-based models are in a sense inverse to each other. With the field-based approach, the first-class entities in the model (fields) are functions from the spatial framework to the other attributes. On the other hand, the object-based approach constructs a population of entities with a spatial embedding, i.e. a function from entities to a spatial framework (see Figure 4.7).

## 4.2   FIELD-BASED MODELS

### 4.2.1   Fields and the field-based approach

The left-hand part of Figure 1.5 in Chapter 1 shows a representation of a digital elevation model of Biddulph Moor, near the Potteries region. The representation has already been discretized by elevations within a small number of bands. The degree of lightness in the grey-scale image indicates the level of the value of the elevations at locations in the spatial framework. The spatial reference is the independent variable and the elevation attribute varies with respect to this reference.

Suppose that we are set the task of providing a geographic database for a given region and we decide to use a field-based model of the information within it. One of the first tasks will be to construct a suitable spatial framework for our model.

A *spatial framework* is a partition of the given region into a finite tessellation of spatial objects. The individual areal components of the partition are called *locations*. For some purposes, it is sufficient to approximate the locations by points.

The tessellation used in a spatial framework is often regular (but this is not a requirement), and will often be a grid of squares (e.g. Ordnance Survey Grid) or triangles (as in triangulated irregular network (TIN) models). The spatial framework is a finite structure – it must be in order to be tractable to computation. Often the application domain will not be finite, or it may be larger than can be practically accommodated in totality. This implies that the phenomena that are to be modelled will be *sampled* and the necessary error introduced in the sampling process must be considered.

It is usual for an information system based upon the field-based approach to contain several fields, so that comparisons and combinations of the constituent fields can be made. Measurements of topographical altitude with respect to this framework constitute a single layer of our model. Another layer may consist of the variation of temperatures over the same region. In general, there will be many layers in our database, each with respect to the same or different underlying spatial frameworks. Figure 4.8 shows two fields having the same spatial framework, relating to Biddulph Moor, as in section 1.2.3.

The field-based spatial model is now defined. For simplicity, assume that each field in the model has the same spatial framework *F*. In general, we would allow the

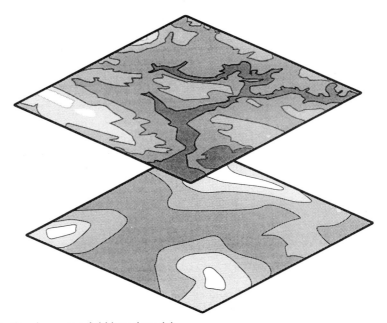

**Figure 4.8**   Two layers in a field-based model.

spatial framework to be based upon any mathematical model of space. In practice, most spatial frameworks are based upon Euclidean space (in two-dimensions, the Euclidean plane), as in Chapter 3, where distance and angle are defined in the usual way. (Later in this chapter we will see an example where the underlying space is not Euclidean.)

A *field-based model* based upon spatial framework $F$ consists of a finite collection of $n$ spatial fields, $\{f_i : 1 \leq i \leq n\}$.

For $1 \leq i \leq n$, each *spatial field* $f_i$ is a computable function from set $F$ to a finite attribute domain $A_i$.

In order that this model will work computationally, the number of fields, the cardinality of the domains $A_i$ and the spatial framework $F$ are all stipulated to be finite. Also, the field functions $f_i$ are defined to be computable (i.e. are capable of evaluation using a digital computer). In the special case where the spatial framework is a Euclidean plane and the attribute domain is a subset of the set of real numbers, a field may be visualized as a surface in a natural way. The Euclidean plane plays the role of the horizontal $xy$-plane and the spatial field values give the $z$-coordinates, or heights above the plane. Assuming that the function is single-valued, then the field may be represented as a surface.

Fields are often visualized using a set of isolines. An *isoline* is the locus of all points in the field with the same attribute value. The left-hand part of Figure 1.5 shows an isoline representation of topographical elevations for the area around Biddulph Moor. The right-hand part of Figure 1.5 shows an isometric projection of the same surface, draped with the same isolines. To sum up the modelling method using a field-based approach:

- Construct, or use a given suitable model of the underlying space to act as the spatial framework, $F$.
- Find suitable domains for the attribute(s) to act as the $A_i (1 \leq i \leq n)$.
- Sample the phenomena under consideration at the locations in the spatial framework, so as to construct the spatial field functions $f_i$, for $1 \leq i \leq n$.
- Perform analyses by computing with the spatial field functions.

*Example 1 (Regional climate variations)*

A simple model would place a square grid over the region and measure aspects of the climate at each node of the grid to a pre-determined level of accuracy. A more realistic model would use an irregular spatial framework to reflect where it is practical to set up weather stations to record such measurements. Fields are associated with variations of each of the climatic attributes. Fields may be smoothed, banded, combined, etc.

*Example 2 (Regional health)*

Population and health measures may be derived from census information, hospital databases, and other sources. Many of the values of attributes may be measured with respect to predefined and differing subdivisions of the region. For example, population measurements may be referenced to UK enumeration districts. In the case that a field function has a domain that is different from the original spatial framework, it would be necessary to supply an algorithm to redistribute the values of the attributes. Having set up the fields so that they are mutually compatible, the model could be used to analyse spatial patterns of specific diseases or relate such patterns to other spatial fields in the model.

It is worth re-emphasizing that the conceptual models under discussion are independent of any implementation or any physical representation of the data, although we shall find that some models lead more naturally to certain types of data structure than others. For example, field-based models may often lend themselves more readily to raster-based data structures, but the two concepts are quite independent.

### 4.2.2 Properties of fields

In this section possible properties of spatial fields are described. These properties may be properties of the spatial framework, attribute domains or spatial field functions.

*Properties of the spatial framework and attribute domains*

The spatial framework may be regular or irregular and will usually be a Euclidean space that allows the measurement of lengths and angles within it. Its resolution and locational error structure are often very important and should be appropriate for the types of data and analyses that it is designed to support.

The attribute domain may contain values that are measurements belonging to one of the following types (Stevens, 1946):

*Nominal*: Qualitative rather than quantitative values, and therefore incapable of any arithmetic operations. For example, the name of a town is a nominal value. The resulting field, if based upon a gridded spatial framework, is often called a *categorical grid*.

*Ordinal*: Quantifies by ordering on a linear scale, but not by magnitude. An example might be a level of accident risk, given as category one, two, three, etc. The lower the category the higher the risk, but just how much higher is not quantified. Ordinal values may be compared for size but they may not be added, subtracted, multiplied, etc.

*Interval*: Quantifies by defining relative position on an interval scale, but without reference to a fixed point. Thus, differences in temperature measurements may be quantified without our being able to quantify ratios of temperature (unless we bring in the concept of absolute zero). In commonsense terms it makes sense to say that Madrid is 10°C hotter than Edinburgh, but not that Madrid is twice as hot as Edinburgh. Interval measurements may be compared for size, with the magnitude of the difference being a meaningful notion.

*Ratio*: Measurements on a ratio scale are defined with respect to a fixed point. Such measurements are capable of supporting a wide range of arithmetical operations, including addition, subtraction, multiplication and division. There are many examples of such attribute domains, amongst which are annual rainfall, topographical altitude above sea level, population density and incidence of particular diseases.

Another property of an attribute domain is its support for *null values*. A null value is registered if the value is unknown or the measurement inappropriate. For example, in a field-based model of biotype variation, tree classification values would be inappropriate in ocean regions. On the other hand, tree classification values may be unknown for certain inaccessible inland regions. A rich domain will be able to support the difference between these two concepts of null value.

### Continuous, differentiable and discrete fields

A spatial field is *continuous* if the underlying spatial field function is continuous. Intuitively, this means that any small change in location results in only a small change in attribute value. Thus, there are no sudden jumps in value, as might occur with a vertical cliff in a digital elevation model. Continuity is only appropriate if the notion of 'small change' is well-defined in the spatial framework and attribute domain, that is if the domains are themselves continuous.

A spatial field is *differentiable* if the underlying spatial field function is differentiable. A differentiable function is one for which its rate of change (slope) is always defined. As with continuity, differentiability only makes sense for a continuous spatial framework and attribute domain. Examples of continuity and differentiability are shown in Figure 4.9. For the sake of simplicity, assume that the spatial framework is one-dimensional, ranged along a horizontal line. Then the variational field may be plotted as a graph of attribute value against spatial framework. In case A of Figure 4.9, the variation is represented as a continuously smooth curve: clearly differentiable as the slope of the curve can be defined at every point. In case B, although the field is continuous (the graph is connected), it is not differentiable everywhere. There is an ambiguity in the slope, with two choices at the articulation point between the two straight line segments. In case C, the graph is not connected

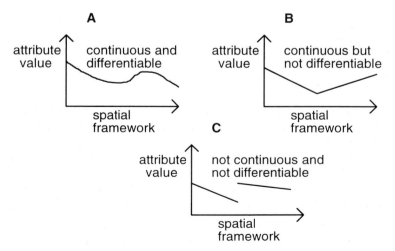

**Figure 4.9** Examples of continuity and differentiability for fields over a one-dimensional spatial framework.

and so the field is not continuous and *a fortiori* not differentiable. In the case where the spatial framework has dimension two (or higher), the slope is dependent not only on the particular location but also on the bearing at that location (see Figure 4.10). There is a logical connection between the concepts of continuity and differentiability. Every differentiable function must be continuous, but not necessarily conversely.

Another important distinction is between *continuous* and *discrete* variation, and this can be well illustrated by examples.

■ A glacier movement is *continuous*. If today a measuring pole is at location *A* and this time next year the pole is at location *B*, then it makes sense to interpolate. We might make a linear interpolation and estimate that in six months the pole will be mid-way between *A* and *B*. The marker pole moves continuously.

■ The change of an administrative boundary is *discrete*. If today the boundary follows line *A* and this time next year the boundary follows line *B*, then it does not make sense to say that in six months' time the boundary will be half-way between *A* and *B*. The boundary jumps discretely.

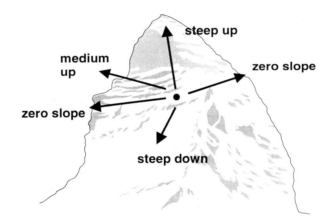

**Figure 4.10**   A differentiable field where the slope at a point depends upon the bearing.

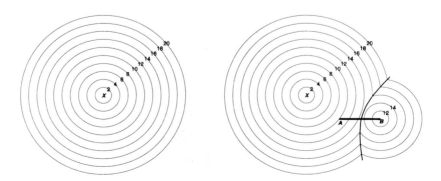

**Figure 4.11**   Travel time conditions in isotropic and anisotropic fields.

*Isotropic and anisotropic fields*

A characteristic feature of a spatial field is whether conditions within it vary with direction. A field whose properties are independent of direction is called an *isotropic* field. Consider travel time in a spatial framework. Let us start with the simplest possible assumption that the time taken to travel between locations is in direct proportion to the Euclidean distance between them. Then, the locus of all points that are a constant time (*isochrone*) from an arbitrary point $X$ is a circle. Figure 4.11 (left) shows concentric circles, being the isochrones for times 2, 4, 6, ... units. This field is isotropic, because the time from $X$ to any point $Y$ is dependent only upon the distance between $X$ and $Y$ and is independent of the bearing of $Y$ from $X$.

Now, let us make life more interesting. Suppose that there is a high-speed link $AB$ in the spatial framework (see Figure 4.11, right). For simplicity, assume that the link is so fast as to make the travel time from $A$ and $B$ negligible compared with the travel times anywhere else in the field. Then, when travelling between two points, there is a choice whether or not to use the high-speed link (worm-hole?). Consider points close to $X$ (say, within 10 time units) in Figure 4.11 (right). Clearly in these cases, a traveller would do better not to use the high-speed link. However, for points near $B$ it would be better for the traveller from $X$ to travel to $A$, take the link (zero-time) and continue on from $B$ to the destination. The isochrones are shown in the figure. The hyperbola marks the boundary between regions where it is better/worse to use the link. In the second case, to determine the travel time, it clearly matters in which direction the destination location is from $X$. The field is said to be *anisotropic*.

Anisotropic fields are common in real-world situations. They are often linked to networks (in our example, the high-speed link was a very simple network). Other examples that introduce anisotropic conditions are natural and artificial barriers to direct accessibility.

*Spatial autocorrelation and other spatial pattern descriptors*

Spatial autocorrelation (Cliff and Ord, 1981) measures the degree of clustering of values in a spatial field. It is a quantitative expression of Tobler's famous proposition (1970), that 'everything is related to everything else, but near things are more related than distant things'. If a spatial field has the property that like values tend to cluster together, then the field exhibits high positive spatial autocorrelation. If there is no apparent relationship between attribute value and location neighbourhoods, then there is zero spatial autocorrelation. If there is the propensity of like values to

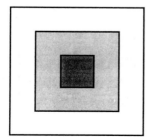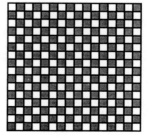

**Figure 4.12**   Patterns with high positive and negative spatial autocorrelation.

be located away from each other, then there is negative spatial autocorrelation. Spatial autocorrelation therefore measures the relationship between attribute values at a location and attribute values in the location's neighbourhood. Figure 4.12 gives examples of a high positive (left-hand bull's eye) and high negative spatial autocorrelation (right-hand chequer).

Spatial autocorrelation is one of many descriptors of patterns in spatial fields. The understanding of fields in geographic space leads into spatial analysis. Openshaw (1991) gives a summary of the methods available for spatial analysis and suggests appropriate spatial analysis functionality for GIS. Patterns and relationships in spatial fields are often interesting, but are not ends in themselves. The aim is to identify relationships and then to use them to further understand the application domain.

### 4.2.3   Operations on fields

A field operation takes as input one or more fields and returns a resultant field: thus, fields form a closed structure under field operations. This section describes some of the typical operations that field-based models allow, dividing them into three main classes. Before these classes are described, some preliminary definitions of neighbourhoods and zones are needed.

> Given a spatial framework $F$, a *neighbourhood function* $n: F \rightarrow \mathscr{P}(F)$ is a function that associates with each location $x$ a set of locations which are 'near' to $x$.

For set $F$, section 3.3.1 defined the power of $F$, $\mathscr{P}(F)$, as the set of subsets of $F$. A member of $\mathscr{P}(F)$ is therefore a subset of $F$. For each location $x$ in $F$, $n(x)$ will then be a subset of $F$, called the neighbourhood of $x$. (Note the parallel with neighbourhood topologies in Chapter 3.) Figure 4.13 illustrates the neighbourhood function idea. Nearness, of course, depends upon the underlying spatial framework. If, as is the case with many practical applications, the space is Euclidean, then the neigh-

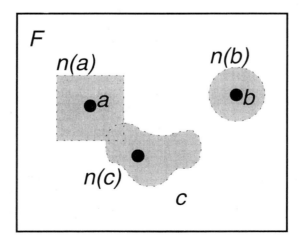

**Figure 4.13**   A neighbourhood function $n$.

bourhood of $x$ might be the set of points within a specified distance and/or bearing of $x$. If the space is metric, then the neighbourhood of $x$ might be the set of points within a specified distance of $x$. If the space is topological, then the neighbourhood could be defined in terms of topological neighbourhood. The neighbourhood of $x$ will usually contain $x$ itself.

The next definition allows us to partition the spatial framework into disjoint subsets (zones), the elements of each zone satisfying some condition based on the field function.

> Given a spatial field function $f$, an *f-zone* is a subset of $F$, members of which satisfy a condition defined using $f$.
>
> An *f-zoning* of $F$ is a partition of $F$ into disjoint *f*-zones.

When the spatial field function is obvious, we omit it. Figure 4.14 shows an example of a zoning of a spatial framework based upon field function $f$. Zones are then created according to the conditions:

for all $x \in L$,    if $f(x) = 0$          then $x \in$ zone $A$

for all $x \in L$,    if $0 < f(x) < 5$    then $x \in$ zone $B$

for all $x \in L$,    if $f(x) \geq 5$        then $x \in$ zone $C$

The algebra of a field-based model is specified as the structure of the possible operations on its constituent spatial field functions. Such operations take as arguments existing spatial field functions and produce as a result a new spatial field function. Tomlin (1990) has provided a typology of these operations, which is summarized below in an amended and slightly generalized context.

### Local operations

A *local operation* acts upon one or more spatial field functions of the field-model to produce a new field. Its distinguishing feature is that the value of the new function

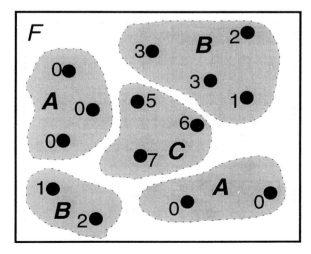

**Figure 4.14** A zoning of a spatial framework.

at any location *is dependent only on the values of the input field functions at that location*. Local operations may be unary (transforming a single field), binary, etc. We give a more formal definition in the binary case. The general case may easily be extrapolated from this.

Formally, suppose given a framework $F$ and spatial field functions $f$ and $g$. Suppose further that $\bullet$ is a binary operation that acts on values in the attribute domains of $f$ and $g$, respectively, to produce a value in another attribute domain. Then, we may pointwise construct a new spatial field function $h$ defined by:

for each location $x$, $h(x) = f(x) \bullet g(x)$.

This binary combination of the two fields $f$ and $g$ is shown in Figure 4.15.

Suppose our earlier field model of regional health (example 2) contains fields of populations and lung cancer mortalities, registered to the same spatial framework. The mortality and population attributes are both real numbers and so can be divided in the case when there is non-zero population at a location (operation $\bullet$ is division in this case). Then we can derive a new field of ratios of lung cancer mortalities per unit population. The derivation of a value at any location depends only upon the values of lung cancer mortality and population at that location. Therefore, the operation that derives the new field is local.

*Focal operations*

For a focal operation, in the creation of the derived field, the attribute value derived at a location $x$ may depend not only on the attributes of the input spatial field functions at $x$, but also on the attributes of these functions in the neighbourhood $n(x)$ of $x$. Thus, the values of the field nearby may influence the value of the derived field at that location. In the unary case, the process is as follows. Suppose given a framework $F$, a neighbourhood function $n$, and a spatial field function $f$. For each location $x$:

1. Compute $n(x)$ as the set of neighbourhood points of $x$ (usually including $x$ itself).

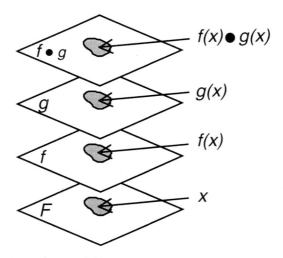

**Figure 4.15**   Local operation on two fields.

2. Compute the values of the field function $f$ applied to appropriate points in $n(x)$.

3. Derive a single value of the new field from the values computed in step 2, possibly taking special account of the value of the field at $x$.

A good example of a focal operation is the operation that computes the gradient of a field of topographical altitudes over a continuous spatial framework. In this case $n(x)$ will be close neighbours of $x$ and $\varphi$ will compute the differences between values of the neighbours of $x$ in different directions to produce a slope vector.

### Zonal operations

A zonal operation aggregates values of a field over each of a set of zones (arising in general from another field function) in the framework. In more precise terms, suppose given a framework $F$, a spatial field function $f$ and set of $k$ zones $\{Z_1, \ldots, Z_k\}$ that partition $F$. A zonal operation results in the following kind of derivation of a new field. For each location $x$:

1. Find the zone $Z_i$ in which $x$ is contained.

2. Compute the values of the field function $f$ applied to each point in $Z_i$.

3. Derive a single value of the new field from the values computed in step 2, possibly taking special account of the value of the field at $x$.

For example, given a layer of temperatures and a zoning into administrative regions, a zonal operation is required to create a layer of average temperatures for each region.

### Summary of field operations

By way of summary, Table 4.1 contains brief notes on a sample of general field operations. For some of these, where the spatial framework is assumed to be a Euclidean plane and the field values are real numbers, it is useful to view the field as a surface. The local operations **lsum**, **ldiff**, **lprod**, **lquot**, **lmax**, **lmin**, **lmean** pointwise compute arithmetic combinations of fields. Focal operations **slope**, **aspect**, **fmean**,

**Table 4.1** A sample of field operations

| Type | Name | Degree | Description |
|------|------|--------|-------------|
| Local | **lsum, ldiff, lprod** | binary | Pointwise sums, differences |
|       | **lquot** |        | products and quotients |
|       | **lmax, lmin** | binary | Pointwise maximums or minimums |
|       | **lmean** | n-ary | Pointwise means |
| Focal | **slope** | unary | Maximum gradients at locations |
|       | **aspect** | unary | Bearings of steepest slopes at each location |
|       | **fmean** | unary | Weighted average based upon the neighbourhood |
|       | **fsum, fprod** | unary | Sum and product of field values in the neighbourhood of each location |
| Zonal | **zmin. zmax** | unary | Minimum and maximum values of the field values in each zone |
|       | **zsum, zprod** | unary | Sum and product of field values in each zone |
|       | **zmean** | unary | Mean of field values in each zone |

**fsum, fprod** are all unary and at each location compute a value dependent upon the field values in its neighbourhood. The zonal operations **zmin, zmax, zsum, zprod, zmean** are also unary and at each location compute a value dependent upon the field values in the zone containing it. A more detailed list may be found in Tomlin (1991).

One of the fundamental cartographic operations is the overlay of one layer upon another, thus producing a composite function of the underlying space. Without using the language of overlay specifically, it should be clear that overlays are implicit in many of the operations performed above. Indeed, the example of a local operation involved the overlay of population and cancer mortalities, one on the other, in order to perform an arithmetic operation (division) upon the individual pairs of elements in the layers. Similarly, a zonal operation may be viewed as an overlay of a layer onto a zoning layer.

### 4.2.4   Digital elevation models

Digital elevation models (DEMs) provide an excellent application of the field model. (There is some confusion in the literature in usage of the terms 'digital terrain model' (DTM) and 'digital elevation model' (DEM). A DEM models topographic elevations. Weibel and Heller (1991) note that a DTM is a fuller model of the surface of the Earth, representing topographic elevations and other terrain features.) In this case, the spatial framework directly represents a portion of the Earth's surface, usually as a piece of Euclidean plane or part of the surface of a sphere embedded in Euclidean 3-space. The field values represent heights, measured at locations on the spatial framework. It is assumed that the spatial field function is single-valued, so cliff overhangs may not be represented. The model may be visualized as a '2.5-dimensional' surface. (A spatial object $O$ is commonly known as 2.5-dimensional if it is embedded in Euclidean 3-space and has the property that if $(x, y, z)$ and $(x, y, z')$ are members of $O$ then $z = z'$.)

The locations in the spatial framework are usually conceived as points rather than areas. The configuration of points may be in the form of a regular grid, leading to an *elevation matrix* or irregularly placed wherever it was convenient and useful to measure the topographic elevation. The advantage of a regular grid is simplicity of modelling and analysis. However, there is no flexibility to take denser sets of readings in areas of interest or difficulty; neither can the structure of the grid parallel the structure of the terrain (following ridges, etc.). The pros and cons of grid *versus* irregular samplings are discussed in (Peucker 1978, Mark 1979).

Interpolation/extrapolation is an important issue, determining the value of the elevation at points on the Euclidean plane that are not part of the spatial framework. There is a rich literature on interpolation, with no one method viewed as 'best'. In the case of an irregular grid, a simple approach is to triangulate and make a linear interpolation within each of the triangular facets. The structure that supports this is the triangulated irregular network (TIN) (discussed in Chapter 3 and later chapters). There is a choice of triangulations, with the Delaunay triangulation being the most popular. Operations on DEMs include:

- *Volume and surface area*: The volume of material under the surface may be calculated, given a depth and a portion of the spatial framework (useful for cut-and-fill engineering works). Surface area refers to the area of the surface facet.

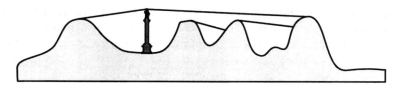

**Figure 4.16**   Section of a visibility map showing terrain seen by Admiral Nelson.

- *Visibility analysis*:   Given a point *p* in space, the *visibility* map depicts the *view-shed* of the surface with respect to *p*. That is, the set of all points on the surface that are visible by direct line of sight from *p* (or, what amounts to the same thing, from which *p* is visible). Figure 1.6 shows a viewshed with respect to the proposed site of an opencast mine on Biddulph Moor. Figure 4.16 shows a section of another visibility map. Nelson's Column is placed in a hilly terrain: the diagram shows which sections of the terrain can be seen by Nelson's one eye (or which areas can see Nelson's head). Visibility analysis is important in many applications, for example in radio transmission, military operations, and visual effects of plant.

- *Slope, aspect, higher derivatives*:   As already discussed, the steepness of a path in a surface at a point depends upon the direction of travel. The direction that produces the greatest slope at a point in a surface is called the *aspect* at that point. The *slope* of a surface at a point is the measurement of the gradient in the direction of the aspect. Higher derivatives, such as measures of concavity/convexity may also be useful, and are also dependent upon direction. Slope and aspect are often important as components of land-use analysis.

- *Route planning*:   This is a higher-order operation that brings in slope, aspect, visibility and other operations.

Isolines and projections are useful aids to visualizing DEMs. Hill-shading may be used to enhance projections. Facets are shaded according to their slope and aspect, and positioning with respect to a notional light source, according to some model of lighting, such as diffuse reflection of the illuminated object. It is also possible to drape isolines or some other surface (e.g. a road map) over the DEM.

### 4.3   OBJECT-BASED MODELS

Object-based models decompose an information space into *objects* or *entities*. As stated in (Mattos *et al.*, 1993), an entity must be:

- identifiable,
- relevant (be of interest),
- describable (have characteristics).

With respect to the last item, the description of an entity is provided by static properties (such as the name of a city), behavioural characteristics (such as a method of plotting a representation of the city at a particular scale) and structural characteristics (placing the object in the overall structure of the information space).

Instead of laying the spatial framework down, like a carpet, and allowing the attribute values (e.g. population densities) to vary with respect to it, the object-based approach populates the *information space* with objects (e.g. cities, towns, villages, districts) that have attributes (e.g. population density, centroid, boundary) given as the characteristics of other objects (e.g. rational number, point, polygon). In this approach, the entire frame of spatial reference does not become distinguished and prescribed as it does in the field-based approach (the spatial framework), but is provided by one or more attributes of the objects in the information space. In a more general context, this is precisely the object-oriented approach to data modelling discussed in Chapter 2.

Entities in such a model will have several categories of dimension along which attributes may be measured. These include spatial, graphical, temporal and textual/numeric (see Figure 4.17). To give a simple example, a land parcel may have a polygon representing its boundary in the real world (spatial object), a polygon and point representing its cartographic form at differing levels of generalization (graphical objects), times when it was created in the real world and in the system (temporal objects), and attributes describing its area, owner and name (textual/numeric objects). In practice, the dimensional categories do not arise in such an unmixed form. For example, a land parcel may change shape at different times and have different owners at different times. Modelling approaches must allow these categories to be unified. The temporal dimension will be discussed further in Chapter 8.

There is a distinction, not always made, between the *spatial* and *graphical* dimensions: spatial objects exist to directly model the application domain, while graphical objects are their presentational form. This resonates with the earlier distinction between modelling at the application domain level and the systems level. Graphical objects are required for systems modelling, and the relationship between spatial and graphical objects has traditionally been the subject of cartography. The transformation from spatial to graphical objects introduces abstraction and generalization processes. Graphical objects themselves are required to hold methods for rendering on appropriate presentation media.

Another useful distinction is between *spatially referenced objects*, such as houses, roads and hospitals, with spatial properties, and *spatial objects*, such as points, arcs and areas, that are referenced by spatially referenced objects. Object type inheritance cannot be used here. It is not the case that a house is a polygon, therefore

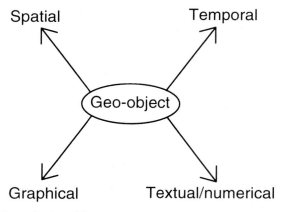

**Figure 4.17**   Dimensions of a geo-object.

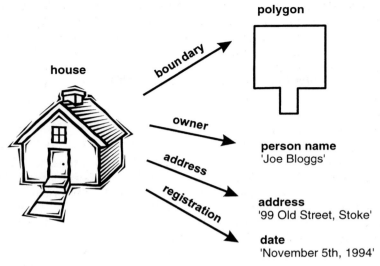

**Figure 4.18**   Spatially-referenced object **house**.

object type **house** does not inherit from object type **polygon**. It *is* the case that a house has a boundary that may be represented as a polygon at a particular scale and resolution on a display (*object type* **house** *references object type* **polygon**). Figure 4.18 shows an instance (more precisely a graphic of an instance) of object type **house**. The arrows show some of the object's attributes, referencing further object types, **polygon, person name**, **address** and **date**.

In summary, an object-based model for a GIS application is mainly populated by spatially referenced objects. (In fact, we really want to say 'spatio-temporally referenced objects' here, but will delay the explicit consideration of time until Chapter 8.) A spatially referenced object has attributes, some of which are spatial objects. There may be more than one spatial object that a spatially referenced object has as an attribute, for example, the house above may reference a polygon as its boundary and a point as its seed. The alphanumeric object types referenced by a spatially referenced object (person name and address in the above example) are well understood and a part of traditional systems modelling: they will not be discussed further. The spatial object types are the subject of the next section.

### 4.3.1   Spatial objects

Spatial objects, are called 'spatial' because they exist inside 'space', called the *embedding space*. The specification of a spatial object depends upon the structure of its embedding space: for example, it would not be easy to define a circle that is embedded in a space, each point of which can only have integer coordinates. The following typology of embedding spaces is often useful (see Chapter 3).

*Euclidean*:   Admitting measurements of distances and bearings between objects. Given a suitable frame, objects in Euclidean space may be represented using sets of coordinate tuples.

*Metric*: Admitting measurements of distances (but not necessarily bearings) between objects. An example of a metric space is a travel time surface (assuming that travel time is symmetric, in the sense that the travel time between $x$ and $y$ is the same as the travel time between $y$ and $x$.

*Topological*: Admitting topological (but not necessarily metric or Euclidean) relationships between objects. Topological relationships are those that are preserved under 'rubber-sheet' transformations, which continuously deform the underlying space. Examples of topological relationships are connectivity and adjacency.

*Set-oriented*: Admitting only general set-based relationships, such as membership, containment, union and intersection. The containment structure of a hierarchy of administrative regions provides an example of a set-oriented space.

The specification of an object requires the definition of its state and behaviour. The most common situation is that the underlying space is Euclidean and each spatial object is specified by a set of coordinate tuples or computable equations. Another approach is to specify a set of primitive objects, out of which all others in the application domain can be constructed, using an agreed set of operations. Primitive spatial object classes that have been suggested include closed half-planes (although these suffer from the non-intuitive property of being infinite in extent), simplicial complexes (the disadvantage here being computational expense in building with such primitive objects) and point-line-polygon primitives (common in existing systems). Consider the following very simple analysis that may be required of a GIS.

Calculate the total length of major roads inside a circle of radius 10 miles and centre a given hospital (Figure 4.19).

The modelling process might proceed as follows. The spatially referenced object classes in the application domain are identified as **road** and **hospital**. The spatial object classes that they reference (after abstraction and generalization) might be **arc** and **point**, respectively. The spatial object class **disc** is also important. Operation **length** will act on **arc**, returning a real number, and **intersect** will apply to form the piece of an **arc** in common with the **disc**. The configuration of spatial objects is shown in Figure 4.20.

The analysis might proceed as with the four stages below. Each stage of the analysis uses objects defined in the model and legal operations that may be applied to those objects.

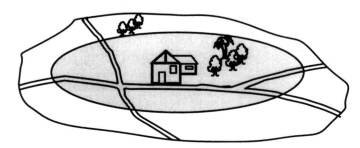

**Figure 4.19**  Hospital and environs.

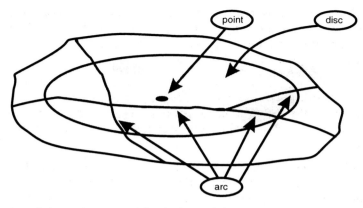

**Figure 4.20**   Spatial objects (continuous) for the hospital domain.

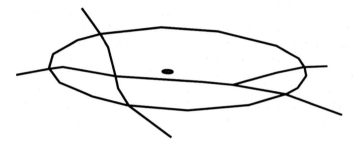

**Figure 4.21**   Spatial objects (discrete) for the hospital domain.

1. Find the disc that has as centre the point representing the hospital and radius 10 miles.
2. Find the intersection of this disc with each of the arcs representing roads. (In practice, not all the roads in the application domain are required; there may be an access method that can retrieve from the database only those arcs that are near to the hospital point.)
3. Find the length of each of the clipped arcs obtained in the previous stage.
4. Total the lengths (an operation on real numbers).

The objects and operations described above are not suitable for computation, because they are continuous and infinite. A process of discretization must convert the objects to types that are computationally tractable. For example, discs may be represented as polygonal areas, arcs by chains of line segments and points will be embedded in some discrete space. Operations such as **intersection** and **length** may then be computed using standard algorithms from computational geometry. Figure 4.21 shows a discretized version of the example.

*Spatial object types in the Euclidean plane*

Figure 4.22 shows a possible inheritance hierarchy for objects in a continuous two-dimensional space, the Euclidean plane with the usual topology. (Names for spatial

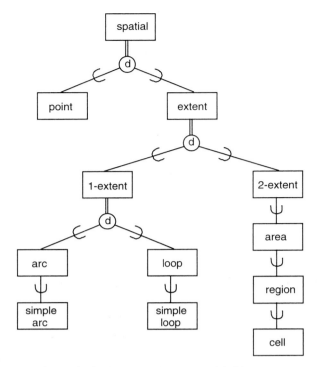

**Figure 4.22**   Inheritance hierarchy for some continuous spatial object types.

object types have never been standard and cause some confusion. We have tried to be as precise and 'standard' as possible.) The most general spatial object type **spatial** is at the top of the hierarchy. This type is the disjoint union of types **point** and **extent**, distinguishing single points and extended objects embedded in two dimensions. Class **extent** may be specialized by dimension into the types **1-extent** and **2-extent**. Two subtypes of the one-dimensional extents have been described in Chapter 3 as **arc** and **loop**, specializing to **simple arc** and **simple loop** when there are no self-crossings. The fundamental areal object was described in Chapter 3 as the regular closed set, called here by the type name **area**. A connected area we name a **region** and a region that is simply-connected (no holes) is a **cell**, homeomorphic to the unit disc.

*Spatial object types in a discretization of the Euclidean plane*

The Euclidean plane is not computable and must be discretized in order that computation can take place. Discrete forms exist for all the continuous types in Figure 4.22. Figure 4.23 shows part of the inheritance hierarchy for discrete one-dimensional extents. The type **line segment** comprises bounded straight line segments that may or may not contain their extreme points. In the case that the segment contains both extreme points, it is *closed*: if it contains neither extreme point, it is *open*. Type **polyline** is a sequence of line segments connecting successive points. A polyline may be *looped* if its extreme points are coincident, or *unlooped* otherwise. A two-dimensional discrete spatial object is a 2-extent that has a discrete

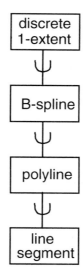

**Figure 4.23**   Inheritance hierarchy for some discrete one-dimensional extents.

1-extent for its boundary. The interior and boundary of a simple looped polyline defines the type **simple polygon**. (This construction is well-defined by the Jordan Curve Theorem (Chapter 3).)

Polygons and polylines provide piecewise first-degree polynomial approximations to two-dimensional and one-dimensional extents, respectively. Higher polynomial approximations are also computationally tractable and may be appropriate in certain circumstances, for example when smoothly contoured lines rather than jagged edges are required. Also a higher-degree curve may be more storage-efficient for storing some shapes, rather than polylines with large numbers of segments. Many higher-degree curves in widespread use, mainly in graphics and CAD systems, are based upon cubic polynomials. A cubic polynomial:

$$ax^3 + bx^2 + cx + d$$

has four coefficients, $a$, $b$, $c$, $d$, and thus four controls (*degrees of freedom*) that may be varied to produce different shapes. The most widely used cubic forms are: *Bézier*, *Hermite* and *B-spline*. Bézier and Hermite curves are each defined by two end-points and two controls, while a B-spline is defined by two end-points and four controls. Two examples of a Bézier cubic curve, with its controls, are shown in Figure 4.24. Points $A$, $B$ are of course the end-points of the curve, the other points $C$, $D$ control the shape of the curve: the vectors $\underline{AC}$ and $\underline{BD}$ represent the tangent vectors at the end-points. The left-hand and right-hand curves show how the shape of the curve can be changed by altering the tangent vectors: the end-points are the same and the new tangent vectors are $\underline{AC'}$ and $\underline{BD'}$. Type *B-spline* in Figure 4.23 is a sequence of B-spline curves connecting successive points.

### 4.3.2   Static spatial operations

Object behaviour is defined by operations that are applied to an object or objects (*operands*) and produce a further object (*result*). Spatial operations on spatial objects

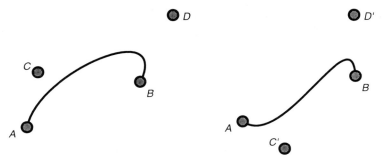

**Figure 4.24**   Two Bézier cubic curves with control points shown.

may be broadly classified as *static* or *dynamic*, static operations defining operations and relationships that do not make essential changes to the operand objects, while dynamic operations change the state of (or even create/delete) one or more of the operands. This section discusses some of the static operations that can be applied to continuous spatial objects embedded in the Euclidean plane. The operations will need modification when acting upon discrete spatial object types. Dynamic operations are discussed in section 4.3.3.

Table 4.2 catalogues operations on continuous spatial objects, embedded in the Euclidean plane, according to their group, operand type(s) and resulting type. Operations in the table are unary, applying to a single operand, or binary, applying to two operands. Operand and result object types are given for each operation. Of course, any subtype of an operand type will also be able to inherit the operation. For example, the **union** operation may be applied to the union of an object of type **arc** with an object of type **region**. The concept of operation polymorphism implies that methods may be different between subtypes and supertypes. For binary operations (relations in the case of binary Boolean operations), we have used the infix form where appropriate, that is $x$ **op** $y$ rather than **op**$(x, y)$.

The operations in the table are grouped into general, set-oriented, topological and Euclidean. In each case, an operation falls into a particular group when its definition requires the structuring of space appropriate for the group. The operation **equals** between **spatial** and **spatial** requires no particular structure, merely that it is possible to tell whether two spatial objects are the same or not. Set-oriented operations require for their specification only the structuring of space into sets of points. Spatial objects of type **extent** have an extension and may be treated as purely sets of points. Operations classified as set-oriented were defined in section 3.3.1 and will not be considered further here, except to note that they are not independent: the usual set-theoretic constraints hold. For example, for any objects $X$ and $Y$ of type **extent**, we have De Morgan's equalities:

$$X \backslash (Y \cap Z) = (X \backslash Y) \cup (X \backslash Z)$$

$$X \backslash (Y \cup Z) = (X \backslash Y) \cap (X \backslash Z)$$

As we saw in section 3.4, spatial object properties and relationships are considerably more complex in a topological setting than the set-oriented context above. Object types with an assumed underlying topology are **arc**, **loop**, **area** (also sometimes **point**). Table 4.2 gives some of the possible topological operations. Operations

**Table 4.2**  A catalogue of static continuous spatial operations

| Group | Operation | Symbol | Operand | Operand | Result |
|---|---|---|---|---|---|
| General | equals | = | spatial | spatial | Boolean |
| Set-oriented | equals | = | extent | extent | Boolean |
| | is a member of | ∈ | point | extent | Boolean |
| | is empty | = ∅ | extent | | Boolean |
| | is a subset of | ⊆ | extent | extent | Boolean |
| | is disjoint from | | extent | extent | Boolean |
| | intersection | ∩ | extent | extent | extent |
| | union | ∪ | extent | extent | extent |
| | difference | \ | extent | extent | extent |
| | cardinality | | extent | extent | cardinal |
| Topological | boundary | ∂ | area | | set(loop) |
| | interior | ° | area | | open area |
| | closure | ‾ | area | | closed area |
| | meets | | area | area | Boolean |
| | overlaps | | area | area | Boolean |
| | is inside | | area | area | Boolean |
| | covers | | area | area | Boolean |
| | connected | | area | | Boolean |
| | components | | area | | set(region) |
| | extremes | | arc | | set(point) |
| | is within | | point | simple loop | Boolean |
| Euclidean | distance | ‖ | point | point | real |
| | bearing/angle | ∠ | point | point | [0,2π) |
| | length | ‖ | arc/loop | | real |
| | area | | area | | real |
| | perimeter | | area | | real |
| | centroid | | area | | point |

**boundary**, **interior**, **closure** and **connected** are defined in the usual manner (section 3.4). The operation **components** returns the set of maximal connected components of an area. Operation **extremes** acts on each object of type **arc** and returns the pair of points of the arc that constitute its end-points. Operation **is within** provides a relationship between a point and a simple loop, returning **true** if the point is enclosed by the simple loop. This relationship is the often used *point-in-polygon* operation.

The Boolean topological operations **meets, overlaps, is inside** and **covers** apply to areas, and are exemplified in Figure 4.25. (Similar topological operators can be defined for arcs and loops.) Informal definitions of these operations are as follows. Let $X$ and $Y$ be objects of type **area**.

$X$ **meets** $Y$ if $X$ and $Y$ touch externally in a common portion of their boundaries.

$X$ **covers** $Y$ if $Y$ is a subset of $X$ and $X$, $Y$ touch internally in a common portion of their boundaries.

$X$ **overlaps** $Y$ if $X$ and $Y$ impinge into each others' interiors.

$X$ **is inside** $Y$ if $X$ is a subset of $Y$ and $X$, $Y$ do not share a common portion of boundary.

Figure 4.25 also shows that these topological operations are more discriminating than set-oriented operations. The figure shows two pairs of objects whose relationships are indistinguishable using only set-oriented operations. From within the structure of pure sets, given two objects $X$ and $Y$ of type **area**, we cannot distinguish the relations $X$ **meets** $Y$ and $X$ **overlaps** $Y$. From the set-oriented viewpoint, the relations are both cases of the relation $X$ **is not disjoint from** $Y$. Similarly, $X$ **is inside** $Y$ and $Y$ **covers** $X$ are both instances of the Boolean set-oriented operation $X$ **is a subset of** $Y$. Also, not all the given topological operations are independent. Some examples of interdependencies between the **boundary, interior** and **closure** operations now follow. Let $X$ be an object of a type that supports a topology, then:

$\partial X$ is the set difference of $X^-$ and $X^\circ$

$X^-$ is the set complement of the interior of the complement of $X$

$X^\circ$ is the set complement of the closure of the complement of $X$

The operations in the topological section of Table 4.2 by no means provide a complete typology of the diversity of spatial relationships in topologically structured spaces. Figure 4.26 shows some of the infinite number of possible topological relationships that are available between objects of type **cell**. In the top-left of the figure, there is a case of a meet relationship, satisfying our informal definition, but the configuration of objects $X$ and $Y$ is not homeomorphic to the objects $X$ and $Y$ shown meeting in Figure 4.25. Similarly, there is a variety of cover and overlap relationships. The spatial relationship between two cells shown in the bottom-right of the figure seems to be a combination of meeting and covering/overlap: in fact, it would be formally described as an overlap (see the later definition of Egenhofer, p. 174).

The operations **distance, bearing/angle, length, area** and **perimeter** available in metric and Euclidean spaces have been covered in Chapter 3. Operation **centroid** returns the centre of gravity of an areal object as an object of type **point**. Distances

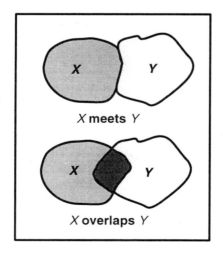

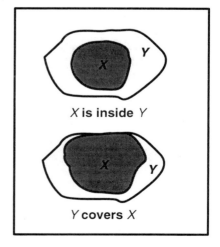

$X$ **is not disjoint from** $Y$                    $X$ **is a subset of** $Y$

**Figure 4.25**   Topological and set-oriented operations.

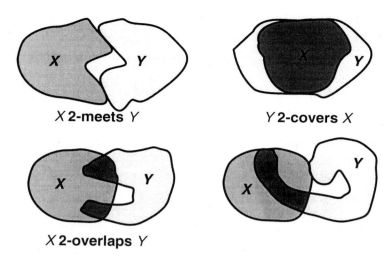

**Figure 4.26** Further topological relationships between cells.

and angles are defined between the **point** elements of the space. In practice, it is often important to measure distances and angles between objects of different dimensions. For example, we might wish to know the distance of a town from a motorway. There are several ambiguities here that must be resolved before the question can be answered properly. Do we mean the town centre or the town as an area? Are we measuring distance along roads, as the crow flies, or by some other means? If, for example, we are measuring as the crow flies, do we want the distance to the nearest point on the motorway or to the nearest intersection? Definitions of distance and bearing/angle may be extended to operations upon objects that are not just points but are subtypes of **extent**. Algorithms have been developed to this end. See, for example, Peuquet and Zhan (1987), where an algorithm is given which determines a directional relationship between polygonal shapes in the plane.

Returning to topology, one characterization of topological relationships between spatial objects (Pullar and Egenhofer, 1988) has been more widely cited in the literature than any other. This method shows that the large number of spatial operations given in Table 4.2 can be defined in terms of **boundary** and **interior** (or other combinations of two from **boundary**, **interior** and **closure**). Although the method can be applied to any object types for which a topology can be defined, we describe it in the context of the type **cell**. The essence of the method is, given two spatial regions, to ask what can be deduced about the topological relationship between them from the set-theoretic relationship of intersection between their interiors and boundaries. To be more precise, let $X$ and $Y$ be the spatial cells, and assume that the boundaries, $\partial X$ and $\partial Y$, and interiors, $X°$ and $Y°$ are known. To find the topological relationship between $X$ and $Y$, compute the set-oriented relationships between $\partial X$, $X°$, $\partial Y$ and $Y°$. In fact, for a first-pass determination of the topological relationships, all that is required is to test whether intersections are empty or non-empty. Thus, consider the following four sets:

$$\partial X \cap \partial Y \qquad \partial X \cap Y°$$

$$X° \cap Y° \qquad X° \cap \partial Y$$

**Table 4.3**  Eight relations between cells in the Euclidean plane

| $\partial X \cap \partial Y$ | $X° \cap Y°$ | $\partial X \cap Y°$ | $X° \cap \partial Y$ | Operation |
|---|---|---|---|---|
| $\varnothing$ | $\varnothing$ | $\varnothing$ | $\varnothing$ | $X$ **disjoint** $Y$ |
| $\neg\varnothing$ | $\varnothing$ | $\varnothing$ | $\varnothing$ | $X$ **meets** $Y$ |
| $\neg\varnothing$ | $\neg\varnothing$ | $\varnothing$ | $\varnothing$ | $X$ **equals** $Y$ |
| $\varnothing$ | $\neg\varnothing$ | $\neg\varnothing$ | $\varnothing$ | $X$ **inside** $Y$ |
| $\neg\varnothing$ | $\neg\varnothing$ | $\neg\varnothing$ | $\varnothing$ | $Y$ **covers** $X$ |
| $\varnothing$ | $\neg\varnothing$ | $\varnothing$ | $\neg\varnothing$ | $Y$ **inside** $X$ |
| $\neg\varnothing$ | $\neg\varnothing$ | $\varnothing$ | $\neg\varnothing$ | $X$ **covers** $Y$ |
| $\neg\varnothing$ | $\neg\varnothing$ | $\neg\varnothing$ | $\neg\varnothing$ | $X$ **overlaps** $Y$ |

Each of these sets may either be empty or non-empty. There are 16 different mutually exclusive combinations of possibilities. As each possibility is a condition on the boundaries and interiors of sets, each possibility will lead to a relationship between the sets that is preserved under topological transformations (homeomorphisms). For general sets in a pointset topology, each of the 16 combinations can exist and leads to a distinct topological relationship between the two sets $X$ and $Y$. We are concerned, however, only with spatial cells embedded in the Euclidean plane. In this case, only eight of the sixteen combinations can occur. Table 4.3 shows the eight possibilities and their correspondence with the spatial operations given earlier in Table 4.2.

It can now be seen that **meets** and **overlaps** are topological refinements of the set-oriented operation **intersection**, while **inside** and **is covered by** are refinements of the **is a subset of** relationship. Because four intersections are tested for empty in this method, it is sometimes known as the *4-intersection*.

Egenhofer (1991c) has extended this work to higher-dimensional spaces, and also takes into account cases where the co-dimension is non-zero (i.e. where the dimension of the spatial objects is less than the dimension of the space in which they are embedded). In this case, three topological operations are used: boundary, interior and set complement. The binary spatial relation between two spatial objects, $X$ and $Y$, is classified by checking for emptiness/non-emptiness of the nine combinations (hence the name *9-intersection* of the method) of the operations applied to $X$ and $Y$. Let $T$ be the topological space in which $X$ and $Y$ are embedded, then the combinations are:

$$\partial X \cap \partial Y \qquad \partial X \cap Y° \qquad \partial X \cap T/Y$$

$$X° \cap \partial Y \qquad X° \cap Y° \qquad X° \cap T/Y$$

$$T/X \cap \partial Y \qquad T/X \cap Y° \qquad T/X \cap T/Y$$

Readers who wish to follow this material further will find references in the bibliographic notes at the end of the chapter.

### 4.3.3  Dynamic spatial operations

Most of the operations that we have looked at so far have been *static* in the sense that the operands are not affected by the application of the operation. For example, calculating the length of an arc has no lasting effect on the arc itself. *Dynamic* operations alter the objects upon which the operations act. This section discusses some of the dynamic operations upon spatial objects. The three fundamental dynamic operations are **create**, **destroy** and **update**. All dynamic operations are variations upon one of these themes.

The creation of objects can either be *dependent* or *independent*. In the case of *independent creation*, an object is created without reference to any other objects in the model. All that is required is specification of its type and attributes. With *dependent creation* an object is created with reference to other objects in the model. Once the new object is created, then it may or may not continue to be dependent on the existence and properties of the parent object. Several types of dependent creation may be distinguished, given here by operations that are parametrized by the referenced objects $X, Y, Z, \ldots$ (see Figure 4.27):

- **reproduce($X$):** acts to produce an identical replica of the referenced object $X$ with regard to type, state and behaviour. (The careful reader might believe that the reproduce operator is not dynamic because it does not alter the dependent object. However, **reproduce** does not act upon the dependent operator, but merely references it. This operator, like other creation operators, takes no arguments.);

- **generate($X$):** acts to create an object that will depend upon but not replicate the referenced object $X$ with regard to type, state and behaviour;

- **split($X$):** acts to create a set of objects whose composition (aggregation) is the referenced object $X$. The original object may or may not be destroyed as a result of this operation;

- **merge($X, Y, Z, \ldots$):** acts to create a single object that is the composition (aggregate) of the referenced objects $X, Y, Z, \ldots$

Destruction of an object using the **destroy** operation is generally taken to be permanent. However, there are cases when a **reincarnate** operation may be appropriate (the rebirth of a nation state, for example). Strictly, reincarnation can only be possible in a temporal system that keeps a record of past history, so that reference to the type, state and behaviour of the dead object can be made.

Another group of dynamic operations are the *transformations*. Transformations are **update** operations that change the spatial attributes of objects. In a Euclidean

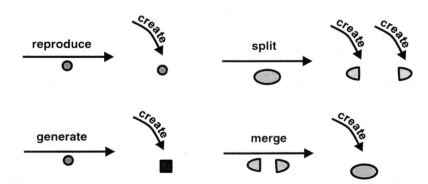

**Figure 4.27**   Some dynamic dependent creation operations.

space, some of the primitive linear transformations are **translate, rotate, scale, reflect** and **shear**. Let $(x, y)$ be an arbitrary point in the Euclidean plane.

- **translate:** changes position of the object in the embedding plane without changing shape, orientation or size of the object. For example, a translation through distance $a$ in the $x$-direction and $b$ in the $y$-direction is effected by the rule:

$$(x, y) \rightarrow (x + a, y + b)$$

- **rotate:** changes orientation of the object in the embedding plane without changing shape, position or size of the object. For example, the rotation through an angle $\theta$ about the origin is effected by the rule:

$$(x, y) \rightarrow (x \cos \theta - y \sin \theta, x \sin \theta + y \cos \theta)$$

- **scale:** changes size of the object in the embedding plane without changing shape, position or orientation of the object. For example, a scaling of $a$ in the $x$-direction and $b$ in the $y$-direction is effected by the rule:

$$(x, y) \rightarrow (ax, by)$$

- **reflect:** mirrors an object in a straight line. For example, the reflection in the line through the origin at angle $\theta$ to the $x$-axis is effected by the rule:

$$(x, y) \rightarrow (x \cos 2\theta + y \sin 2\theta, x \sin 2\theta - y \cos 2\theta)$$

- **shear:** for example, the shear parallel to the $x$-axis is effected by the rule:

$$(x, y) \rightarrow (x + ay, y)$$

All these transformations act on straight line segments to give straight line segments (hence linear). The transformations also act on parallel straight lines to produce parallel straight lines, thus parallelism is an invariant. Any combination of such primitive transformations is an *affine* transformation. Other important groups of dynamic operations are projective transformations, preserving perspective, and topological transformations, preserving topological properties.

### 4.3.4   Formal theories of spatial objects

It is possible to provide formal theories (logical calculi) for spatial relationships that have very general interpretations. Such theories are set in a logical framework, with definitions of terms, well-formed formulae and axioms. An example that we now summarize is Clarke's calculus of individuals (Clarke, 1981, 1985) that has been set in a many-sorted first-order logic by Cui, Cohn and Randell (1993). The focus of the theory is a binary connection relation between regions. (The word 'region' in this context is uninterpreted. Intuitively, a region may be thought to correspond to an object of type **area** in our earlier typology.) Write $C(X, Y)$ as shorthand for 'region $X$ is connected to region $Y$'. The connection relation is reflexive and symmetric, satisfying the following axioms.

1. For each region $X$, $C(X, X)$
2. For each pair of regions $X$, $Y$, if $C(X, Y)$ then $C(Y, X)$

The surprising part of Clarke's calculus is the next step. It turns out that many of the set-oriented and topological relations between spatial objects may be constructed using just the minimal machinery above. Below is a sample of the possible constructions, where the relationships have been named to emphasize the connection with the spatial operation typology of Table 4.2.

$X$ **disjoint from** $Y$:    it is not the case that $C(X, Y)$.

$X$ **a part of** $Y$:    for each region $Z$, if $C(Z, X)$ then $C(Z, Y)$.

$X$ **overlaps** $Y$:    there exists a region $Z$ such that
$Z$ is **a part of** $X$ and $Z$ is **a part of** $Y$
and $X$ is not **a part of** $Y$
and $Y$ is not **a part of** $X$.

$X$ **meets** $Y$:    $C(X, Y)$
and it is not the case that $X$ **overlaps** $Y$.

$X$ **covers** $Y$:    $Y$ is **a part of** $X$
and $X$ is not **a part of** $Y$
and there exists a region $Z$ such that
$Z$ **meets** $X$ and $Z$ **meets** $Y$

$X$ **is inside** $Y$:    $X$ is a part of $Y$
and $Y$ is not **a part of** $X$
and there does not exist a region $Z$ such that
$Z$ **meets** $X$ and $Z$ **meets** $Y$

$X$ **equals** $Y$:    $X$ is **a part of** $Y$
and $Y$ is **a part of** $X$

## 4.4   INTEGRATING FIELD AND OBJECT MODELS

Although the object-oriented approach to systems and object-based spatial data models are superficially similar concepts, they have some distinct differences. It is not necessary to implement object-based models using the object-oriented approach. On the other hand, the object-oriented approach can be used as a framework for describing both field-based and object-based spatial models. For object-based models, the case is clear, but field-based models can also be set in an object-oriented context. In fact, earlier in this chapter fields were described in terms of the properties and behaviour of field function objects, including a set of operations that act upon them: this is precisely the object-oriented approach.

There is some level at which fields and objects can co-exist. We have seen that an object-oriented approach may be taken to the specification of field models. Conversely, some of the types in an object-based approach may be fields. The field and object-based approaches to spatial data modelling are therefore not mutually exclusive. Some applications are more naturally modelled as fields. Example 1 (section 4.2.1), with information on the variation of climatic attributes throughout a region appears to be a natural candidate for a field-based model. However, even this case does not fall completely into the field-based camp. For example, if the data on rainfall are collected at points that are spatially highly dispersed and irregularly spaced, and if these collection points have characteristics of their own, then it might be more natural to model them as objects with position and average annual rainfall as two attributes.

Example 2 (section 4.2.1) on regional health is even less clear-cut. Some aspects of the information, for example population variation over a region, lean more towards field-based modelling. However, data on hospitals, health centres and their attributes are more appropriately modelled as object types (hospital, health centre, town, region, etc.), each with static and dynamic properties, unique identity and relationships with other classes.

A system handling traffic flows along road segments stretches the field-based approach to the limit, because the underlying space is no longer a collection of areal/point locations but a network of arcs, and it is not easy to see how the field-based approach may naturally be applied. In this case, it seems more natural to take the roads as the first-class objects of the model, with references to spatial objects that have properties such as position and length of arc, and other non-spatial attributes such as traffic flow level and road classification.

Example 2 shows that both field and object models are often needed together. There needs to be a rapprochement between the extremes so that both can work together as models and systems that can support the dual approach. This is not easy, due to the nature of their opposition. The problem manifests itself at each level, from high-level modelling discussed in this chapter, through data structures that will support each type, to systems implementations.

### BIBLIOGRAPHIC NOTES

Good discussions of the relative merits of field-based against object-based models (although sometimes not using that terminology) are contained in work by Peuquet (1984), Morehouse (1990), Goodchild (1991) and Couclelis (1992). The quality issue for data and data models is discussed by Burrough (1992, 1994).

The field-based approach is described generally and in great detail by Dana Tomlin (1990). Tests for spatial autocorrelation originate from the work of Cliff and Ord, described in their text (1981): an introductory and accessible account is provided by Unwin (1981). Openshaw (1991) gives a summary of the methods available for spatial analysis and suggests what spatial analysis functionality is relevant for GIS. A parallel development to Tomlin's algebra for image processing is the image algebra of Ritter et al. (1990). Takeyama (forthcoming) and Takeyama and Couclelis (forthcoming) describe work that generalizes Tomlin's algebra, taking ideas from Ritter's image algebra and recognizing the importance of neighbourhood as a foundational concept. Applications of the object-oriented approach to field-based environmental models are discussed in Kemp (1992).

Digital terrain modelling (DTM) has been applied in geographic contexts since the late 1950s. For an early paper in the area, see (Miller and Laflamme 1958). Burrough (1986, Chapter 3) contains a useful survey of principles, techniques and applications. Weibel and Heller (1991) provide a comprehensive coverage of digital terrain models, including grid arrays and TINs, interpolation, manipulation, interpretation and visualization. Foley et al. (1990) has material on graphics relating to the visualization of DEMs.

The literature on visibility analysis is extensive. Burrough (1986) and Weibel and Heller (1991) give overviews. Yoeli (1985) gives a detailed treatment of the basic problem of computing of visibility maps for grid arrays. Mark (1987) trades some accuracy for a fast algorithm. De Floriani *et al.* (1986) provide an algorithm for visibility maps in TIN structures. Fisher (1994a) considers the issue of uncertainty in visibility analysis, distinguishing between probable and fuzzy viewsheds. Franklin and Ray (1994) construct high-performance algorithms for viewsheds and discuss how study of visibility characteristics of the field can give some insights into the field itself. Fisher (1994b) provides amendments to the viewshed operation to take account of more realistic visibility properties: for example, a silhouette against a sky background is more noticeable than the same object against a landscape background.

Work on the formal classification of spatial object types is a fundamental part of GIS research. Egenhofer, Frank and Jackson (1989), Worboys (1992), De Floriani, Marzano and Puppo (1993) and Güting and Schneider (1993b) provide frameworks in which the specification of spatial object types can be rigorously developed. Kirby *et al.* (1989) treat the problem of non-simply-connected areas. First-degree (polyline) and higher-degree polynomial approximations to arcs and loops are well-discussed in Foley *et al.* (1990). Bézier cubic curves were introduced in Bézier (1970).

There has been a good body of research on spatial operations (relationships, in the case of Boolean operations). An early and oft-quoted paper is that by Freeman (1975), which declares a set of 13 operations. Feuchtwanger (1989), Egenhofer and Frank (1987) and Peuquet (1986, 1988) provide further typologies. Egenhofer and co-workers have written a series of papers exploring some of the background and connections between these operations. The 4-intersection relationships are discussed in Pullar and Egenhofer (1988), Egenhofer (1989b), Egenhofer and Herring (1990) and Egenhofer (1991b), while the extension to 9-intersection relationships is given in Egenhofer (1991c) and Egenhofer and Herring (1991). Egenhofer and Franzosa (1995) place models of spatial relationships based upon intersection relationships in a general framework, founded on the idea of *topological similarity*. The relevance of Clarke's calculus of individuals (Clarke, 1981, 1985) to spatial databases is discussed by Cui, Cohn and Randell (1993).

# CHAPTER FIVE

# Representation and algorithms

The crooked paths go every way
Upon the hill – they wind about
Through the heather in and out
Of the quiet sunniness.

James Stephens, *The Goat Paths*

The art of constructing any computer-based system is to construct a bridge between individual and group conceptions of the application domain and the computational processes that support the system. If the human aspects are neglected, then the system may not be usable, or at least people will have to waste energy translating their views of the application into ways understandable by the computer. If the computational aspects are not tackled, then the system risks being inefficient, performing slowly and taking an unnecessarily large amount of data store, if it is able to perform at all. To keep the appropriate balance, a widely adopted approach (universally adopted for large systems) is to divide the process of system development into a collection of tasks, each with a different emphasis.

In the previous chapter, the focus was on high-level conceptual models of spatial information, thus paying special attention to the human views of applications that feature spatial information. We now consider questions that are closer to computational processing: ways of representing spatial information that facilitate computation. Closely related to this is the discussion of how the various spatial operations are performed, that is the algorithms used. In the next chapter, the emphasis changes to ways in which spatial data may be structured and indexed in a database so that storage and retrieval are efficient. However, the distinction between representation, data structures and data access methods sometimes becomes blurred, therefore some topics covered in this chapter could equally have been in the next, and *vice versa*.

The first section of the chapter discusses some general matters concerning computation in a GIS, in particular the multi-dimensionality of the data, distinctions between field-based and object-based representations, and the question of computational efficiency. Discretization emerges as a big issue for object-based representations, and an approach that handles the resultant problems in a self-contained module, called here the geometric layer, is discussed in section 5.2. Section 5.3 describes some of the representations of spatial objects that have been introduced and section 5.4 discusses representations in field-based systems using tessellations.

Basic geometric algorithms (section 5.5) form the engine room of a GIS, without which higher level processing may not take place. Conversions between field and object representations are covered in section 5.6. The final section discusses representations and algorithms for network spaces.

In summary, this chapter considers representations of information commonly found in GIS, usually embedded in the Euclidean plane (or at least in a surface), and algorithms for performing operations upon these objects.

### 5.1  COMPUTING WITH GEOGRAPHIC DATA

Much traditional computing is based upon one-dimensional data representations. The following example is intended to give a feel for the jump from one to two dimensions. Figure 5.1 shows some of the different possible relationships that may exist between two straight line segments, each embedded in the same one-dimensional Euclidean space (the real line). To categorize the relationships, we only distinguish those that are topologically different. Thus, two segments may be disjoint, touching internally or externally, overlapping, nested or equal. These six relationships exhaust all the possibilities (see the set-oriented and topological operators of Chapter 4).

Let us now see what happens when we extend the investigation to two dimensions. In a sense, the simplest finite one-dimensional object is the line segment (1-simplex): in two dimensions, assuming discreteness, the corresponding object is the triangle (2-simplex). The extended problem is now to enumerate the possible topologically distinct relationships that may exist between two triangles that are embedded in the same plane. Figure 5.2 shows some of the many possibilities. As triangles are, in a sense, the simplest of the two-dimensional objects, it is clear that the step from one to two dimensions introduces many more possibilities. It is then natural to assume that the operations underlying two-dimensional spatial data models will have a correspondingly more elaborate algorithmic support. The further jump to three dimensions is even greater and is not yet part of the mainstream of GIS. Three-dimensional GIS are considered in the final chapter.

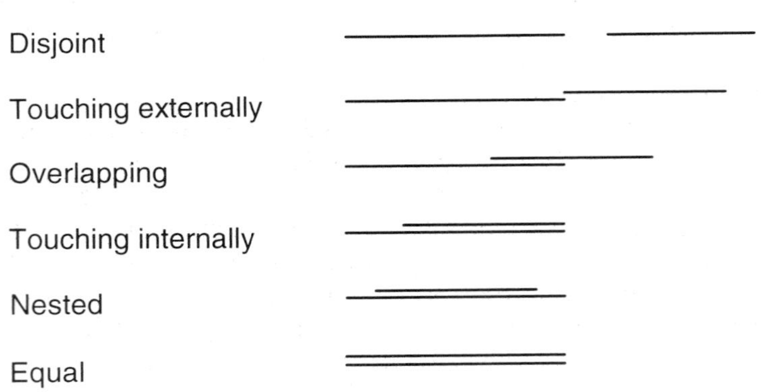

Disjoint

Touching externally

Overlapping

Touching internally

Nested

Equal

**Figure 5.1**    Relationships between two line segments embedded in a one-dimensional Euclidean space.

**Figure 5.2**   Nine topologically distinct relationships between two triangles embedded in a Euclidean plane.

### 5.1.1   Geometric algorithms and computational geometry

As the name implies, a geometric algorithm provides for the execution of an operation acting upon geometric (or spatial) objects. An algorithm is a specification of the computational processes required to perform an operation. Thus, an algorithm for the addition of two numbers in a particular base will specify precisely the steps required to compute their sum. Algorithms may be translated into machine instructions (usually by means of a high-level programming language) that a computer is able to process.

The notion of a geometric algorithm goes back at least as far as the geometer Euclid, living and teaching in the city of Alexandria *c.* 300 BC, in whose works geometric constructions are important features. In Euclid's geometry, the constructions are executed using a specified collection of instruments, straight-edge and compass, and proceeding according to a given set of basic moves. At that time, the issue was not the efficiency of the constructions but the class of objects resulting from them. A typical problem for investigation was the possibility of trisecting an angle using straight-edge and compass alone. It is only in this century that efficiency has become important, expressed by the technical concept of *algorithmic complexity*, and measured in terms of the amount of computation time or space the algorithm takes as a function of the input size.

Geometric algorithms have some interesting special features. Consider the problem of determining whether a point is inside a closed loop or not. Unless the loop is extremely convoluted, the human eye and brain are able to decide this instantly and without much discernible effort. By contrast, we shall see that this classical 'point in polygon' decision procedure, while not being all that hard, is nevertheless not trivial. This property of geometric algorithms contrasts with arithmetic, where many complex calculations requiring paper, pencil and much head-scratching from a person may be solved with ease by a computer.

Another characteristic of geometric algorithms is that they are frequently plagued by special cases. What seems at first a simple problem often becomes more complex as special cases are considered. This feature has been vividly described by Douglas

(1974), where what looks initially like a trivial algorithm to determine whether two straight lines intersect becomes beset by a multitude of special cases.

Computational geometry is the study of the properties of algorithms for solving geometrical problems and includes consideration of questions such as:

1. Is it possible to find *any* algorithm to solve this geometrical problem?

2. What is the most efficient algorithm to solve this geometrical problem?

3. What is the best way of structuring the geometrical data to make the solving of this problem most efficient?

The first question is about computability, where a function or problem is *computable* if it is possible to find some algorithm to compute the function or solve the problem. As noted above, geometric computability questions and the possibilities of particular geometric constructions date back to ancient Greece. General computability issues have been at the forefront of twentieth-century investigations, notable results about limitations of formal systems having been found by Kurt Gödel and Alan Turing.

Finding an algorithm to solve a problem in theory is quite different from solving it in practice. This is the issue raised by questions 2 and 3. Efficiency may be measured in terms of the time that an algorithm consumes (*time complexity*) or storage space used (*space complexity*). The time taken for an algorithm to execute sequentially is the sum of the times of all the constituent operations within the algorithm.

It is almost always the case that the time and space taken to solve a problem increase with the size of the input data. For example, we would expect the time taken to compute the area of a simple polygon to increase as the number of edges of the polygon increases. The question is, 'What is the relationship between the input size and the time or space required by the algorithm?' This relationship may be expressed as a function, called the *complexity function* from input size to amount of time (or space). In the case of the area of a simple polygon, it is not too hard to show that the time taken to compute the area is directly proportional to the number of its edges, and so the complexity function is $f: n \rightarrow kn$, where $k$ is the constant of proportionality and $n$ is the number of edges. If we are interested in the general way that the computation time is related to input size, then in a sense the value of the constant $k$ does not matter: in any case it will depend upon details of the speed of the computer system and other factors. Also, the input size is not $n$ (the number of edges) but is proportional to $n$. It is the linearity of the relationship that is important. This sense of priorities is expressed using the notation $O(n)$ for the set of functions that are at most as large as the product of some constant and $n$.

**Table 5.1** Approximate values of some common functions

| $n$ | $\log_e n$ | $\sqrt{n}$ | $n^3$ | $e^n$ |
|-----|-----------|-----------|-------|-------|
| 1   | 0         | 1         | $10^0$ | $10^0$ |
| 25  | 3         | 5         | $10^5$ | $10^{11}$ |
| 50  | 4         | 7         | $10^6$ | $10^{21}$ |
| 75  | 4         | 9         | $10^6$ | $10^{32}$ |
| 100 | 5         | 10        | $10^7$ | $10^{43}$ |

In general $O(f(n))$ is the set of functions that are at most as large as the product of some constant and $f(n)$. The 'big oh' notation allows us to describe the behaviour of the relationship between computational time (or space) and input size, without being precise about the scalar multipliers. The most important functions that are used as yardsticks to measure the complexity of algorithms are, in increasing order of size, logarithms, fractional powers (e.g. square roots), polynomials and exponentials. To give an idea about the rate of increase of these functions, Table 5.1 gives their approximate values for some values of the input between 1 and 100. Note that it is not just the static value of the function that is important, but the way that it changes as the input size increases. It is often of interest to consider the *asymptotic* nature of the function, that is how it behaves for very large input sizes.

A function order may then be associated with an algorithm to give some indication of how the algorithm will perform relative to the size of input. Of course, different inputs, even though they may be of the same size, will usually result in different performance of the algorithm. It is usual to take a pessimistic view and give functions that measure the *worst-case* performance of the algorithm, that is the performance on the input of each size that results in the most time or space consumption. It may be possible to determine *average-case* performance also: however, it is often difficult to know how to calculate this average. (Consider the problem of generating a truly random sample of 10-sided polygons in the plane!) Returning to worst-case complexity (the usual measure), Table 5.2 gives a very rough idea about the performance of algorithms with different complexity functions.

For each geometric operation, we may now compare the performances of algorithms that implement that operation. The comparison may be experimental, generating sets of inputs and measuring the performance of a particular computer system on the input sets for each of the algorithms. Drawbacks with this approach are:

- Dependency of the results on the particular computer system being used as the platform for the experiments.

- Problems with the generation of representative sample sets of inputs.

Alternatively, we can take the theoretical approach discussed above, finding and comparing the function class of each algorithm. Drawbacks here are:

- Calculating the complexity of the algorithms may be difficult.

- Worst-case analysis may not be a true reflection of experience, for example when the algorithm performs very badly in some cases that hardly ever occur.

- Comparing the function class may be misleading if the scalar multipliers are very different.

Overall, the theoretical approach is more general and independent of any particular computational platform.

**Table 5.2**  A rough guide to complexity

| | |
|---|---|
| $O(1)$ | very fast, constant time, independent of input |
| $O(\log n)$ | fast (e.g. binary search) |
| $O(n)$ | linear time (e.g. linear search) |
| $O(n \log n)$ | (e.g. optimal sorting algorithms) |
| $O(2^n)$ | slow, exponential |

### 5.1.2   Field-based and object-based representations

Issues arising from the representation of field-based and object-based information are quite distinct. In field-based models, the data in the form of field functions have already been discretized to the locations in a fixed spatial framework. On the other hand, with an object-based model discretization has wider ramifications because there is no explicit spatial framework. Unless discretization is properly handled, this can lead to uncontrolled propagation of rounding errors.

A spatial field is often represented as a raster on a regular array, each element of which is a pixel. Such a representation can be extremely greedy, with areas containing only limited variation in the field values taking as much computer storage as areas of greater activity. It follows that there is an important role for data structures that compress the data and facilitate fast retrieval. Such raster structures are discussed in the next chapter. There is scope for further work on representations for spatial fields other than rasters, but unless the data can be approximated to simple analytic functions of two variables it is difficult to see a way forward. An example of an innovative approach (Pratt, 1991) is the representation of convex planar regions by periodic wave forms.

Representations of the object-model must tackle the discretization problem. In this text, discretization issues are separated out by the development of a two-tier structure that handles discretization separately from spatial object representation. This separation is discussed in detail in the next sections (5.1.3 and 5.2). The representations are various, from minimal spaghetti structures that hold little explicit information about object topology but are efficient for a range of applications, including graphics, to more expressive representations.

Just as a *rapprochement* between field and object must be reached at the model level, so also must it at the representation level. The conversion from field-based to object-based, raster-to-vector, and conversely from vector-to-raster, are the subject of section 5.6 of this chapter.

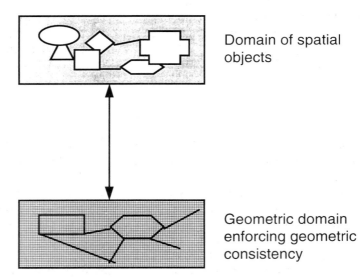

Domain of spatial objects

Geometric domain enforcing geometric consistency

**Figure 5.3**   Geometric and spatial object domains.

### 5.1.3   Geometric and spatial object domains

It is conceptually useful to separate the handling of planar spatial objects into two domains, as shown in Figure 5.3. The primary purpose of this dichotomy is to separate the construction of the classes of spatial objects from the more low-level concerns of accuracy and topological integrity arising from the discretization of the real Euclidean plane. Euclidean geometry assumes an underlying continuum which can only be finitely approximated using a digital computational model. Also, real-world geometric objects may need to be approximated to fit into the resolution allowed by the digital model. This approximation can lead to a large and non-trivial accumulation of small errors. If errors are not controlled, they may alter the geometrical embedding of objects in the Euclidean plane and even accumulate to influence and distort the topology. It is important to keep the error accumulation within bounds, and this is the purpose of the lower geometric level.

## 5.2   A REPRESENTATION OF THE DISCRETE EUCLIDEAN PLANE

An item of surficial geodata is referenced by coordinatization to the surface of the Earth. The more common coordinatizations are:

1. Geographic latitude and longitude.
2. Three-dimensional Cartesian coordinates (X, Y, Z), setting the origin of coordinates at the terrestrial centre, Z-axis through a pole and X- and Y-axes through the equator.
3. Two-dimensional Cartesian coordinates (x, y), achieved by a transformation of a portion of the Earth's surface into part of the Euclidean plane.

While it is immediately more appealing to hold the locational references as directly related to their position on the Earth, as in (1) and (2), it is not usually the most efficient due to the high volume of spatial data. It is more efficient not to store the data with their terrestrial references, but to provide a planar reference (maybe referencing a grid cell) and convert to terrestrial coordinates if required (3). Therefore, most of the following discussion will assume a planar referencing, suitably discretized for computational purposes.

### 5.2.1   Geometric domains

Assume without loss of generality that the number pairs chosen to represent points in discretized Euclidean space are integers. Set up a coordinate frame consisting of a fixed, distinguished point (*origin*) and a pair of orthogonal lines (*axes*), intersecting in the origin. A *point* in the plane of the axes has now associated with it a unique pair of integers (x, y) measuring its distance from the origin in the direction of each axis. The collection of all such points is the *discrete Euclidean plane*, $Z^2$. Straight-line segments may be defined by giving a quadruple of integers, the two pairs of which represent the end points of the line segment.

We are now ready to define more precisely the notion of *geometric domain* as a triple $\langle G, P, S \rangle$, where:

$G$, the domain grid, is a finite connected portion of the discrete Euclidean plane, $Z^2$

$P$ is a set of points in $Z^2$.

$S$ is a set of line segments in $Z^2$

subject to the following closure conditions:

Each point of $P$ is a point in the domain grid $G$

Any line segment in $S$ must have its end points as members of $P$.

Any point in $P$ that is incident with a line segment in $S$ must be one of its end points.

If any two line segments in $S$ intersect in a point, then that point must be a member of $P$.

The idea is that higher-level spatial objects may be constructed from objects in the geometric domain. Figure 5.4(a) shows a small and simple example of a geometric domain. Here the grid $G$ is 10 by 10 square, the set $P = \{a, b, c, d, e, f, g, h, i\}$ and the set $S = \{ab, ac, dh, eh, fh, gh\}$. Note that only points on the grid are allowed, and since line segments must intersect in a point in $P$, the configuration in Figure 5.4(b) is not a geometric domain.

### 5.2.2  Problems arising from discretization and the Greene-Yao algorithm

Euclidean space, based upon real numbers, is topologically dense and capable of a potentially infinite amount of precision. Discretized space, based upon the integers, is capable of only a finite precision. Thus, in the process of discretization, some precision will inevitably be lost. This is an unfortunate fact of life, but it can be controlled if we can minimize the loss of precision by having appropriate discretization processes and by estimating the errors involved. This has a simple analogue in the rounding of numbers during numerical applications. Some of the errors that can arise in the geometric domain are now described.

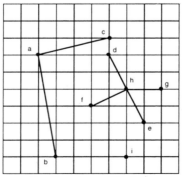

a.   A structure that forms a geometric domain.

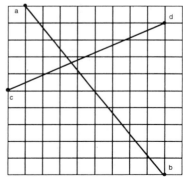

b.  A structure that does not form a geometric domain.

**Figure 5.4**   Grid structures.

Figure 5.5(a) showed a configuration that was not classed as a geometric domain. The problem is that the line segments *ab* and *cd* intersect in a point that is not a member of the domain's pointset. Indeed, it is not possible to add this point of intersection to the pointset, because it is not a domain grid point. So what should we do? A solution is to introduce a grid point *x* to the domain that is the nearest to the point of intersection (or use a convention to choose if several points are equally near). This is the two-dimensional equivalent of numerical rounding and is illustrated in Figure 5.5(a). Of course, we still do not have a configuration that satisfies the domain axioms, but maybe we could extend our notion of geometric domain. However, this solution on its own is unsatisfactory, because topological constraints are broken. For example, the incidence relationship that the point of intersection of two lines is on both the lines, does not hold here. Also, as Figure 5.5(b) shows, it is possible for a point to 'shift' from one side of a line to the other. The 'real' point of intersection of lines *ab* and *cd* lies below line *ef*, but the rounded point of intersection *x* lies above line *ef*.

A first shot at a modification that gets around the problems so far mentioned is to split the line segments so as to join at the rounded intersection point, as shown in

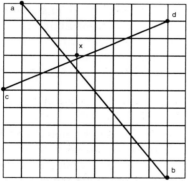

a. Introducing the nearest grid point to the point of intersection to the domain

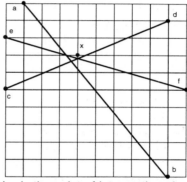

b. Is the point of intersection above or below line *ef*?

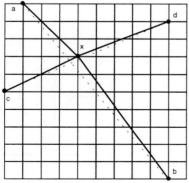

c. New split lines meet at rounded intersection point *x*.

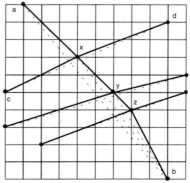

d. More intersection leads to further shifts of line *ab*.

**Figure 5.5** Grid structures.

Figure 5.5(c). Thus lines *ab* and *cd* have been split into lines *ax*, *bx*, *cx* and *dx*. The configuration is now a geometric domain. The problem with this solution is that we lose control of the shift of the line segments as more and more intersections occur in a dynamic situation. Figure 5.5(d) shows some further line segments that have been added, and we notice that the chain *axyzb* has strayed well away from the original line segment *ab*.

A solution has been proposed by Greene and Yao (1986) to handle the problem of drifting lines. Imagine that the grid points are pegs on a peg-board and that the lines *ab* and *cd* are elastic bands stretching between their end points. We require to move their point of intersection to the nearest point on the grid, and we do it in such a way that the bands cannot pass over the pegs but may come to rest against some of them. In our example point *f* forms such a barrier to the movement of segment *xb*. This gives rise to two segments *xf* and *fb*. Thus, segment *ab* is no longer split into two segments, but in this case three, namely *ax*, *xf* and *fb*. Similarly, *cd* splits into *ce*, *ex* and *xd*. The details are shown in Figure 5.6(a).

We have tried to give an appreciation of the algorithm by means of an example. In fact, the formal expression of the algorithm requires some knowledge of the mathematics of continued fractions. It is omitted from this account, but may be found in (Greene and Yao, 1986). The important result is that this algorithm places a limit on the drift of a line segment to its neighbouring points (the so-called envelope of the segment), no matter how many line segment intersections are involved (see Figure 5.6(b)). Therefore, we now have a discretization process that is both well-defined and results in a bounded error accumulation.

### 5.2.3   Discretizing arcs

This section has so far concentrated upon problems caused by points and straight-line segments in a discretized plane. However, many spatial objects are composed of smooth curves, so here we describe approaches to the approximation of curvilinear arcs in a discrete space. The usual technique is to approximate the curve by a poly-

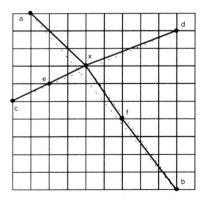

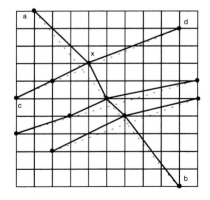

a. Greene Yao algorithm applied to *ab* and *cd*.     b. Further intersections using the Greene Yao algorithm

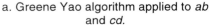

**Figure 5.6**   Greene-Yao handling of segment intersections.

line, but, in order to be computable, curves do not have to be reduced to a finite sequence of line segments. For example, a quadratic curve, given as the set $\{(x, y) : y = ax^2 + bx + c\}$ is computationally tractable, provided that the coefficients $a$, $b$ and $c$ are computable numbers. It is possible to produce an algorithm that will determine the points of intersection of two quadratics. A whole class of line simplification algorithms, based upon Bézier curve segments (cubic polynomials), may be found in the literature (e.g. Foley *et al.*, 1990).

The algorithm described in this section approximates a curve by a polyline. There are of course an unlimited number of ways that a curve may be approximated in this way, due to the number and positioning of the segments. In general, the more segments that are used, the better the approximation, providing that the segments have been chosen appropriately. Related to this discussion is the process of polyline simplification, often needed in cartographic generalization, where it is required to approximate a polyline with a simplified polyline (usually having fewer points) that provides a 'sufficiently good' approximation.

Consider the curve in Figure 5.7(i). A technique related to the Douglas-Peucker algorithm for arc simplification is shown. First, a point is assigned to each end of the arc. A straight-line segment $ab$ is constructed between these points. The point $c$ on the arc furthest from the line (in terms of length of the dropped perpendicular) is added to the pointset. The arc is split into two parts, $ac$ and $cb$, and the algorithm applied recursively to each of the pieces of arc. In Figure 5.7(ii), the arc is split into

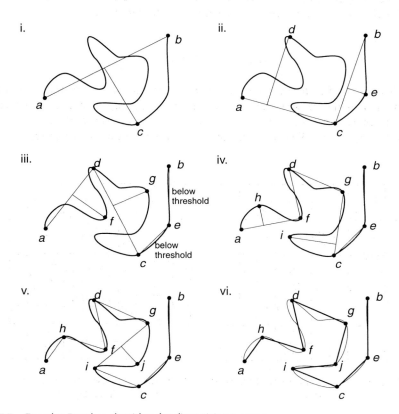

**Figure 5.7** Douglas-Peucker algorithm for discretizing a curve.

*ad, dc, ce* and *eb*. The process stops for any arc segment when the distance of the furthest point on the arc segment from the straight line between the arc segment's end points is below a nominated threshold distance. In the example *ce* and *eb* have this property. The smaller the threshold, the more points will be added and the better will be the approximation. The algorithm continues (Figure 5.7(iii) and (iv)) until there are no more arcs to split. In our example this stage is reached in Figure 5.7(v) with arcs *ah, hf, fd, dg, gj, ji, ic, ce, eb*. The approximating polyline is shown in Figure 5.7(vi).

### 5.3  THE SPATIAL OBJECT DOMAIN

Separating out the geometric from the spatial object domains allows us to divorce topological infrastructure from embedding. This section considers some approaches to representing the structure of spatial objects, often called 'capturing the topology'. The first representation – spaghetti – provides only a minimal level of topological information. We then progress on to representations that handle topology more explicitly. Some of these representations mix topology with the embedding.

#### 5.3.1  Spaghetti

Spaghetti (or maybe spaghetti rings) is a highly expressive metaphor for the representation that is now to be described. The *spaghetti data structure* represents a planar configuration of points, arcs and areas. Geometry is represented as a set of lists of straight-line segments. Each such list is the discretization of an arc that might exist independently, or as part of the boundary of an area. There is no explicit representation of the topological interrelationships of the configuration, such as adjacency relationships between constituent areas.

An example is shown below, where the planar configuration of areas shown in Figure 5.8 is represented as spaghetti. Each of the polygonal areas is represented by its boundary loop. Each loop is discretized (using one of the line discretization algorithms such as the Douglas-Peuker algorithm of section 5.2.3) as a closed polyline. Each such polyline is represented as a list of points, each point being an extreme of

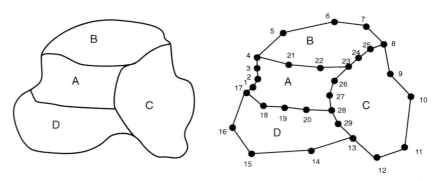

**Figure 5.8**  A planar configuration (continuous and discrete) that partitions a subset of the plane into areas.

a line segment in the polyline. (Here and later, although sequences of arcs are used to represent boundary loops, these sequences are actually structured as cycles. Thus, for example, [a, b, c, d] = [b, c, d, a] = [c, d, a, b] = [d, a, b, c].)

A: [1, 2, 3, 4, 21, 22, 23, 26, 27, 28, 20, 19, 18, 17]

B: [4, 5, 6, 7, 8, 25, 24, 23, 22, 21]

C: [8, 9, 10, 11, 12, 13, 29, 28, 27, 26, 23, 24, 25]

D: [17, 18, 19, 20, 28, 29, 13, 14, 15, 16]

To represent the geometric embedding in the plane, each point would be specified by its coordinate pair. The spatial structure has thus been translated into a set of lists.

Any geometrical configuration based upon points and lines embedded in the Euclidean plane can be represented in this way. Spaghetti provides the basic connectivity within a spatial configuration, but few other spatial relationships are explicitly represented. For example, the fact that areas A and B are adjacent must be calculated by noting a common point sequence in their point lists. Thus, the representation is inefficient for classes of spatial analysis where spatial relationships are important. Also, the representation is inefficient in space utilization since there is redundant duplication of data (point sequences being recorded as parts of the lists for both areas that they bound).

Spaghetti representations are useful in situations involving large numbers of simple geometric operations where structural relationships are not required. This is the case when all that is required is the presentation of the graphic on the screen or other output device, and so suitable applications include simpler forms of computer cartography. The spaghetti representation stems from a conceptually simple data model and has been popular in GIS. It is possible to take the basic spaghetti structure and enhance it to provide a richer representation. The enhancement might be the attachment of additional attributes to the lists (maybe giving feature type or line style), or the attachment of pointers to and from non-spatial databases associated with the spatial data.

### 5.3.2   Representing more topology

This and the following sections introduce representations that capture explicitly some of the spatial relationships not inherent in the spaghetti representation. Such representations are often called *topological*, to show that they can represent topological relationships such as adjacency. However, this terminology is not very useful as even spaghetti is capable of expressing the topological connectivity relationship between edges.

The next representation, we will call it NAA for Node-Arc-Area, represents explicitly the adjacency relationships between areas in a subdivision of a surface. NAA shows very clearly in its symmetry the duality between nodes and areas in a surficial subdivision. It is convenient to describe it in the form of a set of relation schemes for a relational database: in fact, this representation was used as an example for E-R modelling in section 2.3.1. The primary constituent entities are **directed arc**, **node** and **area** (where here, **area** is taken to be a face enclosed by arcs, homeomorphic to a cell, except for the external area) and system rules are:

- Each directed arc has exactly one start and one end node.
- Each node must be the start node or end node (maybe both) of at least one directed arc.
- Each area is bounded by one or more directed arcs.
- Directed arcs may intersect only at their end nodes.
- Each directed arc has exactly one area on its right and one area on its left.
- Each area must be the left area or right area (maybe both) of at least one directed arc.

These rules may be expressed as an extended entity-relationship diagram as shown earlier in Figure 2.18. Entity type **area** has overlapping subtypes **left area** and **right area**. Each instance of these subtypes enters into a one-to-many relationship with the directed arcs that bound it on the left and right, respectively. Entity type **node** has overlapping subtypes **begin node** and **end node**. Each instance of these subtypes enters into a one-to-many relationship with the arcs for which it is the begin or end node, respectively.

Figure 5.9 shows a decomposition of the configuration in Figure 5.8 into nodes, directed arcs and areas ready for NAA representation. The constituent areas are

**Table 5.3** Relation corresponding to areal configuration in Figure 5.9

| Arc ID | Begin node | End node | Left area | Right area |
|--------|-----------|----------|-----------|------------|
| a | 1 | 2 | A | X |
| b | 4 | 1 | B | X |
| c | 3 | 4 | C | X |
| d | 2 | 3 | D | X |
| e | 5 | 1 | A | B |
| f | 4 | 5 | C | B |
| g | 6 | 2 | D | A |
| h | 5 | 6 | C | A |
| i | 3 | 6 | D | C |

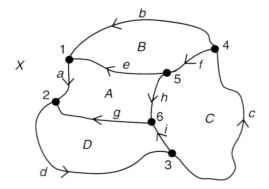

**Figure 5.9** A decomposition of the planar configuration in Figure 5.8 into nodes, directed arcs and areas.

labelled *A*, *B*, *C*, *D* and *X* is the external area, necessary so that the system rules are satisfied. Arcs are labelled *a*, ..., *i* and nodes labelled 1, ..., 6. Area *A* is bounded by four arcs *a*, *e*, *h* and *g*. Arcs *a* and *e* have *A* as a left area, while arcs *g* and *h* have *A* as a right area. Node 1 has incident arcs *a*, *b* and *e*. Arcs *e* and *b* have node 1 as an end node, while arc *a* has node 1 as a begin node.

Readers who enjoy patterns will observe a pleasing symmetry in the NAA representation, realizing the duality between node and area. The symmetry is only marred by the existence of a specially distinguished area that is the external: there is no such distinguished node. However, if we change our embedding space to be the surface of the sphere (the perfect 2-space?), then the need for an external area no longer exists and we have perfect symmetry! So much for mysticism, back to computation.

Following the usual database design procedures, the EER diagram forms the basis for a normalized relational database scheme as a single relation scheme:

ARC (<u>ARC ID</u>, BEGIN_NODE, END_NODE, LEFT_AREA, RIGHT_AREA)

The corresponding relation for the example in Figure 5.9 is shown in Table 5.3.

The NAA representation up to now has only given the connectivity and adjacency relationships between planar objects. It can be extended to include details of the embedding. New entity types **point, polyline** and **polygon** are required for the discretized embedded spatial objects. We also require an entity **coordinate** to pin down positions to the surface. An EER diagram for the extended NAA representation is given in Figure 5.10. Note that every polygon is an embedding of an area and is constituted as a sequence (cycle) of chains. Each chain is an embedding of an arc

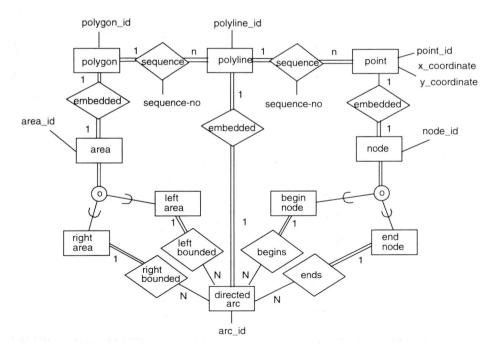

**Figure 5.10**  EER diagram for the extended NAA representation.

and is constituted as a sequence of points. Every node is a point, but not every point is a node, because points trace the embedding of arcs between nodes.

As before, this diagram may be used as the design of a normalized relational database scheme. To simplify matters, since some entities are in one-to-one correspondence with no added attributes, we may make identifications between **area–polygon** and **arc–polyline**. In this case, the tables are:

ARC (<u>ARC</u>, BEGIN_NODE, END_NODE, LEFT_AREA, RIGHT_AREA)

POLYGON (<u>AREA ID, ARC ID</u>, SEQUENCE_NO)

POLYLINE (<u>ARC ID, POINT ID</u>, SEQUENCE_NO)

POINT (<u>POINT ID</u>, X_COORDINATE, Y_COORDINATE)

NODE (<u>NODE ID</u>, POINT_ID)

Relation POLYGON gives the embedding of each area as a sequence of arcs (polylines): each sequence of arcs should be interpreted as a cycle. Relation POLYLINE gives the embedding of each arc as a sequence of points. Relation POINT gives the embedding of each point and relation NODE gives the point corresponding to each node.

### 5.3.3   Doubly-connected-edge-list (DCEL)

The next representation, the doubly-connected-edge-list (DCEL) discussed by Muller and Preparata (1978), allows further capture of topology in surficial configurations. The DCEL provides a complete representation of the topology of a connected planar graph. It omits the details of the actual embedding (i.e. polygons, chains and coordinates of points), but focuses upon the topological relationships embodied in the entities node, arc (edge) and area (face). The DCEL provides in a single table the information necessary to construct:

For each node in the configuration, the sequence (cycle) of arcs around the node.

For each area in the configuration, the sequence (cycle) of arcs around the area.

The word 'sequence' is important here. The earlier NAA representation will give the sets of arcs that bound an area or meet at a vertex, but the sequencing may not be determined (it could only be determined from the NAA representation if the embedding was also given).

The EER diagram for the DCEL, shown in Figure 5.11, is an extension of that for the NAA representation. Two additional relationships have been added, every arc has a unique next arc and a unique previous arc. Given an arc $a$, to find the previous arc to $a$, select the begin node, say $n$, of $a$ and starting at $a$ proceed around $n$ in an anticlockwise direction until the first arc is found. To find the next arc to $a$, proceed around the end node of $a$ in an anticlockwise direction starting at $a$ until the first arc is found. The relationships **previous** and **next** are shown in Figure 5.12. An example of a DCEL representation is given in Table 5.4, representing the configuration shown in Figure 5.9.

There now follows an algorithm that uses a DCEL to compute the anticlockwise sequence (cycle) of arcs surrounding node $n$. We assume that we have scanned the columns begin_node and end_node in the DCEL to find some arc $a$ that is incident

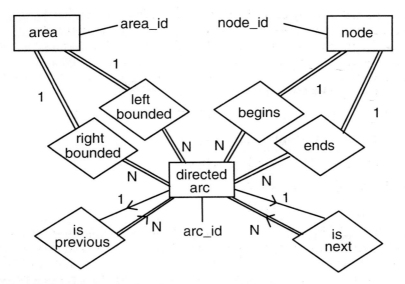

**Figure 5.11** EER diagram showing the information structure of the DCEL.

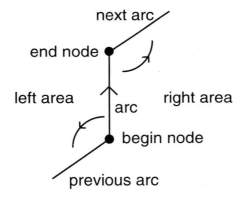

**Figure 5.12** Relationships to a single arc in the DCEL.

**Table 5.4** DCEL table representing the configuration in Figure 5.9

| Arc | Begin node | End node | Left area | Right area | Previous arc | Next arc |
|-----|-----------|----------|-----------|------------|--------------|----------|
| *a* | 1 | 2 | *A* | *X* | *e* | *d* |
| *b* | 4 | 1 | *B* | *X* | *f* | *a* |
| *c* | 3 | 4 | *C* | *X* | *i* | *b* |
| *d* | 2 | 3 | *D* | *X* | *g* | *c* |
| *e* | 5 | 1 | *A* | *B* | *h* | *b* |
| *f* | 4 | 5 | *C* | *B* | *c* | *e* |
| *g* | 6 | 2 | *D* | *A* | *i* | *a* |
| *h* | 5 | 6 | *C* | *A* | *f* | *g* |
| *i* | 3 | 6 | *D* | *C* | *d* | *h* |

with $n$; and if no such arc exists, return that the sequence is empty. To generate the sequence (cycle) of arcs around $n$, apply the method **new_arc** below, initially to arc $a$ and then to the results of previous applications, in order to generate each member of the sequence in turn. When **new_arc** returns the initial arc $a$, then the sequence is complete.

```
method new_arc (arc)
begin
if begin_node (arc)=n
   then new_arc : =previous_arc (arc)
   else new_arc : =next_arc (arc);
return new_arc
end.
```

In the example above, node 6 is the end node for arc $i$. Applying **new_arc** to $i$ gives arc $h$, applying **new_arc** to $h$ gives $g$, then back to $i$, which completes the sequence.

In a similar way, the clockwise sequence (cycle) of arcs around an area $A$ may be computed by modifying method new_arc to new_arc1, given below. Assume that we have scanned the columns left area and right area in the DCEL to find some arc $a$ that bounds area $A$; and if no such arc exists, return that the sequence is empty. To generate the sequence (cycle) of arcs around $A$, apply the method **new_arc1** below, initially to arc $a$ to generate each member of the sequence in turn. When **new_arc1** returns the initial arc $a$, then the sequence is complete.

```
method new_arc1 (arc)
begin
if left_area (arc)=A
   then new_arc : =previous_arc (arc)
   else new_arc : =next_arc (arc);
return new_arc
end.
```

In the above example, area $C$ is the left area for arc $c$. Applying **new_arc1** to $c$ gives arc $i$, applying **new_arc1** to $i$ gives $h$, then $f$ and back to $c$, which completes the sequence.

### 5.3.4   Representations of strongly connected areal planar objects (Object-DCEL)

The same kind of structure can be used to describe aggregations of strongly connected areal objects (section 3.4.3). Assume that the objects under consideration in this section are regular closed in the Euclidean plane, that is they are 'pure area'. The formal concept of the *combinatorial map* was briefly introduced in section 3.4.4. The representation now described, which we call the *object-DCEL*, is based upon the combinatorial map, and has the important characteristic that it is *faithful*, up to homeomorphism and cyclic reordering of the arcs round a polygon. That is, two such areal aggregations that are not homeomorphic must have different representations, subject to cyclic reordering of arcs round the polygon.

The areal object $A$ shown shaded in Figure 3.39 seems to have two equally natural representations, as either the union of two cresents $B$ and $C$, or the difference of circular and elliptical discs $D$ and $E$. It does not really matter which one is chosen, as long as we are consistent: in fact, we will adopt the former. The keys to the object-DECL representation are the notions of strong and weak connectivity. Object *A* is not itself strongly connected, there being two articulation points at each end of its vertical diameter. *A* is weakly connected and has the lunes *B* and *C* as strongly connected components. The object-DCEL representation has the effect of decomposing an areal object into its strongly connected components.

The requirement is to provide a method that will allow the specification of unique and faithful representations of complex, weakly connected areal objects of the kind shown in Figure 5.13. Such a method is now presented. First, overlay upon the object a collection of arcs and nodes, as shown in Figure 5.14. As usual in this section, the arcs are directed. Construct the direction of the arcs so that the object's area is always to the right of each arc. As for the DCEL representation, the weakly connected areal object may be represented in object-DCEL form as a table (see Table 5.5). In this case, no area identifiers need be added, since it is assumed that the whole structure is the spatial reference of a single areal object. Also, we have been sparse with the node information, because an arc's end node may be retrieved as the begin node of the next arc.

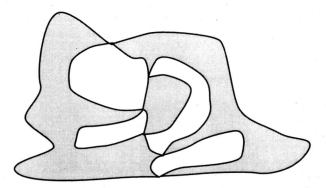

**Figure 5.13**   A complex, weakly connected areal object.

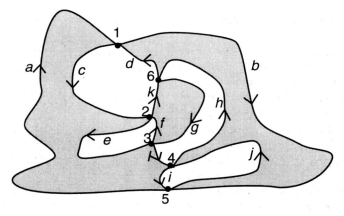

**Figure 5.14**   The object in Figure 5.13 with arcs and nodes defined.

**Table 5.5** Object-DCEL table representing the weakly connected, aeral object in Figure 5.13

| Arc | Begin node | Next arc |
|-----|-----------|----------|
| $a$ | 5 | $c$ |
| $b$ | 1 | $j$ |
| $c$ | 1 | $e$ |
| $d$ | 6 | $b$ |
| $e$ | 2 | $l$ |
| $f$ | 3 | $k$ |
| $g$ | 6 | $f$ |
| $h$ | 4 | $d$ |
| $i$ | 4 | $a$ |
| $j$ | 5 | $h$ |
| $k$ | 2 | $g$ |
| $l$ | 3 | $i$ |

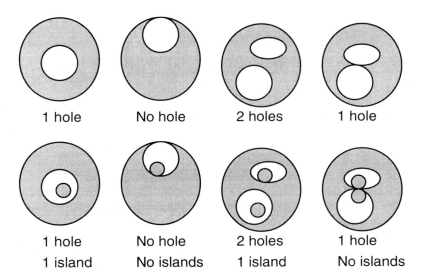

**Figure 5.15**   Areal objects with varying numbers of holes and islands.

The arc boundaries of the strongly connected cells that are components of the weakly connected object may now easily be retrieved by following the sequences of arc to next arc until we arrive back at the starting arcs. For the above example these arc boundaries are:

[$a, c, e, l, i$]

[$f, k, g$]

[$b, j, h, d$]

In summary, this section has described the object-DCEL representation of regular closed, weakly connected areal objects. This representation comprises the lists of arcs bounding the strongly connected components of the object. The representation provides a complete and faithful description of the topology of the object. Later in the chapter, we will describe operations that can be performed on areal objects using this representation.

Before leaving this section on the representations of planar areal objects, it is worth dwelling for a short time on the problem of holes and islands. In the context of regular closed areal objects, a hole (or island) is defined as a subregion within the main region but whose boundary is disjoint from the main region's boundary. Figure 5.15 gives examples of combinations of holes and islands. Note that a hole cannot touch the boundary of the main region or another hole. Areal objects with holes can be represented using extensions of the structures above, although care must be taken if faithful representations are still required. One possibility is to provide a second structure that is a tree of containment, showing how holes, islands, etc., are contained within each other. This is the approach adopted in David *et al.* (1993), for example. Extensions to combinatorial maps are proposed in Bryant and Singerman (1985).

### 5.3.5  Other boundary representations of geometric objects

Boundary representations represent geometric objects in terms of their boundaries. Much of the work on boundary representations has been done by geometric modellers for computer-aided design (CAD) applications, where the geometric objects are usually bounded, reasonably regular and in three dimensions. Spatial objects commonly occurring in GIS, on the other hand, are irregular and, in most present systems, two-dimensional. This section reviews some of this work where it is relevant to GIS, looking only at two-dimensional representations. Some of the structures in the sections above will be revisited here in a more general context.

Recalling the earlier material on topological foundations, the *boundary* of a geometric object is that portion of it which is neither in its interior nor in the interior of its complement. In the case of an areal object, the boundary is a collection of arcs. For solid three-dimensional objects, the boundary is a surface or collection of surfaces. For discretized geometric objects, the boundary is divided into a collection of components, for example, nodes and edges. A *boundary representation* is a representation of that subdivision explicitly holding information on structural relationships, such as incidence and adjacency, between its component parts.

In a similar way to earlier sections, a geometric object is defined by a boundary representation at two levels: *topological level* and *embedding level*. The topological model describes the topological infrastructure of the object; the embedding model (see geometric level above) describes the embedding of the object into geometric space. Usually, this space has been the Euclidean plane. However, in general, the space may be any manifold or manifold-like object, such as the plane, sphere, torus or more exotic object. The space may itself be bounded or unbounded. Separating the embedding from the topology allows a distinction to be made between characteristics of objects that are due to their topology and those that arise out of their embedding. It also allows the work on topological relationships to be applied in a variety of embeddings. Thus, much of the earlier work on DCELs, for example, can be applied to embeddings in a range of spaces.

The work on representations of subdivisions of two-dimensional spaces (surfaces) will be presented by studying the winged-edge representation of Baumgart (1975), its extension by Weiler (1985), and the edge algebra and quad-edge data structure of Guibas and Stolfi (1985). We will follow the terminology of these authors by referring to the primitive 0-, 1- and 2-dimensional objects as *vertex*, *edge* and *face*, respectively. These terms are consistent with the separation of the connectivity relationships from the embeddings.

### Winged-edge representation

The winged-edge representation is a variant of the DCEL structure discussed earlier. It may only be used to represent subdivisions of orientable surfaces (i.e. surfaces for which the notion of 'clockwise' may be well-defined). The fundamental spatial object is the edge. With each edge are associated its two incident faces: *p-face* and *n-face*, and two incident vertices: *p-vertex* and *n-vertex*. With each edge are also associated four other edges: *nc-edge* (next edge clockwise), *pc-edge* (previous edge clockwise), *na-edge* (next edge anti-clockwise), *pa-edge* (previous edge anti-clockwise). This is shown schematically in Figure 5.16.

At first sight, there seems to be ambiguity regarding the decision of whether a related face, edge or vertex is *n* or *p*, since the edges have no direction. Although there are no explicit orderings, an implicit direction on each edge constrains the representation uniquely. By way of illustration, a winged-edge representation is given in Table 5.6 for the subdivision of an orientable surface (indicated by the orientation symbol) shown in Figure 5.17.

The winged-edge representation provides a complete characterization of the topology of the subdivision in terms of incidence, adjacency and orderings of edges around vertices and faces. The original structure represented subdivisions of orientable surfaces without boundaries, where each constituent edge is incident with two distinct vertices and two distinct faces: thus loops and isthmuses were disallowed. Weiler (1985) has extended the representation to handle loops and isthmuses.

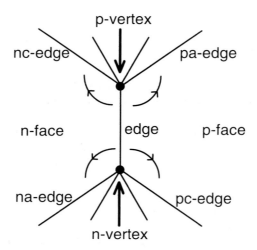

**Figure 5.16**   Winged-edge representation showing the vertex-edge-face relationships.

**Table 5.6**   Entries for a winged-edge structure representing the subdivision in Figure 5.17

| edge | n vertex | p vertex | n face | p face | nc edge | na edge | pc edge | pa edge |
|------|----------|----------|--------|--------|---------|---------|---------|---------|
| a | 1 | 2 | A | X | g | e | b | d |
| b | 4 | 1 | B | X | e | f | c | a |
| c | 3 | 4 | C | X | f | i | d | b |
| d | 2 | 3 | D | X | i | g | a | c |
| e | 5 | 1 | A | B | a | h | f | b |
| f | 4 | 5 | C | B | h | c | b | e |
| g | 6 | 2 | D | A | d | i | h | a |
| h | 5 | 6 | C | A | i | f | e | g |
| i | 3 | 6 | D | C | g | d | c | h |

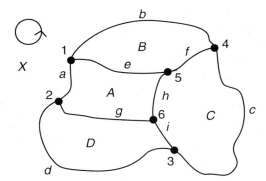

**Figure 5.17**   Subdivision of an orientable surface (as Figure 5.9).

*Edge algebra and quad-edge data structure*

The edge algebra of Guibas and Stolfi (1985) provides representations of subdivisions of surfaces without boundaries. The surfaces may be orientable or non-orientable (i.e. the notion of 'clockwise', invariant over the surface, is not required). In order to understand this representation a few preliminaries are required.

It may be recalled from Chapter 3 that the *dual* $G^*$ of a planar graph $G$ is obtained by associating with each face in $G$ a vertex in $G^*$. Two vertices in $G^*$ are connected by an edge if and only if their corresponding faces in $G$ are adjacent. Given an edge $e$ in $G$, the *dual edge* $e^*$ to $e$ joins the vertices in $G^*$ corresponding to the two faces incident with $e$ and $G$.

A *directed* edge has a direction defined from its start vertex to its end vertex. An *oriented* edge has a distinguished side defined. Direction is indicated as usual with an arrow along the edge. Orientation is indicated by an arrow emitting from one side (the distinguished side) of the edge. Figure 5.18 shows a directed, oriented edge. Note that an edge may itself be directed and oriented even when the embedding surface is not oriented. Directed, oriented edges (called *do-edges*, here) are the basic elements of this representation.

Coming now to the quad-edge representation itself, each edge of the surficial subdivision is represented by eight do-edges: four do-edges associated with the edge

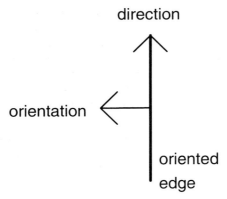

**Figure 5.18**   A directed, oriented edge.

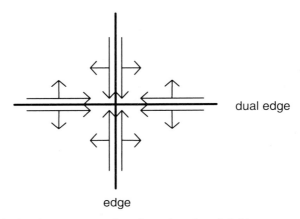

**Figure 5.19**   Quad-edge data representation of an edge of a subdivision.

and four do-edges associated with its dual edge (see Figure 5.19). Three functions act upon the do-edges, namely *Flip*, *Onext* and *Rot*. Further details may be found in Guibas and Stolfi (1985).

The reason for the added complexity of this model is the ability to handle embeddings in non-orientable surfaces. If the surface is orientable, then the algebra may be simplified. The edge-algebra representation has an equivalent quad-edge data structure (so-called because four do-edges are required to represent a single topological edge). As with other representations, the implementation may either be in relational form (each edge having an identifier and eight further attributes) or by means of pointers.

## 5.4   REPRESENTATIONS OF FIELD-BASED MODELS: REGULAR AND IRREGULAR TESSELLATIONS

The overall development of the last section steered a path in the direction of ever more comprehensive representations of spatial objects embedded in the plane. These representations enable more information about topological relationships between constituent parts of spatial objects to be held explicitly. We now change course so as to describe an important class of representations based upon tessellated structures. A *tessellation* is a partition of the plane or portion of the plane as the union of a set of disjoint areal objects. If these constituent objects are all exact copies of the same regular polygon, and each of the vertices is also regular (in a sense shortly to be explained) then the tessellation is called *regular*, otherwise the tessellation is *irregular*. By far the most important class of regular tessellated representations is the grid or raster representation, based upon a tessellation of squares. The irregular tessellated representations of most interest are the triangulated irregular networks (TINs). Triangles are the simplest discrete areal objects that populate the Euclidean plane, and so it is interesting to use them as the primitive areal elements in representations.

### 5.4.1 Regular tessellated representations

In terms of discrete areal objects, a *tessellation* of a surface is a covering of the surface with an arrangement of non-overlapping polygons. A *regular polygon* is a polygon for which all edges have the same length and all internal angles are equal. At each vertex of a tessellation, the *vertex figure* is the polygon formed by joining in order the mid-points of all edges incident with the vertex. A *regular tessellation* of a surface is a tessellation for which all the participating polygons are regular and equal, and all the vertex figures are also regular and equal. For example, a regular tessellation of equilateral triangles has a regular hexagon for its vertex figure.

It may have been Kepler (1571–1630) who first investigated the regular tessellations of the plane. In the case of the Euclidean plane, only three tessellations are possible, based upon equilateral triangles, squares and regular hexagons, and shown in Figure 5.20.

Of the regular tessellations, by far the most commonly used for spatial representations is the square grid. This provides the very common *raster* representation of spatial data, where planar spatial configurations are decomposed into a pattern of squares (*pixels*) on a grid. Conversion to a raster grid (*rasterization*) is a particular example of discretization. Regular triangular and hexagonal tessellations are rarely used for planar data representation. However, nested regular triangular tessellations have been suggested for representing spherical data (see Chapter 6). The regular tessellation representation based upon squares fits well with standard and well-supported programming data types, such as two-dimensional arrays. However, this representation does not often accord with the way in which the data have been

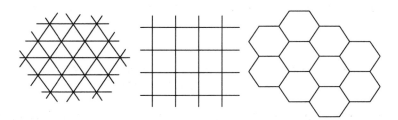

**Figure 5.20** Regular triangular, square and hexagonal tessellations.

collected, often with measurement locations arranged irregularly through the region. In this case, irregular tessellations are more appropriate, and such representations are now considered.

### 5.4.2 Irregular tessellated representations

The most commonly used irregular tessellation is the triangulated irregular network (TIN). A TIN may represent the variation of a field function over a spatial framework. Its irregularity allows the resolution to vary over the surface, capturing finer details where required. A useful concept in what follows is again the notion of duality of planar graphs, discussed in Chapter 3 and in section 5.3.5 above, where faces become nodes and nodes become faces. If we label each triangular face of the TIN with a node and join with edges those faces that are adjacent in the TIN, then the planar mesh that results is the *dual* of the original mesh. If $T^*$ is the dual of the TIN $T$, then the degree of each node (the number of edges incident with that node) in $T^*$ must be three.

TIN models admit linear and higher-order interpolation methods. For example, imagine that a representation is required for a surface of topographical elevations over a region. Suppose that the heights have been measured at points irregularly throughout the region in such a way that there is more sampling in areas where there is more variation. The problem is to calculate a good estimate using interpolation of the elevation of the surface at a non-measured point in the region. A solution is to construct a TIN over the region, where the measurement points become the nodes of the TIN. For the moment, suppose that this triangulation has already been achieved (see below): then the constituent triangles positioned in the surface have discretized the surface into a collection of planar facets. These facets may be used to construct a linear interpolation of height throughout the region.

Figure 5.21 shows the linear interpolation process working for a single general triangular facet of the surface. Assume that triangle $ABC$ is part of a TIN. The coordinates (in vector form) $\boldsymbol{a}$, $\boldsymbol{b}$ and $\boldsymbol{c}$ of the vertices are assumed known, as are the

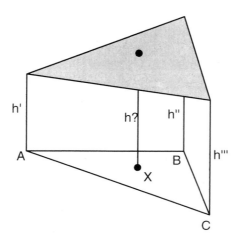

**Figure 5.21**  Interpolation is required to find the height of point $X$.

heights $h'$, $h''$ and $h'''$, at $A$, $B$ and $C$, respectively. The problem is to estimate the height $h$ of the surface at point $X$, whose coordinates are known to be $\boldsymbol{x}$. Since point $X$ is inside or on the boundary of the triangle $ABC$, we have the following relationship:

$$x = \alpha \times \boldsymbol{a} + \beta \times \boldsymbol{b} + \gamma \times \boldsymbol{c}$$

where $\alpha$, $\beta$ and $\gamma$ are scalar coefficients that can be uniquely determined, such that

$$\alpha + \beta + \gamma \leq 1.$$

The height $h$ at the point $X$ can now be found by using the equation:

$$h = \alpha \times h' + \beta \times h'' + \gamma \times h'''$$

The preceding equation gives an estimate of the height of an intermediate point between measured points, based on the assumption that the discretization is into planar facets, that is first-order (linear) surfaces. It is quite possible to base the interpolation upon higher-order surfaces. For details, see for example Akima (1978).

Given a set of points in a region, there remains the problem of tessellating the region with triangles whose vertices are the given set. Such tessellations are not unique and some tessellations may be judged better than others. Figure 5.22 shows two possible triangulations based upon the same set of vertices. The right-hand triangulation has a higher proportion of long thin triangles and so may be judged inferior for some purposes.

### 5.4.3 Delaunay triangulations and Voronoi diagrams

A triangulation that has many desirable properties, both for itself and for its dual, is the *Delaunay triangulation*. Loosely, it can be said that the constituent triangles in a Delaunay triangulation are 'as equilateral as possible'. The dual of a Delaunay triangulation is a *Voronoi diagram* (or *Thiessen polygons*). These constructions are long-standing, as discussed in Okabe *et al.* (1992): Voronoi diagrams are clearly used in the work of the mathematicians Dirichlet (1850) and Voronoi (1908), although similar ideas appeared much earlier, for example in writings of Descarte in 1644. Delaunay triangulations appear in Voronoi (1908) as the dual of Voronoi diagrams.

Delaunay triangulations may be approached by considering the dual Voronoi diagrams. Suppose we are given a set of points in a planar region (maybe they

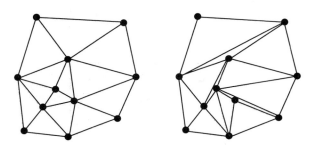

**Figure 5.22**  Two triangulations based upon the same vertex set.

represent emergency accident units) with the property that no three are collinear (to avoid degenerate situations). Imagine that we wish to surround each point by an area that contains all the points in the region which have that accident unit as their nearest. Suppose, for simplicity, that distance is measured 'as the crow flies' using the usual Euclidean metric. It turns out that these areas of closest proximity (or *proximal areas*) are polygons, and constitute a Voronoi diagram. The Delaunay triangulation is now simply formed as the dual of the Voronoi diagram, using the nodes in that diagram. Figure 5.23(b) shows that nodes in the Delaunay triangulation are joined if they are in adjacent proximal polygons. A Delaunay triangulation has the property that each circumcircle of a constituent triangle does not include any other triangulation point within it. This property is illustrated in Figure 5.24.

To understand the intricate details of Delaunay triangulations and Voronoi diagrams, a considerable amount of geometry outside the context of this book would be required. Such details are fascinating, almost endless and well covered in Okabe *et al.* (1992). Below is a list of some of the more striking properties of the Delaunay construction. Suppose given an initial pointset *P* that has no sets of three points collinear:

1. The Delaunay triangulation is unique.
2. The external edges of the triangulation form the convex hull of *P* (i.e. the smallest convex set containing *P*).
3. The circumcircles of the triangles contain no members of *P* in their interior. Note that this is a defining property, in that if it holds, then the triangulation must be a Delaunay triangulation.
4. The triangles in a Delaunay triangulation are best-possible with respect to regularity (closest to equilateral).

Property 4 is vague and needs some explanation. Suppose that we have a triangulation (not necessarily Delaunay) and make a list of the minimum angles in each of the constituent triangles in such a way that the list starts with the smallest angle and moves along in sequence to the largest angle. Call this list the *ordered minimum angle vector* for the triangulation. We may compare two such vectors *lexicographically* as if they were ordered in a telephone directory. Now Property 4 can be made more precise as follows:

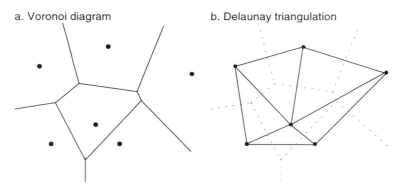

**Figure 5.23**    A Voronoi diagram of proximal polygons and its dual Delaunay triangulation.

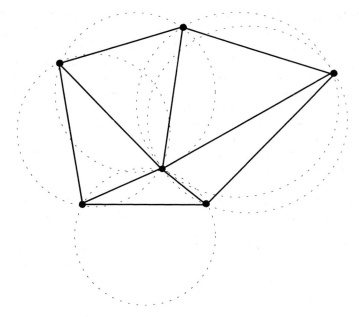

**Figure 5.24** The circumcircles of a Delaunay triangulation.

4'. The triangles in a Delaunay triangulation have the greatest ordered minimum angle vector of any triangulation of $P$.

Property 4' has the intuitive consequence that, since a Delaunay triangulation has the greatest possible ordered minimum angle vector, it will contain the smallest collection of long thin triangles. In a sense, this implies that the Delaunay triangulation is 'as regular as possible'.

Delaunay triangulations and their dual Voronoi diagrams have been proposed and used for GIS applications. A recent paper by Gold (1994) describes their use to quickly and efficiently build topology in forestry applications.

### 5.4.4  Triangulations of polygons

In general, a triangulation tessellates an unbounded region with triangles based upon a prescribed set of vertices. This section describes several approaches to the triangulation of a bounded polygonal region. It would be nice to modify the Delaunay triangulation for an unbounded region, but if we take the collection of points comprising the vertices of a polygon and form the Delaunay triangulation based upon them, this will not necessarily be the triangulation of the polygon, because some of the edges of the polygon may not be edges in the triangulation. However, it is possible to use the so-called *constrained Delaunay triangulation*, which is constrained to follow a given set of edges. Constrained triangulations have been used in some GIS applications: for example, Christensen (1987) discusses applications where edges of the triangles must follow along ridges or contour lines in a digital terrain model. In our case, the constraint is that the edges of the polygon must be edges in the triangulation. Figure 5.25 shows a very simple case with five vertices. The Delaunay triangulation is as in Figure 5.25(a), but when an edge is added, the triangulation is constrained as in Figure 5.25(b).

Apart from Delaunay methods, other triangulations may be found in the literature. The so-called 'greedy triangulation' has the objective of minimizing the total edge length in the triangulation. This is achieved by introducing the shortest possible internal diagonal at each stage. The algorithm is 'greedy' for computational time. Figure 5.26(a) shows the greedy triangulation of a sample polygon. The regularity of the subdivision may be observed.

Other triangulation methods abound in the literature. Much of the effort has been to improve efficiency of the algorithms. One further example is the Garey *et al.* (1978) algorithm for triangulating a simple polygon. The method depends on a preliminary decomposition into monotone polygons (see Chapter 3). The resulting polygons may be triangulated efficiently. Figure 5.26(b) shows the results of triangulating the sample polygon using this method.

There are many other approaches to the triangulation of sets of points and of polygons. A summary and appraisal of the polygonal triangulations may be found in Jayawardena (1994) with a briefer review in Jayawardena and Worboys (1995). Algorithms are considered later in the chapter.

### 5.4.5   Tessellations of the sphere

The representations described so far have all been referenced to planar objects. However, of course the Earth is topologically a sphere and so it should be our

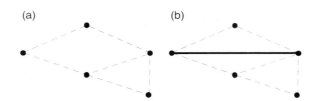

**Figure 5.25**   Unconstrained and constrained Delaunay triangulations.

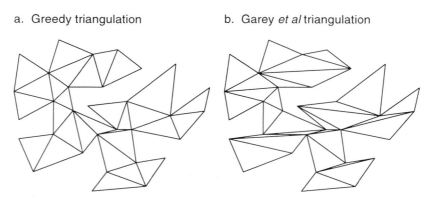

**Figure 5.26**   Greedy and Garey *et al.* (1978) triangulation of a polygon.

concern to find representations appropriate for spherically referenced data. The reasons that so much emphasis has been placed upon the plane are:

- Planar data are computationally simpler than spherical data.
- Small areas of the Earth's surface may be approximated by planes.

Regular tessellations of the sphere correspond to the Platonic solids of antiquity: tetrahedron, cube, octahedron, dodecahedron and icosahedron. For example, the tetrahedral tessellation comprises four spherical triangles (called *triangular facets*) bounded by parts of great circles. Each spherical triangle has an internal angle of 120°. Three triangles meet at each vertex. In the case of the octahedral tessellation, there are initially eight triangular facets, each having an internal angle of 90°. Four triangles meet at each vertex.

Figure 5.27 shows a possible positioning of the octahedral tessellation on the sphere, having triangular facets *nab, nbc, ncd, nda, sab, sbc, scd, sda.* Unfortunately, unlike the plane, regular tessellations of arbitrary fine resolution are not possible on the sphere. However, it is possible to recursively nest regular polygons within the facets of the Platonic solids to produce the resolutions required. This will be discussed in the next section on nested tessellations.

### 5.4.6   Recursively tessellated representations

A *nested tessellation* is a tessellation in which the cells themselves are partitioned using a further and finer tessellation. This process can be recursive, providing a graded granularity to which spatial phenomena may be referenced. *Regular nested tessellations* occur where the regular constituents are themselves partitioned using the same regular figure. Examples of planar regular and irregular nested tessellations are shown in Figure 5.28. The first two (triangle and square) are regular, but the third (hexagon) is not since hexagons cannot fit together to make a larger hexagon. (The idea for this diagram came from Peuquet (1984).)

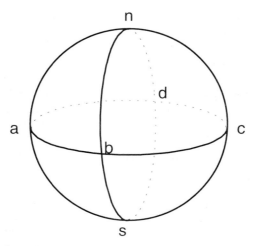

**Figure 5.27**   Octahedral tessellation of the sphere.

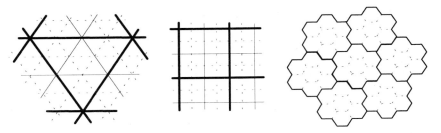

**Figure 5.28**    Regular and irregular nested tessellations of the plane.

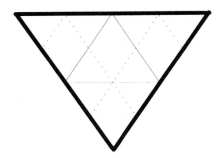

**Figure 5.29**    Recursive subdivision of one facet of the octahedral tessellation of the sphere.

The nested square tessellation of the plane leads to the quadtree data structure, discussed in the next chapter in the context of spatial data structures and indexes. The nested triangular tessellation may be used to provide the nested octahedral tessellation of the sphere. This has been investigated by several authors, notably Dutton (1984, 1989). The idea is to take each of the eight triangular facets *nab, nbc, ncd, nda, sab, sbc, scd, sda,* shown in Figure 5.27 and recursively subdivide them. Figure 5.29 shows a schematic of such a subdivision, where a single facet is subdivided into four triangles, each itself further subdivided into four. This process can continue recursively until the required granularity is obtained. This structure, when indexed in an appropriate way, can act as the spherical equivalent of the quadtree, and is further considered in the next chapter.

Returning to the plane, there is a growing body of work on nested irregular tessellations. A distinction is made between *hierarchical* models, based upon recursive regular or irregular tessellations and *stratified* models in which each stratum is defined to have appropriate properties. A survey of hierarchical models is given by De Floriani *et al.* (1994) with particular emphasis on hierarchical TIN's in (De Floriani and Puppo, 1992). A stratified Delaunay triangulation, where the circumcircle property is satisfied at each level, is defined in the *Delaunay pyramid* of De Floriani (1989). Stratified models are discussed in general by Bertolotto *et al.* (1994).

## 5.5    FUNDAMENTAL GEOMETRIC ALGORITHMS

It would be impossible to describe in this single section all the spatial algorithms that have been devised. The aim is to provide a representative sample, giving preference to commonly used algorithms and those that show an example of a generic approach. Also, the author's own interests have coloured the choice, for example emphasizing triangulation algorithms. The representations introduced in earlier sections will be used in the algorithms.

All of the algorithms described below assume an embedding in the Euclidean plane. Thus, point objects have a position that may be given by a coordinate pair $(x, y)$. The set of points on a straight-line in the Euclidean plane may be constrained by a linear equation to give $\{(x, y) | ax + by + c = 0\}$, where $a$, $b$, $c$ are constants. The set of points on a straight-line segment between two distinct points $p$ and $q$ may also be represented in parametric form as $\{\lambda p + (1 - \lambda)q | \lambda \in [0, 1]\}$ (see Chapter 3).

### 5.5.1 Metric and Euclidean algorithms

*Distance and angle between points*

For completeness, we restate these formulas (from Chapter 3). The (*Euclidean*) *distance* $|ab|$ between points $a = (a_1, a_2)$ and $b = (b_1, b_2)$ is given by the formula

$$|ab| = \sqrt{((b_1 - a_1)^2 + (b_2 - a_2)^2)}$$

The bearing $\theta$ of $b$ from $a$ is given by the unique solution in the interval $[0, 360[$ of the simultaneous equations:

$$\sin \theta = (b_1 - a_1)/|ab|$$
$$\cos \theta = (b_2 - a_2)/|ab|$$

*Distance from point to line*

By distance, the minimum distance between spatial objects is meant. To compute the distance between a point and a straight line, the most compact formula arises when the straight line $l$ is given in the form $\{(x, y) | ax + by + c = 0\}$. Suppose that the point $p$ is given by the coordinate pair $(X, Y)$, then the distance from $p$ to $l$, measured by the length of the line segment through $p$ and orthogonal to $l$ is given by the formula:

**distance**$(p, l) = |aX + bY + c|/\sqrt{a^2 + b^2}$

For the distance between a point $p$ and a straight line *segment* $l$, the calculation is no longer quite so straightforward. Figure 5.30 shows that the distance will be measured as the distance from $p$ to one of the end points of $l$ or from $p$ to the line incident with the end points of $l$, depending on the relative positions of $p$ and $l$. The line segment $l$ defines a partition of the plane into two pointsets, a connected set that we have called **middle**($l$) and a disconnected set called **end**($l$). The calculation to determine the distance of $p$ from $l$ depends upon whether $p$ is in **middle**($l$) or **end**($l$).

The problem becomes more complex for a polyline, as can be seen from Figure 5.31. For each segment $l$, it must be checked whether $p$ lies in **middle**($l$) and if so the distance calculated. The distance from $p$ to each vertex is also calculated. The distance of $p$ from the chain will be the minimum of all these distances. If the number of segments in the chain is $n$, then the time complexity of the computation is O($n$). An approximation to the distance may be achieved by considering only the

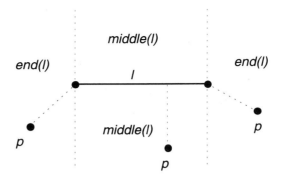

**Figure 5.30**   Distance between a point and line segment.

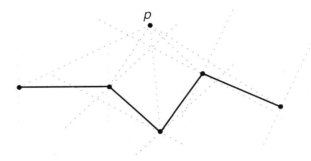

**Figure 5.31**   Distance between a point and a polyline.

distances to vertices of the chain. This approximation will be good in general if the segment lengths are small compared with the distance of the point from the chain.

Distances from a point to a polygon or between polygons may be required in terms of their boundaries. Thus, the distance between two polygons may be interpreted as the distance between their closest points. In that case, a calculation involving boundary polylines may be necessary. More usually, the distance between two polygons is interpreted as the distance between their centroids. This leads to the next section, where the calculation of the centroid of a polygonal area is discussed.

*Centroid of a polygon*

The *centroid* or *centre of gravity* of an areal object is the point at which it would balance if it were cut out of a sheet of material of uniform density. If the area is a polygon $P$ with vertex vectors $p_1, \ldots, p_n$, then the calculation is straightforward as the mean of the vertex vectors:

**centroid** $(P) = (p_1 + \ldots + p_n)/n$

*Area of simple polygon*

Let $P$ be a polygon with vertex vectors $p_1, \ldots, p_n$, then a formula for the area is:

**area** $(P) = (p_1 \times p_2 + p_2 \times p_3 + \ldots + p_{n-1} \times p_n + p_n \times p_1)/2$

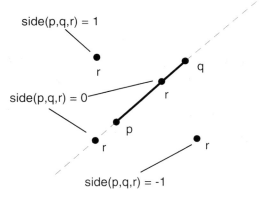

**Figure 5.32** The *side* operation.

where the product symbol × indicates the vector product defined by

$$p \times q = (x\text{-coord of } p) \cdot (y\text{-coord of } q) - (y\text{-coord of } p) \cdot (x\text{-coord of } q)$$

In the case of a triangle *pqr*, the formula gives:

**area** $(pqr) = (p \times q + q \times r + r \times p)/2$

It will be noticed that the formulas give a sign (positive or negative) to the area. Also, according to our formulas, **area** $(pqr) = -$**area** $(qpr)$. If point *p* is to the left-hand side of the directed segment *qr*, then **area** $(pqr)$ will be positive; if point *p* is to the right-hand side of the directed segment *qr*, then **area** $(pqr)$ will be negative; and if *p*, *q*, *r* are collinear, then **area** $(pqr)$ will be zero. Thus, a useful spin-off from our work on area is an algorithm for the *side* operation (Figure 5.32):

$$\textbf{side } (p, q, r) = \begin{cases} 1 & \text{if point } p \text{ is to the left-hand side of directed} \\ & \text{segment } qr(\textbf{area } (pqr) > 0) \\ 0 & \text{if } p, q, r \text{ are collinear } (\textbf{area } (pqr) = 0) \\ -1 & \text{if point } p \text{ is to the right-hand side of directed} \\ & \text{segment } qr \ (\textbf{area } (pqr) < 0) \end{cases}$$

### 5.5.2   Topological algorithms

*Point-in-polygon*

The Boolean topological algorithm **point_in_polygon** that we now consider is one of the most common operations in GIS. Given a point *p* and a polygon *P*, **point_in_polygon** $(p, P) = $ **true** if and only if *p* is in the interior of *P*.

If the polygon is convex, then the operation *side* discussed in the previous section can be used. Assume that the vertices of the polygon are $p_1, \ldots, p_n$, ordered going anti-clockwise around the polygon. The **point_in_polygon**$(p, P) = $ **true** if and only if

$$\textbf{side } (p, p_1, p_2) = \textbf{side } (p, p_2, p_3) = \ldots = \textbf{side } (p, p_n, p_1) = 1$$

The more usual and interesting case is where the polygon is not necessarily convex, and two algorithms are considered: the semi-line algorithm and the winding number algorithm.

The underlying principle behind the semi-line algorithm is very simple. Given a point $p$ and polygon $P$, for simplicity assume that $p$ is not on the boundary of $P$. Draw an infinite semi-line (or *ray*) out from $p$. Count the number of times that the ray intersects with the boundary of $P$. If the number is even, then $p$ is outside $P$. If the number is odd, then $p$ is inside $P$ (see Figure 5.33).

However, as with many of these algorithms, several special cases can arise. For example, if a ray intersects with a vertex of the polygon, how should it be counted? Figure 5.34 shows two such cases. At vertex $a$, the intersection should not be counted while at vertex $b$, the intersection should be counted. It is important that

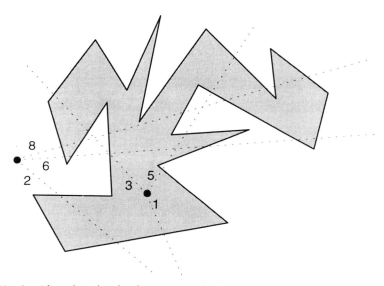

**Figure 5.33**   Semi-line algorithm for determining whether a point is inside a polygon.

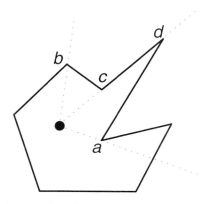

**Figure 5.34**   Special cases in the semi-line algorithm.

the polygon actually crosses the ray for the intersection to be counted, and this does not happen at *a*. Figure 5.34 also shows the intersection of the entire segment *cd* with the ray. The same principle may be applied: it is only the *crossing* that is counted.

With regard to implementation, it simplifies matters to choose the semi-line to be horizontal and directed to the right, then to translate the polygon so that the semi-line is the positive *x*-axis with the point *p* as the origin. The polygon is then traversed and, for each edge, a simple calculation with coordinates determines whether the edge crosses the positive *x*-axis. The algorithm has time complexity $O(n)$, where *n* is the number of edges of the polygon.

The winding number algorithm provides an alternative and quite distinct approach. It has the same complexity as the semi-line algorithm, but in practice is much less efficient due to the trigonometric calculations. Imagine that an observer is moving anti-clockwise along the boundary of *P*, always facing the point *p*. If after having completed the journey round *P*, the observer has turned through one complete circle, then *p* is within *P*. Otherwise, *p* is without *P*. Figure 5.35 shows examples of this. In each case a path is walked visiting the vertices in the order 1-2-3-4-5-6-1. In the left-hand configuration, where the point is inside the polygon, the

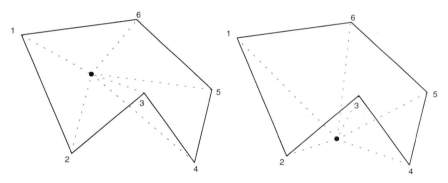

**Figure 5.35** Traversing the boundary of the polygon 123456 always facing the point.

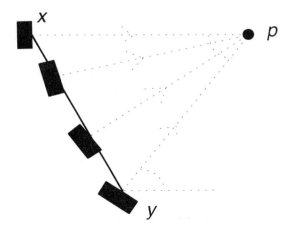

**Figure 5.36** Turn experienced while moving along a segment.

observer will rotate through a full turn; in the right-hand configuration, where the point is outside the polygon, the net turning effect will be zero.

As the observer moves along a single edge $xy$, facing the point $p$, the total angle turned will be the angle $\angle xpy$ (see Figure 5.36). Given the coordinates of $x$, $y$ and $p$, this angle may be calculated using elementary trigonometry. The angles are then summed over the entire walk round the polygon to find the total turn.

### 5.5.3  Set based algorithms

Throughout this section spatial objects are treated as sets of points. It is assumed, unless the contrary is indicated, that all the sets are topologically closed, that is they contain their boundary points and edges.

*Collinearity*

Given three points $a$, $b$ and $c$, the Boolean operation **collinear** $(a, b, c)$ determines whether $a$, $b$ and $c$ are collinear (lie on the same straight line). The algorithm for this operation needs no new ideas. From section 5.5.1, **collinear** $(a, b, c) = $ **true** if and only if **side** $(a, b, c) = 0$.

*Point on segment*

Let straight-line segment $l$ have end points, $p = (p_1, p_2)$ and $q = (q_1, q_2)$, say. For any point $x$, the operation **point_on_segment**$(x, l)$ returns the Boolean value **true** if $x \in l$.

The first step is to use the previous algorithm to decide whether $x$, $p$ and $q$ are collinear. If not, the determination is negative and there is no need to proceed further. Otherwise, consider the minimum bounding box of $l$, **mbb**$(l)$, which is the smallest rectangle with sides parallel to the coordinate axes that contains $l$. Then, $x \in l$ if and only if $x \in$ **mbb**$(l)$. Finally $x \in$ **mbb**$(l)$ if and only if both the following inequations hold:

$$\textbf{min } (p_1, q_1) \leq x_1 \leq \textbf{max } (p_1, q_1)$$

$$\textbf{min } (p_2, q_2) \leq x_2 \leq \textbf{max } (p_2, q_2)$$

where **min** and **max** return the minimum and maximum, respectively, of the arguments.

*Segment intersection*

There are two related operations for straight-line segment intersection: one Boolean, returning **true** if, and only if, the segments have at least one point in common; the other returning the point of intersection if it exists. Clearly, the result of the latter operation immediately gives a result for the former. Assume, as usual, that the segments are closed and therefore contain their end points.

The intersection detection operation is a decision method, returning **true** if and only if the segments have at least one point in common. It may be defined in terms of previous operations, in particular the *side* operation. Figure 5.37 shows an example of how the *side* operation is used. Segments $ab$ and $cd$ will intersect if

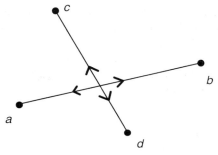

**Figure 5.37** Using the **side** operation for intersection detection.

points $a$ and $b$ are on the opposite sides of the line through $cd$ and points $c$ and $d$ are on opposite sides of the line through $ab$. (There are also special cases where **side** takes the value zero, covered in the definition.) More formally, $ab$ and $cd$ intersect if both the inequations below hold:

side $(a, b, c) \neq$ side $(a, b, d)$

side $(c, d, a) \neq$ side $(c, d, b)$

The converse of this statement is not true, however. It is possible for $ab$, $cd$ to intersect when one or both of these inequations is false. This might occur when one of the segments partially lies on the other. Thus, also to be checked are the possibilities:

**point_on_segment** $(a, cd) = $ **true**

**point_on_segment** $(b, cd) = $ **true**

**point_on_segment** $(c, ab) = $ **true**

**point_on_segment** $(d, ab) = $ **true**

If two segments are shown to intersect by the above decision procedure, unfortunately it does not go on to determine their point of intersection. The calculation of a point of intersection of two line segments, if it exists, is at first sight a trivial exercise in coordinate geometry, reducing to the simultaneous solution of two simple linear equations. However, as with many geometric algorithms, special cases (e.g. vertical segments) and errors due to the discretization process provide some difficulties. (David Douglas (1974) provides an amusing account of this problem in his paper 'It makes me so CROSS'.)

Suppose that we are given segments $l = pq$ and $l' = p'q'$, with respective parametric forms for their pointsets:

$\{\lambda p + (1 - \lambda)q \mid \lambda \in [0, 1]\}$

$\{\lambda p' + (1 - \lambda)q' \mid \lambda \in [0, 1]\}$

Let $r$ be a point of intersection of these sets. Then there are $\alpha, \beta \in [0, 1]$ such that:

$$r = \alpha p + (1 - \alpha)q = \beta p' + (1 - \beta)q' \tag{1}$$

(1) may be rearranged to give $\alpha(p - q) + \beta(p' - q') = q' - q$. This vector equation may be decomposed into two simultaneous equations that will have a unique

solution provided that $l$ and $l'$ are not parallel. It remains to check that the solutions for $\lambda$ and $\lambda'$ are in the specified bounds and then to substitute for one of them in (1).

### Half-plane intersection

The problem here is to form the intersection of a finite number of half-planes, embedded in the Euclidean plane. A half-plane is the set of solutions to a linear inequality: $ax + by + c \leq 0$. Thus, it comprises all points on and to one side of the straight line $ax + by + c = 0$. The result of such an intersection will be a convex (but not necessarily bounded) polygon. The most simple-minded approach to the implementation of this operation is given by the following algorithm

> **Basic half-plane intersection algorithm**
> **Input:** Given $n$ half-planes, $H_1, \ldots, H_n$, defined by $n$ linear inequalities
> **Output:** Winged-edge representation of the convex (possibly unbounded) polygon $P$ formed as $H_1 \cap \ldots, \cap H_n$.
> **Procedure:**
> Step 1.  $P \leftarrow H_1$
> Step 2.  For $i = 2$ to $n$ do steps 2.1 to 2.2
>    2.1  $P' \leftarrow H_i \cap P$
>    2.2  $P \leftarrow P'$

For step 2.1, the intersection of $H_i$ and convex polygon $P$ is constructed, by computing the intersection of $H_i$ with each of the $i - 1$ boundary line segments of $P$. (There can be at most two such intersections.) If there are two intersections, $P$ is sliced by $H$ and the appropriate slice is retained, otherwise no action is taken. The complexity of the algorithm is given by, $O(1) + O(2) + \ldots + O(n) = O(n^2)$.

The basic algorithm may be speeded up by a divide-and-conquer approach. The set of $n$ half-planes is partitioned into two roughly equal-sized parts. Recursively use this divide-and-conquer approach on each part to form two convex polygons $P$ and $Q$, say, then form the intersection of $P$ and $Q$.

> **Divide-and-conquer half-plane intersection algorithm**
> **Input:** Given $n$ half-planes, $H_1, \ldots, H_n$, defined by $n$ linear inequalities
> **Output:** Winged-edge representation of the convex (possibly unbounded) polygon $P$ formed as $H_1 \cap \ldots, \cap H_n$.
> **Procedure:**
> Step 1.  $N \leftarrow$ integer part of $n/2$
> Step 2.  $X \leftarrow \{H_1, \ldots, H_N\}$, $Y \leftarrow \{H_{N+1}, \ldots, H_n\}$
> Step 3.  Form the intersection of $X$ by calling the algorithm recursively, with resulting convex polygon $P_X$
> Step 4.  Form the intersection of $Y$ by calling the algorithm recursively, with resulting convex polygon $P_Y$
> Step 5.  Result $\leftarrow P_X \cap P_Y$

Let $t(n)$ be the time taken to execute the divide-and-conquer algorithm. Then, because step 5 takes linear time, we have the recurrence relation:

$$t(n) = \text{time taken for step 3} + \text{time taken for step 4} + O(n)$$

$$= t(n/2) + t(n/2) + O(n)$$

$$= 2t(n/2) + O(n)$$

Thus, the time complexity of the algorithm is $O(n \log n)$, in fact, optimal for this operation (Theorem 7.10 of Preparata and Shamos (1985)).

*Intersection, union and overlay of polygons*

This section discusses the Boolean operations of intersection and union of two simple polygons. Let the two simple polygons be $P$ and $P'$, with segment cycles $[l_1, \ldots, l_m]$ and $[l'_1, \ldots, l'_n]$. The key is to find the points of intersection of line segments from $P$ with line segments from $P'$. In the general case, we must examine each pair $(l_i, l'_j)$ $1 \leq i \leq m$, $1 \leq j \leq n$) and determine the point of intersection, if it exists. (If the segments overlap or meet at their end points, then special cases must be constructed – not considered here). There are $mn$ such pairs to examine and so the time complexity of the algorithm is $O(mn)$. In the special case where both polygons are convex, then we may traverse each cycle in order to determine intersections, thus the time complexity in this case can be reduced to $O(m + n)$ (see Figure 5.38).

Let us assume that $P$ and $P'$ are embedded in larger planar partitions and are represented as winged-edged structures. The algorithm takes the edge information of the polygons represented by the winged-edged structures, along with the new

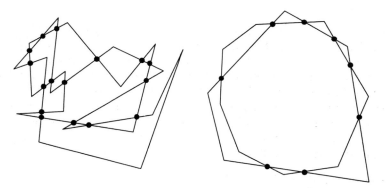

**Figure 5.38**    Intersection points of a general simple and a convex polygon.

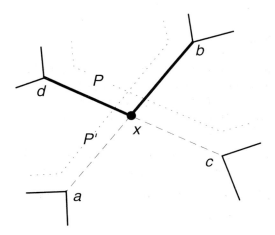

**Figure 5.39**    Constructing the intersection of two polygons using the winged-edge representations.

intersection points, and constructs a representation of the polygon intersection (or union). This is not described in detail here (the interested reader is referred to the bibliographic notes for some references on this). Figure 5.39 shows some of the considerations involved. Two winged-edges $ab$, $cd$ are shown, along with their point of intersection $x$. The polygons $P$, $P'$ are assumed to be along the right of directed edges $ba$ and $cd$, respectively. The directed edges, $bx$, $xd$ form part of the edge structure of the intersection polygon. The edge portions $ax$, $xc$ are not required. For the union, the other pair of edges $cx$, $xa$ will be kept and $dx$, $xb$ discarded. Polygon overlay will require essentially the same information.

### 5.5.4   Raster algorithms

This section discusses some of the operations that are often particularly associated with raster-based data, or arising from the field-based model. Many algorithms developed for raster-based data stem from the graphics and image processing communities. Some are of great use for GIS applications. Space precludes a detailed discussion, but a brief survey is given and pointers to further reading are provided in the bibliography at the end of the chapter.

*Planar geometric transformations:*   Functions from the plane, or at least a finite portion of the plane stored in the computer, to itself. Such transformations were discussed in Chapter 3, where formulas were provided in some cases. Thus, to transform a raster image, apply the transformation to each pixel of the image, resulting in a collection of pixels for the transformed image. Unfortunately, there are at least two problems with this. First, if the image is enlarged, then it will have only as many pixels as the original and thus be less dense. A further problem arises from the discrete computer arithmetic used for the coordinates. This relates to aliasing effects (jagged edges and staircases) caused by discrete sampling of a rapidly changing continuous variation. Algorithms have been developed that can partially overcome these problems. See Foley *et al.* (1990) for a general discussion and Weiman (1980) for scaling and shearing algorithms.

*Graphics algorithms:*   A very large number of algorithms come into this category, and most are outside the scope of this text and well-covered by specialist graphics texts, for example Foley *et al.* (1990). Examples include planar transformations (as above), shape filling, clipping, anti-aliasing and scan-conversion. Required for viewing three-dimensional shapes are projective transformations, hidden line determination and hidden surface elimination.

*Edge-detection and enhancement:*   High values for the derivatives of fields (large changes) may imply the existence of an edge. For example, a change in vegetation may result in a sudden change in spectral frequencies recorded in a satellite image. It is often important to detect and record the positions of such edges. It may also be appropriate to enhance the edge by exaggerating the difference between neighbouring values.

*Smoothing:*   The removal of random distortions, often in the form of a speckled effect, may be caused by noisy communication channels.

*Thresholding:*   Converts the intensity of each element of an image to one of two levels, to produce a bilevel image where pixels with values in a given range are distinguished. For example, in a grey-scale image, we might want to change the values of all pixels below a given level of greyness to black and those above to white. Thus, a threshold is created between black and white areas. For *simple thresholding*, a fixed value is chosen for the threshold value throughout the raster; in *adaptive thresholding*, the value varies according to the local conditions in part of the image. Thus, in a generally dark area, it might be necessary to choose a different value to a generally light area.

### 5.5.5    Triangulation algorithms

*Polygon triangulation algorithms*

This section describes two of the simpler non-Delaunay algorithms for triangulation, taking triangulation of a polygon as an exemplar. The triangulations will not introduce any extra interior vertices (Steiner points). The objective is to arrive at reasonably regular triangulations, that is with few long thin triangles.

*Greedy triangulation algorithms*    The greedy triangulation has been briefly considered in section 5.4.4 and illustrated in Figure 5.26(a). This algorithm is appropriate for both unconstrained and constrained triangulations. We frame it as a triangulation of a simple polygon. The objective is to minimize the total edge length in the triangulation. This is achieved by an iterative process that selects the shortest available internal diagonal at each stage. A simple implementation (Düppe and Gottschalk, 1970) of this algorithm is given below.

**Greedy triangulation algorithm**
**Input:** Winged-edge representation of a simple polygon $P$ with $n$ vertices
**Output:** Winged-edge representation $T$ of the triangulation of $P$.
**Procedure:**
Step 1.    Form the set $S$ of the $N = n(n - 3)/2$ diagonals of $P$.
Step 2.    Sort $S$ in ascending order, placing the diagonals in list $D = [d_1, \ldots, d_N]$.
Step 3.    $T \leftarrow P$
Step 4.    For $i = 1$ **to** $N$ do steps 4.1 to 4.2
     4.1    $d \leftarrow d_i$
     4.2    **if** $d$ is an internal diagonal of $T$
          **and** $d$ does not properly intersect any diagonal in $T$
          **then** $T \leftarrow$ winged-edge representation of $T \cup d$.

In the worst case, all $O(n^2)$ diagonals will be considered. Each of these edges must be tested for intersection with the other edges in the triangulation. The triangulation will contain $O(n)$ edges. Assuming a naive test for intersection of two edges, running in constant time, the algorithm is of worst-case complexity $O(n^3)$.

We can make some immediate improvements on the efficiency of this algorithm. For example, keep a count of the edges added to the triangulation. As the number of such edges must be $n - 3$, when we arrive at this point, abort Step 4. An alternative and more efficient method due to Gilbert (1979), and given in Preparata and Shamos (1985), balances the scanning of list $D$ against deciding whether an edge is in the triangulation. The method uses the segment tree data structure (see Chapter 6) and results in an algorithm of complexity $O(n^2 \log n)$.

*The triangulation algorithm of Garey et al.*   The algorithm of Garey *et al.* (1978) to be discussed now is suitable for the triangulation of simple polygons. With respect to time complexity, it improves on the greedy method above. The algorithm proceeds by first decomposing the polygon into monotone polygons (see section 3.2.3) and then triangulating each of the monotone polygons. Let us assume that monotonicity is with respect to the *y*-axis and that no two different vertices of the polygon have the same *y*-coordinate. Stage 1 of the algorithm decomposes the polygon into monotone polygons. This stage is essentially the regularization of the polygon as described in Preparata and Shamos (1985). Extra edges are added between vertices of the polygon so that:

1.  Edges only intersect at polygon vertices.

2.  Apart from the highest vertex (with respect to the *y*-axis), every vertex is joined by a single edge to a higher vertex.

3.  Apart from the lowest vertex (with respect to the *y*-axis), every vertex is joined by a single edge to a lower vertex.

The algorithm proceeds by making two plane sweeps of the polygon. The first sweep is from top to bottom and regularizes each vertex *v* (except for the bottom vertex) that has no direct edge to a vertex below. Such a non-regular vertex *v* is shown in Figure 5.40. To regularize vertex *v*, let edges *e* and *f* be the two closest edges of the polygon to *v*, intersecting the plane sweep through *v* (at least one of these edges must exist). From the lower vertices of edges *e* and *f*, choose the upper, *v'*. The new edge *vv'* is added to the structure. It is important to note that *vv'* cannot intersect any existing edge (except at a vertex).

This process is repeated for each non-regular vertex in the sweep down the polygon. A symmetrically opposite process is applied in the second sweep up the polygon. We claim without proof that the resulting structure is a decomposition of the polygon into monotone polygons. Some of these polygons may be external to the original polygon and may be discarded. Readers who are interested in a more detailed discussion of the regularization process may consult Preparata and Shamos (1985).

The complexity of the regularization stage is $O(n \log n)$, where $n$ is the number of vertices of the polygon. An example of the procedure is given in Figure 5.41. Figure 5.41(b) shows the regularization of the polygon in Figure 5.41(a). Figure 5.41(c) shows the polygon with the new external edges removed. It may be observed that the polygons $A, \ldots, I$, into which the original polygon has been decomposed, are all monotone.

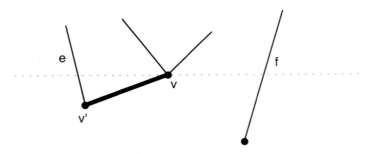

**Figure 5.40**   Regularization of non-regular vertex *v*.

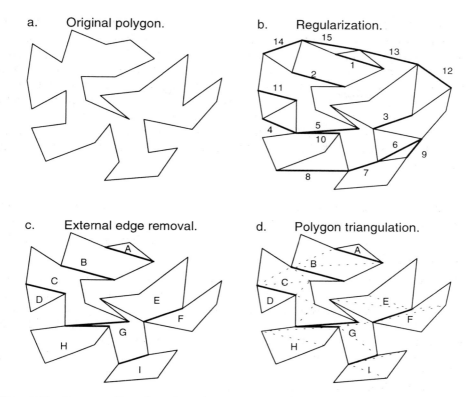

**Figure 5.41** Decomposition of a polygon into monotone polygons.

The problem has now been reduced to the triangulation of a monotone polygon. Ingeniously, it is possible to triangulate a monotone polygon in time $O(n)$. Let $P$ be a monotone polygon with respect to the $y$-axis. Begin by performing a sort-merge procedure on the vertices of each of the chains to produce a list $V = [v_1, \ldots, v_n]$, from top to bottom, of the $n$ vertices of $P$. This initial sorting can be done in linear time. Now move down list $V$, adding diagonals for the triangulation as we go and keeping a second list $X$ of all the vertices that have been visited but not yet fully processed. Initially $X = [v_2, v_1]$ and the process starts with vertex $v_3$, move through $V$. Suppose that the process has reached the vertex $v$, and the list $X = [x_m, \ldots, x_1]$. New diagonals of $P$ are generated according to the rules:

1. If $v$ is adjacent to $x_1$ but not to $x_m$, as in Figure 5.42(a), then add diagonals $vx_1$, $vx_2, \ldots, vx_m$. Reset list $X = [v, x_m]$.

2. If $v$ is adjacent to $x_m$ but not to $x_1$, and $x_m$ is a convex vertex (angle $vx_mx_{m-1}$ is less than 180°), as in Figure 5.42(b), then add diagonal $vx_{m-1}$ and remove $x_m$ from the head of $X$. Repeat this process until the condition for the rule no longer holds. Add $v$ to the head of the list.

3. If $v$ is adjacent to both $x_m$ and $x_1$, as in Figure 5.42(c), then add diagonals $vx_2$, $\ldots, vx_{m-1}$ and halt.

To illustrate this procedure, consider the monotone polygon shown in Figure 5.43(a). The two monotone chains are [1, 2, 3, 4, 5, 6, 7] and [1, 12, 11, 10, 9, 8, 7],

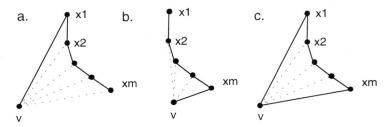

**Figure 5.42**    Three cases occurring in the triangulation of a monotone polygon.

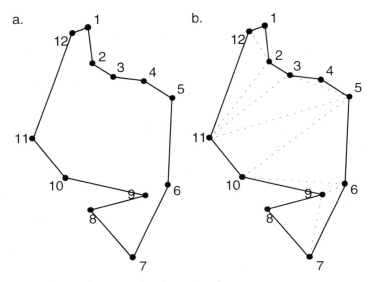

**Figure 5.43**    Triangulation of an example monotone polygon.

and sort-merging the chains gives the vertex list V = [1, 12, 2, 3, 4, 5, 11, 10, 6, 9, 8, 7].
List $X$ is initialized to [12, 1]. The first vertex to be considered is 2, to which rule 1
applies. Diagonal 2–12 is formed and $X$ reset to [2, 12]. Rule 2 applies to vertex 3,
but because angle 3–2–12 is reflex, no diagonal is added and $X$ reset to [3, 2, 12].
Vertex 4 leads to a similar case, but for vertex 5, rule 2 applies and angle 5–4–3 is
convex, so diagonal 5–3 is added. An attempt is made to add diagonal 5–2, but this
proves impossible because angle 5–3–2 is reflex. List $X$ is reset to [5, 3, 2, 12]. Rule
2 is applied to vertex 11 and several diagonals are added. The procedure continues
through the vertex list. Finally, vertex 7 is reached, to which rule 3 is applied.
Diagonal 7–9 is added and the procedure terminates. The result is shown in Figure
5.43(b).

The triangulation procedure iterates over the vertices of the monotone polygon
and runs in linear time. Combining this with the $O(n \log n)$ monotone partitioning
algorithm results in an algorithm of time complexity $O(n \log n)$ for triangulating an
arbitrary simple polygon. The resulting triangulation of the simple polygon in
Figure 5.41(a) is shown in Figure 5.41(d).

*Further speed-ups*    The history of triangulations is interesting. Algorithms of time
complexity $O(n^2)$ have been around for a long time. Garey *et al.* published their

$O(n \log n)$-time algorithm in 1978, this being an achievement of the new-born field of computational geometry. Although a lower bound on the time-complexity of a general planar triangulation was known to be $O(n \log n)$-time, it was not known whether this still held for triangulations of simple polygons or whether an even lower bound was possible. It was not until nearly 10 years later that Tarjan and Van Wyk (1988) published an $O(n \log \log n)$-time algorithm. Of course, the best that can be hoped for is a linear time algorithm, and this was actually achieved by Chazelle (1991), thus providing a climax to a large research effort.

### Delaunay triangulation algorithms

Algorithms for the Delaunay triangulation and its dual Voronoi diagram are abundant in the literature. The worst-case time complexity of any Delaunay triangulation algorithm is at least $O(n \log n)$. This follows from the observation that a triangulation for $n$ points along the parabola $y = x^2$, say, provides a means of sorting the $x$-coordinates of these points in increasing order of magnitude. As it is well known (Aho et al., 1974) that sorting algorithms must be at least of time complexity $O(n \log n)$, any Delaunay triangulation algorithm is at least $O(n \log n)$. In fact, divide-and-conquer $O(n \log n)$ algorithms have been achieved (Shamos and Hoey, 1975; Lee and Schachter, 1980) that are theoretically best possible. In practice, an incremental method enhanced by an efficient data structure, due to Ohya, Iri and Murota (1984a, 1984b), is often more useful, because it is conceptually simple and has linear average time complexity (even though worst-case time-complexity $O(n^2)$). Dwyer (1987) published a variant of the divide-and-conquer approach that led to an algorithm with optimal worst-case time complexity and average time complexity of $O(n \log \log n)$. In this section, we outline some of the ideas behind these constructions.

Assume throughout that there are $n$ points in the Euclidean plane, with the requirement to construct a Delaunay triangulation of the plane with the $n$ points as vertices. Some of the following are constructions for Voronoi diagrams, but it is a straightforward, linear-time, process to convert the Voronoi diagram to its dual triangulation. The first, brute force, approach is to construct each of the $n$ polygons separately as follows.

**Brute-force Delaunay triangulation algorithm**
**Input:** $n$ points $p_1, \ldots, p_n$ in the Euclidean plane
**Output:** Winged-edge representation $T$ of the triangulation of $P$.
**Procedure:**
Step 1.　For $i = 1$ to $n$ do steps 1.1 to 1.2
　　1.1　Form the $n - 1$ half-planes containing $p_i$ and bounded by the perpendicular bisectors of the line segments
　　　　$p_i p_1, p_i p_2, \ldots, p_i p_{i-1}, p_i p_{i+1}, \ldots, p_i p_n$
　　1.2　apply the divide-and-conquer algorithm for half-plane intersection (section 5.5.3) to construct the Voronoi polygon around $p_i$
Step 2.　Form the collection of all the polygons computed in step 1
Step 3.　Convert resulting Voronoi diagram to Delaunay triangulation

As the divide-and-conquer algorithm for half-plane intersection is of time complexity $O(n \log n)$, this approach is of time complexity $O(n^2 \log n)$.

The algorithms of Shamos and Hoey (1975) and Lee and Schachter (1980) use the divide and conquer approach. We describe an example of the use of divide-and-conquer in this context, similar to the Lee-Schachter algorithm.

**Divide and conquer Delaunay triangulation algorithm** (Lee-Schachter)
**Input**: $n$ points $p_1, \ldots, p_n$ in the Euclidean plane
**Output**: Winged-edge representation $T$ of the triangulation of $P$.
**Procedure**:
Step 1.   Sort the $n$ input points in lexicographically ascending order
          (i.e. order of ascending $x$-coordinates, but if $x$-coordinates
          are equal, then by ascending $y$-coordinates)
Step 2.   Divide the sorted list into two roughly equal halves
Step 3.   Recursively apply this algorithm to triangulate each half
Step 4.   Merge the triangulations of the two halves together to
          triangulate the entire set of points

Of course, this description glosses over some crucial details, especially how the merger is conducted. Interested readers are referred to the literature for details: here is provided a brief description and an example. To merge the two triangulations, form their respective convex hulls and construct upper and lower common tangents. Figure 5.44 shows the merging process by means of an example. Starting from the lower common tangent, proceed upwards between the triangulations, adding and deleting as appropriate. In the figure, one edge from the right-hand partial triangulation is deleted and nine edges are added. The time complexity of the merger operation is linear, and so the whole algorithm runs in $O(n \log n)$ time.

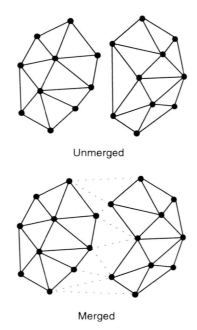

Unmerged

Merged

**Figure 5.44**   Merging two Delaunay triangulations.

For a final example of a Delaunay triangulation algorithm, we consider an application of an incremental method. This is described for a Voronoi diagram, but can be adapted directly to triangulations (see Tsai (1993) for a description). The general idea is to construct the Voronoi diagram incrementally, modifying it as new vertices from the vertex set are considered. Figure 5.45 shows the method in action. At each stage, the newly introduced vertex is shown unfilled in the figure. The polygon in which the new vertex lies is calculated, and a trail (shown as a dotted line) is made around the new vertex, creating its proximal polygon as it goes. The other edges are then modified appropriately and the process iterates until no more new vertices are available.

## Conclusions

Computational geometry has traditionally been concerned with measures of efficiency derived from theory – especially complexity theory. As usual, theory and practice are not always in accord. Methods provided by computational geometers for computing triangulations that are theoretically efficient may not be practical to implement. However, computational geometry provides a basis from which practical approaches may be constructed and evaluated. An additional difficulty is caused by the usual problem of discretization and approximation necessitated by the fixed

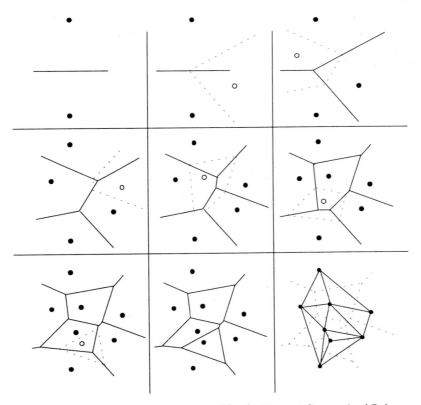

**Figure 5.45**  Some steps in the incremental method for the Voronoi diagram (and Delaunay triangulation).

finite precision of numeric calculations in a digital computer. This problem is addressed in, for example, Milenkovic (1993), where a robust construction of Voronoi diagrams is discussed.

## 5.6   RASTER-VECTOR AND VECTOR-RASTER CONVERSION

### 5.6.1   Raster-to-vector (vectorization)

As system hardware and software become cheaper and provide more functionality, the cost of spatial data capture increasingly dominates. Manual digitization, relying on large amounts of human operator time, remains prohibitively expensive for large data-driven applications. Scanning technology has improved dramatically in the last decade and the possibility arises to improve data capture by making the raster-to-vector conversion process, called *vectorization*, more efficient. Computerized approaches to raster-to-vector conversion can be divided into fully automatic and semi-automatic processes. Although there has been a substantial research effort on fully automatic raster-to-vector conversion processes, they have yet to prove really practical. The difficulty is that the data themselves contain within the pure spatial configurations a degree of information that is difficult to fully extract automatically. Therefore, semi-automatic processes may be more appropriate and feasible at this stage.

Musavi *et al.* (1988) have set out a series of steps that need to be taken in the course of converting a raster image to vector form. These steps are more or less closely followed by most commercial systems. The form in which they are reproduced here is very similar to that of Flanagan *et al.* (1994).

*Threshold:*   Form binary (black-and-white) image from a multi-scale raster (see section 5.5.4).

*Smooth:*   Remove random noise, usually in the form of speckles (see section 5.5.4).

*Thin:*   Thin lines so that they are one pixel in width. Thinning algorithms are often derived from the Medial Axis Transform of Montanari (1969).

*Chain code:*   Transform the thinned raster image into a collection of chains of pixels, each representing an arc.

*Vector-reduction:*   Each chain of pixels is transformed into a sequence of vectors. The number of vectors in the sequence depends upon wiggliness of the line and the accuracy required.

A commonly used erosion algorithm for thinning is that due to Zhang and Suen (1984). Assume that the binary raster has 0 to represent white and 1 to represent black. Assume that we want to thin the black (or grey, in the case of our diagrams). For each pixel position $p$, let $N(p)$ be an integer that is the sum of the values of the eight neighbours of $p$. Also, $p_N, p_S, p_E, p_W$ denote the values at pixels positioned above, below, right and left, respectively, of $p$. Finally, let $T(p)$ be a count of the number of transitions from 0 to 1, visiting the immediate neighbours of $p$ in rotation. In Figure 5.46, $N(p) = 5$, $p_N = p_W = 1$, $p_S = p_E = 0$, and $T(p) = 2$. The algorithm now follows, and an example showing the successive stages of an erosion is given in Figure 5.47.

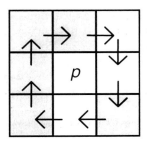

**Figure 5.46**   Pixel neighbourhood.

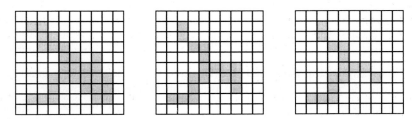

**Figure 5.47**   Zhang-Suen erosion example.


**Zhang-Suen erosion algorithm for raster thinning**
**Input:** $m \times n$ binary raster (0 represents white, 1 represents black)
**Output:** $m \times n$ thinned binary raster (black is thinned)
**Procedure:**
Step 1.   For each point $p$ in the raster do step 1.1
   1.1   If $2 \le N(p) \le 6$ and $T(p) = 1$ and $p_N.p_S.p_E = 0$ and $p_W.p_E.p_S = 0$
      then mark $p$
Step 2.   For each point $p$ in the raster do step 2.1
   2.1   If there are any marked points
      then set all marked points to value 0
      else halt the process
Step 3.   For each point $p$ in the raster do step 3.1
   3.1   If $2 \le N(p) \le 6$ and $T(p) = 1$ and $p_N.p_S.p_W = 0$ and $p_W.p_E.p_N = 0$
      then mark $p$
Step 4.   For each point $p$ in the raster do step 4.1
   4.1   If there are any marked points
      then set all marked points to value 0
      else halt the process
Step 5.   Go to Step 1

Chain-coding was originally presented in Freeman (1961). Chains are formed
from sequences of pixels in a thinned raster image. For each pixel, we need to
determine whether it is on a line segment, the end point of a line segment, or forms a
point of intersection between segments. A chain-coding algorithm will find a pixel
that constitutes the end-point of a strand, and then follow the pixels along the line,
stopping at an end-point or an intersection point. Thus a sequence of pixels, the
*chain*, is generated. The algorithm continues until all pixels in the raster image have

been processed. The algorithm is refined a little to handle loops (with no end-points). A fuller description may be found in Gonzales and Wintz (1987).

The final stage in the vectorization process is vector-reduction, the conversion of each chain to a set of vectors. A long, wiggly line will of course need more vectors than a straighter line, depending upon the level of accuracy required. The Douglas-Peucker algorithm (section 5.2.3) is often used for the vector-reduction process.

Automatic vectorization is by no means completely reliable. However, in practice even small gains in automatic vectorization processes will be significant, due to the very large amount of spatial data to be captured and the consequent reduced operator time that results from a non-automatic process. A compromise is to allow an automatic vectorization to be checked at known points of difficulty by a human expert.

### 5.6.2   Vector-to-raster (rasterization)

Vector-raster conversion (so-called *rasterization*) is not at present a great issue for GIS and will be discussed only very briefly in this book. However, in graphics processing generally it is a key issue. Many geometric primitives are much more compactly stored and efficiently communicated if given in vector rather than raster form. Thus, a straight line segment may be given in terms of its end-points, rather than a large number of pixels that constitute its image. Graphics processors must have the capability to convert these vector or parametrized forms (e.g. centre and radius of a circle) into pixels for display purposes. This topic is known in the graphics literature as *scan-conversion* and is fully discussed in graphics texts such as that by Foley *et al.* (1990).

### 5.7   NETWORK REPRESENTATION AND ALGORITHMS

Network models are fundamental to many GIS applications, for example transport and cable applications. This section discusses the representation of networks in a computer and goes on to consider algorithms for some of the most common and generic network operations.

### 5.7.1   Network representation

There are essentially two straightforward ways of representing a general network for computational purposes. Both assume an ordered labelling of the $n$ nodes (for example by the integers 1, 2, ... $n$). The first approach is to represent the network as a set of node pairs, each pair representing an edge of the network. If multiple edges between two nodes are allowed, then a multiset will be required. If the graph is directed, then the order of the nodes in the node pair representing an edge will be significant.

The other approach follows similar lines, except that the network is represented as a square $n$ by $n$ *adjacency matrix*, where each row and column is labelled by a node. A '1' is placed in a matrix cell if, and only if, the nodes corresponding to the appropriate row and column are connected by an edge, otherwise a '0' is placed in the cell. If multiple edges between two nodes are allowed, then the number in the cell can represent the number of edges. If the graph is directed, then the matrix may not be symmetric.

Often it is important to label the edges. For example, in a traffic flow application where the nodes represent cities and the edges represent flows between cities, the edges will be labelled with the appropriate flows. Edge labels may be accommodated by adding an extra field to the node pair in the edge set representation, and for the adjacency matrix representation, placing the label in the matrix. Figure 5.48 shows an example of a network of hypothetical average travel times (in minutes) for tram routes between places of interest in the Potteries.

The network may be represented as the set of labelled edges:

{(AB, 20), (AG, 15), (BC, 8) (BD, 9), (CD, 6), (CE, 15), (CH, 10), (DE, 7), (EF, 22), (EG, 18)}

In this case, each edge has associated with it an appropriate travel time. The adjacency matrix representing the Potteries network is given in Table 5.7. A non-zero entry in the matrix indicates the existence of an edge between the appropriate row and column nodes and gives the average tram travel time. A zero entry indicates no edge between the appropriate row and column nodes.

**Table 5.7**  Adjacency matrix for the network in Figure 5.48

|   | A | B | C | D | E | F | G | H |
|---|---|---|---|---|---|---|---|---|
| A | 0 | 20 | 0 | 0 | 0 | 0 | 15 | 0 |
| B | 20 | 0 | 8 | 9 | 0 | 0 | 0 | 0 |
| C | 0 | 8 | 0 | 6 | 15 | 0 | 0 | 10 |
| D | 0 | 9 | 6 | 0 | 7 | 0 | 0 | 0 |
| E | 0 | 0 | 15 | 7 | 0 | 22 | 18 | 0 |
| F | 0 | 0 | 0 | 0 | 22 | 0 | 0 | 0 |
| G | 15 | 0 | 0 | 0 | 18 | 0 | 0 | 0 |
| H | 0 | 0 | 10 | 0 | 0 | 0 | 0 | 0 |

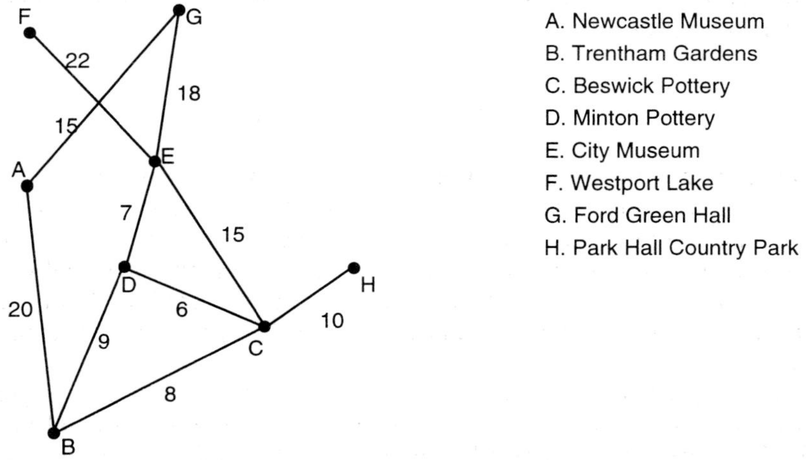

A. Newcastle Museum
B. Trentham Gardens
C. Beswick Pottery
D. Minton Pottery
E. City Museum
F. Westport Lake
G. Ford Green Hall
H. Park Hall Country Park

**Figure 5.48**  A network of hypothetical tram routes between some Potteries locations, labelled by average travel times.

### 5.7.2   Depth-first and breadth-first traversals

One of the most fundamental operations on any connected network is a systematic
traversal of the nodes within it. Such traversals have many applications, for
example, searching a problem space (future possible positions in a chess game) or
traversing a cable network to locate the source of a fault. There are two main alternative approaches: *depth-first* or *breadth-first* traversal. We may either search
through the network by probing ever deeper and only coming up when all alternatives below are exhausted (depth-first), or search through the network by examining
all nodes connected to a given node before going deeper (breadth-first). Consider the
depth-first and breadth-first traversals of the network in Figure 5.48, starting at
node *B*.

*Depth-first traversal of the Potteries network*

Begin at a given node *B* and find a node *A* adjacent to *B*.
Find an unvisited node *G* adjacent to *A*.
Find an unvisited node *E* adjacent to *G*.
Find an unvisited node *F* adjacent to *E*.
Since there are no unvisited nodes adjacent to *F*, backtrack to *E* and continue
the traversal.
Find an unvisited node *C* adjacent to *E*.
Find an unvisited node *H* adjacent to *C*.
Since there are no unvisited nodes adjacent to *H*, backtrack to *C* and continue
the traversal.
Find an unvisited node *D* adjacent to *C*.
Backtrack to *C*, then *E*, then *G*, then *A*, then *B*.
There is now no further option to backtrack, therefore halt the traversal.

This depth-first traversal can be represented as a tree as in Figure 5.49. There is
some indeterminacy in the traversal due to the choices at some nodes, therefore, in
general, several depth-first traversals are possible for a given network and starting

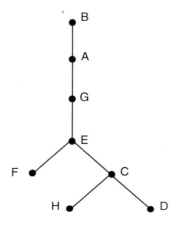

**Figure 5.49**   Depth-first traversal of the Potteries network.

node. Any depth-first traversal of a connected network must visit all the nodes of the network.

*Breadth-first traversal of the Potteries network*

> Begin at a given node *B*.
> Find a node *A* adjacent to *B*.
> Find other nodes *C*, *D* adjacent to *B*.
> As there are no further unvisited nodes adjacent to *B*, look for adjacent nodes to *A*.
> Find a node *G* adjacent to *A*.
> As there are no further unvisited nodes adjacent to *A*, look for adjacent nodes to *C*.
> Find unvisited nodes *E*, *H* adjacent to *C*.
> Continue this process to find node *F* adjacent to *E*.
> Eventually, halt the traversal.

This breadth-first traversal can be represented as a tree as in Figure 5.50. Again, the choice possible at some points of an adjacent unvisited node results in several possible breadth-first traversals of a network.

### 5.7.3 Shortest path

Another operation on a network that is useful in many applications is the computation of the shortest path between two given points. When the edge weighting is non-Euclidean, then it is often the case that the shortest/fastest path between two points is not necessarily the most direct. Dijkstra's algorithm (Dijkstra, 1959) for computing the shortest path is based upon the calculation of values in three one-dimensional arrays, each of size equal to the number of nodes in the network. Each row of each array corresponds to one of the nodes of the network. As the algorithm proceeds, paths are calculated from the start node to other nodes in the network, paths are compared and best (minimum weight) paths are chosen, given the state of

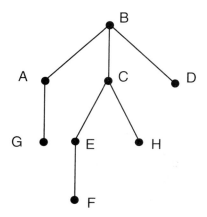

**Figure 5.50**  Breadth-first traversal of the Potteries network.

knowledge of the network at that stage in the progress of the algorithm. At each
stage in the computation:

- The array *distance* keeps track of the current minimal distances from the start
  node to the array nodes. (As the algorithm proceeds, these distances become
  refined to closer approximations to the shortest distance.)
- The array *path* keeps a record of the preceding nodes on the current best paths
  from the start node to the array nodes.
- The array *included* marks off the nodes as they are used in the calculation of the
  minimal distance path from start node to end node.

The algorithm initializes arrays as follows:

- Cells in the *distance* array are initialized to 0 for the start node cell; infinity for the
  cells whose nodes are not directly connected to the start node; and to their direct
  distances from the start node for all other cells.
- Cells in the *path* array are initialized to the start node for cells whose nodes are
  adjacent to start, otherwise undefined.
- Cells in the *included* array are initialized to 'no' for all cells except the start node
  cell.

The algorithm then iterates the following procedure until the end node is marked
'yes' in the *included* array.

1. Find the node (say *n*) whose distance from the start node is smallest amongst all
   those nodes not yet marked 'yes' in *included*.
2. Mark *n* as 'yes' in *included*.
3. For each node *m* not yet marked 'yes' in *included*:
   if *n* and *m* are adjacent
   and if the current distance (in the *distance* array) from start to *n* plus
   the edge weight of *nm* is less than the current distance from start to *m*
   then update the distance array to the new and smaller distance
   from start to *m*; also update the *path* matrix to show the
   preceding node in the path to *m* to be *n*.

Eventually, the end node will be included in the computation and the algorithm
halts, returning the shortest path from the start node to the end node in the end
node cell of *distance*.

The example below uses Dijkstra's algorithm to calculate the shortest path from
*B* to *G* in the Potteries network. The initial arrays are shown in Table 5.8(a). The
nodes directly adjacent to *B* have their edge-weights given in the distance table and
the preceding node on paths from *B* to them is *B* itself. Node *B* is the only node
marked as included. Of the non-included nodes, *C* has the minimum distance
entered at this stage. *C* is marked as included and distances from *B* to other non-
included nodes on paths passing through *C* as the last-but-one node are calculated
and compared with the already calculated distances. For example, the distance of
path *BCD* is $8 + 6 = 14$, longer than the current distance of 9 from *B* to *D*, and no
gain is made there. On the other hand, the path *BCH* has length 18, clearly better
than $\infty$. After processing of node *C*, the arrays are those given in Table 5.8(b).
Processing of node *D* is next and makes only one change. The route from *B* to *E* can

**Table 5.8** Changing states of the arrays for the example of Dijkstra'a algorithm

| | distance | path | included | | distance | path | included |
|---|---|---|---|---|---|---|---|
| A | 20 | B | no | A | 20 | B | no |
| B | 0 | ⊥ | yes | B | 0 | ⊥ | yes |
| C | 8 | B | no | C | 8 | B | yes |
| D | 9 | B | no | D | 9 | B | no |
| E | ∞ | ⊥ | no | E | 23 | C | no |
| F | ∞ | ⊥ | no | F | ∞ | ⊥ | no |
| G | ∞ | ⊥ | no | G | ∞ | ⊥ | no |
| H | ∞ | ⊥ | no | H | 18 | C | no |

a. Initial state of the arrrays.      b. Arrays after processing node C.

| | distance | path | included | | distance | path | included |
|---|---|---|---|---|---|---|---|
| A | 20 | B | no | A | 20 | B | yes |
| B | 0 | ⊥ | yes | B | 0 | ⊥ | yes |
| C | 8 | B | yes | C | 8 | B | yes |
| D | 9 | B | yes | D | 9 | B | yes |
| E | 16 | D | no | E | 16 | D | yes |
| F | ∞ | ⊥ | no | F | 38 | E | no |
| G | ∞ | ⊥ | no | G | 34 | E | yes |
| H | 18 | C | no | H | 18 | C | yes |

c. Arrays after processing node D.      d. Final state of the arrays.

be shortened by going from B to D (9) and then proceeding directly to E (7), giving a total of 16. The revised arrays are given in Table 5.8(c). The nodes not included continue to be processed in this fashion until node G (the end node) is included. At this point the algorithm halts. The final arrays are shown in Table 5.8(d) and the shortest path has been found to be 34.

It is not hard to see that the time complexity of Dijkstra's algorithm is $O(n^2)$, where $n$ is the number of nodes. If we wanted to find the shortest paths between all pairs of nodes in the network, then using an algorithm given by Floyd (1962) yields a time complexity of $O(n^3)$.

Shortest path algorithms may be applied more generally. For example, given a digital elevation model, it may be useful to calculate optimal routes between points in the terrain. Several measures of optimality, such as shortest Euclidean length, avoiding steep paths, and minimum sum of height differences, are discussed by van Kreveld (1994), who considers shortest path in the context of polyhedral terrains.

### 5.7.4 Other network operations

In this last section, we briefly mention some other network operations and their algorithms. The previous operation determined the shortest path between two points of a labelled graph: a more fundamental operation is deciding whether any path at all exists between the points. If the network is not connected, then there will be nodes between which no path exists. A decision about connectivity between two points can be made by using the shortest path algorithm, because if the shortest path length returned is ∞ then there is no path between them. However, this is not the most efficient method.

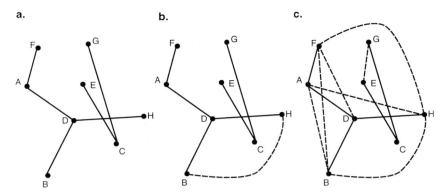

**Figure 5.51**   Transitive closure of the Potteries tram network.

The *transitive closure* operation augments the edge set of a network by placing an edge between two nodes if they are connected by some path. Standard algorithms exist for computing the transitive closure of a network. The algorithm of Warshall (1962) acts iteratively, adding edges between all nodes for which there is a two edge path but not direct edge (completing the triangle). Figure 5.51(b) shows the completion of the triangle *BDH* for the network of Figure 5.51(a), and Figure 5.51(c) shows the full transitive closure.

Another common operation is the *travelling salesperson* operation, that requires the round-trip traversal of an edge-weighted graph, visiting all the nodes, in such a way that the sum of the weights along the edges is minimized. A simple-minded solution, calculating all possible routes and choosing the best, has exponential time complexity and is impractical for any but the smallest examples. Heuristic methods, such as at each stage visiting the nearest unvisited node, sometimes allow good approximations in reasonable time. The travelling salesperson operation is an example of an *NP-complete* operation, meaning that the construction of an algorithm for it that is of polynomial time complexity will lead to polynomial time algorithms for a very large class of other problems. No such algorithm has yet been found. NP-completeness has been the subject of extensive research by the theoretical computer-science community (Brookshear, 1989).

## BIBLIOGRAPHIC NOTES

Computational geometry is sometimes said to have begun with the publication of Shamos (1975). The classic book in the field is by Preparata and Shamos (1985). More recent work (late 1980s) is nicely surveyed in an article by Toussaint (1992), which forms the introduction to a special issue of the *Proceedings of the IEEE* that is devoted to the subject. An even more recent and more elementary text by O'Rourke (1994) introduces computational geometry in a practical way, using many examples and implementations in the C programming language. A more advanced text introducing randomized algorithms is Mulmuley (1994). Mehlhorn and Näher (1995) describe a library of combinatorial and geometric algorithms, including many discussed in this chapter, obtainable by anonymous ftp from ftp.mpi-sb.mpg.de, directory pub/LEDA, and freely usable for teaching and research. Work on the specification of the geometric domain may be found in Güting and Schneider

(1993a) and an algorithm for computing line intersections in a discrete domain in Greene and Yao (1986). For coordinate systems and map projections, see Maling (1991). Waugh and Hopkins (1992) give a parallel algorithm for polygon overlay.

Many of the topological structures described in this chapter have had practical application. In 1979, the United States Bureau of Census adopted the Dual Independent Map Encoding (DIME) scheme for representing the spatial data associated with the US Census. DIME files held the topological relationships between streets in urban areas. The earliest DIME encoding did not consider the embedding (coordinates) but gave attention to the topology (Schweizer, 1973). DIME led to the US Bureau of Census Topological Integrated Geographic Encoding Reference (TIGER) format (Broome, 1986). The name of each street is held, along with coordinate embeddings of its start and end-points and identifiers of the blocks to the left and right of it. This representation is closely related to the winged-edge structure and uses 0-cells, 1-cells and 2-cells to capture explicitly the topological infrastructure of the spatial data. It is well described in a practical context in Marx (1986).

The literature on triangulated irregular networks (TINs) is extensive. A seminal work is by Peucker et al. (1978). Samples of the literature describing GIS applications of TINs are Burrough (1986), Gold and Cormack (1987) and Lee (1991). Delaunay triangulations and their dual Voronoi diagrams also have a very large literature. Voronoi diagrams are clearly used in the work of the mathematicians Dirichlet (1850) and Voronoi (1908), although similar ideas appear much earlier, for example in Descartes' work (1644) (translated by Mahoney (1979)). Delaunay triangulations appear in Voronoi (1908) as the dual of Voronoi diagrams. A most extensive and scholarly recent work in this area is by Okabe et al. (1992).

Algorithms for the Delaunay triangulation and its dual Voronoi diagram are abundant in the literature. Divide-and-conquer algorithms are described in Shamos and Hoey (1975) and Lee and Schachter (1980). Incremental methods are common, and may be found in Lawson (1977), Lee and Schachter (1980) and Bowyer (1981). Incremental methods enhanced by an efficient data structure are often useful in practice (Ohya et al., 1984a, 1984b). Dwyer (1987) published a variant of the divide-and-conquer approach that led to an algorithm with optimal worst-case time complexity and very good average time complexity. The plane sweep method uses a 'sweep line' to range across the vertices and construct a triangulation as it goes: some ingenuity is required for this construction, described in its Voronoi form in Fortune (1987). Puppo et al. (1994) present a parallel algorithm for Delaunay triangulating a terrain. For a survey of approaches to Delaunay triangulation, see Tsai (1993).

For non-Delaunay triangulations, particularly triangulations of polygons, the following provide a representative sample of methods: Garey et al. (1978), Kirkpatrick et al. (1992), Tarjan and Wyck (1988), Fournier and Montuno (1984) and Chazelle (1991). The text of O'Rourke (1994) covers the topic well. For a survey of non-Delaunay triangulations, see also Jayawardena and Worboys (1995).

A survey of hierarchical terrain models is given by De Floriani et al. (1994), and with particular emphasis on hierarchical TINs in De Floriani and Puppo (1992). A stratified Delaunay triangulation, where the circumcircle property is satisfied at each level, is defined in the *Delaunay pyramid* of De Floriani (1989). Stratified models are discussed in general by Bertolotto et al. (1994).

The application of the arc-node structure in a cartographic setting was realized by Peucker and Chrisman (1975). The mathematics of OODCELS goes back to the work on planar maps of Edmunds. Reference to Edmunds (1960) is usually cited, but in fact is a very short (less than half a page) summary that gives no great amount of information. The paper by Cori and Vauquelin (1981), giving much more detail but couched in the language of permutation functions, is readable only by a person with a good background in modern algebra. A recent treatment, with special reference to GIS, is the paper by Stell and Worboys (1994): also relevant is the work of David *et al.* (1993) and Worboys and Bofakos (1993).

For a treatment of boundary representations in general, with the flavour of solid modelling, Mäntylä (1988) and Lienhardt (1991) are excellent. The winged-edge representation and its extensions are presented by Baumgart (1975) and Weiler (1985). Edge-algebras and the quad-edge data structure is given in Guibas and Stolfi (1985). A more abstract treatment of such representations that deals with any number of dimensions is Pigot (1994), where a large number of references to related work may be found. Boolean operations on boundary representations are discussed by Requicha and Voelcker (1985).

Raster algorithms and algorithms related to rasterization and vectorization have a large literature that does not generally form part of the GIS corpus. The two main areas of relevance are graphics and image processing. A highly recommended and comprehensive introduction to graphics is provided by Foley *et al.* (1990). Image processing is introduced by Gonzales and Wintz (1987). For more specialist research material on the automatic vectorization process, the reader could consult Musavi *et al.* (1988) and Flanagan *et al.* (1994).

# CHAPTER SIX

# Structures and access methods

All words,
And no performance!

Philip Massinger

In this chapter there is a further movement from high-level, conceptually indepen-dent spatial models towards the machine level. In the last chapter, the emphasis was upon computationally appropriate ways of representing and manipulating different kinds of spatial data. We now move on to consider storage questions, and in partic-ular storage structures that allow acceptable *performance*. In this chapter, the per-formance of an information system is taken to be measured in terms of response times to queries and database size. Other, and arguably more critical, aspects of performance, such as the suitability and usability of the interface, are considered in the next chapter. The discussion here begins with an introduction to the main issues of performance for general purpose databases and then takes up these issues in the special context of spatial data. We will see that while the fundamental principles of good database indexing practice still apply, spatial data access has its own special problems demanding solution.

## 6.1 GENERAL DATABASE STRUCTURES AND ACCESS METHODS

The typical organization of a database is a collection of files, each containing a collection of records, stored on a set of disks. We noted the main physical character-istics of a computer disk in Chapter 1. The atomic unit of data held on a disk is referred to as the *disk block*. The time taken to transfer a disk block to or from a disk has three components:

- *Seek time*: Time taken for the mechanical movement of the disk heads across the disk to the correct track.
- *Latency*: Time taken for the disk to rotate to the correct position.
- *CPU transfer time*: Required to transfer the block into the CPU.

Seek time is the dominant performance factor in data retrieval from secondary storage because it is based upon mechanical movement. The considerable conse-quences of this fact for data structures may be summarized:

Structuring of data in secondary storage that seeks to lessen mechanical movement of the disk heads will result in improved database performance. So as to
minimize disk head movement, data are ideally placed in such a way that blocks
often accessed together are close by on the disk. Lessening disk head movement
will also be a consequence of the construction of appropriate indexes, so that
unnecessary searches, maybe of entire files, are avoided.

Databases, by their very nature, are usually designed to be flexible and to
respond well to a wide variety of queries. In general, queries are unknown in
advance, even though the requirements and analysis phases of database construction should provide a good statistical basis for the prediction of query patterns.
Another difficulty is caused if the database is highly dynamic, with its content
changing considerably in the course of its lifetime. Therefore, the physical placement
of data on the disk can only provide a partial answer to good performance. Indexes
and access methods are traditionally the main tool for the database developer.

### 6.1.1  File organization and access methods

Before getting to the main discussion, we review the fundamental concepts of file,
record and field, and techniques for physical file placement on disks.

*Record*:  A sequence of data items related to a single logical entity (e.g. student
record).

*Field*:  A place for a data item in a record (e.g. first-name field in student record).
A field whose length varies according to the record that contains it is a *variable-
length* field (e.g. a field for comments); if it contains multiple data items, it is a
*repeating* field (e.g. student grades); otherwise it is a *fixed-length* field. A field that
serves to identify each record unambiguously (e.g. student identifier) is called a
*key* field.

*File*:  A sequence of records, usually all of the same type (e.g. student file, comprising records for students in the institution). If all records are of the same size
(measured in bytes), then the file is composed of *fixed-length* records, otherwise
the file is composed of *variable-length* records.

Files are physically placed upon a disk by assigning disk blocks to hold records. If
the disk block is smaller than the record size, then records will be spread across
blocks, otherwise (and maybe more usually) each block contains several records.
*Contiguous allocation* occurs when successive groups of records are located on adjacent disk blocks. However, such an allocation will be disrupted in a dynamic situation with records coming and going. *Linked allocation* provides a pointer link
between blocks. Other allocations are also in common usage (Elmasri and Navathe,
1994).

The term *file organization* is used to include the physical organization of the
records on the disk (we assume throughout that the disk is the form of secondary
storage used) and the manner in which the blocks of records are linked. File organization also includes the way that the records of the file are inserted into storage.
Having allocated our files an area of secondary storage, the next step is to describe

the *access methods*. Access methods determine the way that the files may be manipulated by user programs. They are described in terms of a set of primitive operations, the most important of which are listed below:

*Open*: Prepares a file for access operations.

*Find*: Locates the disk block containing the next record in the file that satisfies the search condition. Transfers the block to main memory. Marks the found record as the *current record*.

*Read*: Copies the current record in main memory into the program that requires it.

*Delete*: Updates the file by deleting the current record in main memory and writing back to disk.

*Modify*: Updates the file by altering a value of a field of the current record and writing back to disk.

*Insert*: Updates the file by reading the appropriate block into main memory, inserting a record and writing back to disk.

*Close*: Releases a file from access operations.

A complex access method to a file is composed of a sequence of primitive access operations. For example, a change of address for a student with a known identifier is effected by opening the student file, finding the record for the student with the given identifier, modifying the file and finally the file is closed.

A file is either mainly *static* (with almost no updates once the records of the file are in place) or *dynamic* (with records coming, being altered and going during the file's lifetime). The static-dynamic nature of a file is determined by the access operations upon it and will influence the method of its organization. We will consider three basic types of file organizations: unordered, ordered and hashed.

### 6.1.2   Unordered files and linear searches

In an *unordered* file organization, as new records are inserted into the file, they are placed in the next physical position on the disk, either in the last used disk block or in a new disk block. Insertion of a new record is therefore very efficient. However, the file has no structure beyond entry order, so retrievals will require a search through each record of the file in sequence. If all that is required is that the file is accessed sequentially, for example to print out labels of students' names and addresses, then an unordered file organization is acceptable. However, problems arise with a *direct* and *random* access, where specific information is required from targeted records. For example, retrieving the grades of a student with a given last-name will require an examination of the values of the last-name fields of each record in turn until a match is found. Such a search is called a *linear* search, or a *brute-force* approach, and is to be avoided if possible. For *n* records, we can expect to retrieve *n/2* records, on average, therefore a linear search has linear time complexity. We will see shortly how indexes can help to solve this problem.

Another difficulty arises for a highly dynamic file where, as records are deleted, holes are created. It is possible to modify the organization to mark holes and insert new records into these positions, but this will slightly slow down the insertion time.

### 6.1.3  Ordered (sequential) files and binary searches

In an ordered file organization, each record is inserted into the file in the order of values in one or more of its fields. For example, we may decide to order the student file by student last-name. In this case the last-name field is called the *ordering field*. The values in the ordering field must of course be capable of being totally ordered (e.g. integers with the usual ordering, or character strings with the lexicographic ordering). The great advantage of an ordered file organization is that binary searches on ordered fields become possible and the binary search algorithm is now given. Note that the operation **div** forms the integer part of the quotient of two integers (thus 15 **div** 2 is 7).

**Binary search algorithm**
**Input:**   Given an ordered file with an ordering field, placed on $n$ disk blocks (labelled 1 to $n$), and a search value $V$
**Output:**   Records from the file with value $V$ in their ordering field
**Procedure:**
Step 1.   $low \leftarrow 1$
Step 2.   $high \leftarrow n$
Step 3.   while $high \geq low$ do steps 3.1 to 3.7
    3.1       $mid \leftarrow (low + high)$ **div** 2
    3.2       read block $mid$ into memory
    3.3       if $V <$ value of ordering field in the first record of block $mid$
    3.4          then $high \leftarrow mid\text{-}1$
    3.5          else if $V <$ value of ordering field in the last record of block $mid$
    3.6             then $low \leftarrow mid + 1$
    3.7             else search block $mid$ for records with value $V$ in their ordering field (proceeding to next block(s) if appropriate)

It can be seen that the binary search algorithm takes advantage of the ordering in the file to successively chop it in half until the targeted records are found and retrieved or no matching records are found. With a file utilizing $n$ disk blocks, the number of chops can be at most $\log_2(n)$. Thus, the time complexity of the binary search method is logarithmic for retrieval of a single record – a huge improvement on linear search. For our previous example, retrieving the grades of students with a given last-name, suppose that the file is placed upon 1000 blocks. Linear search would require on average 500 block accesses, while if the records were in order of last-name, binary search would require approximately $\log_2(1000) \approx 10$ accesses. Note that if multiple records match the search condition, then the time complexity may no longer be logarithmic (in the worst case, where all records match the search condition, the time is clearly linear). However, in all these cases where the number of records retrieved is small compared with the size of the data file, the complexity is approximately logarithmic.

Many retrievals are based on range search conditions, rather than equality conditions. For example, the search might involve finding all students whose last name begins with 'C' or whose grades are in the range A–C. As records close by in the range will be close by in the ordering, a binary search will also effect a significant speed up, and it is straightforward to modify the binary search algorithm to handle range searching.

Although ordering a file makes a large difference to performance with searches on the ordering fields, retrievals on other fields are once again reduced to linear search.

For a dynamic file, insertion of a record is very expensive as all succeeding records must be shunted along. The same problem arises with deletion, although we could just mark records that are deleted and clean the file up at longer intervals. Insertion can also be improved by placing records in another file, with pointers, and reorganizing periodically.

### 6.1.4   Hashed files

A hashed file is organized using a *hash function*. For one or more fields (*hash fields*), the hash function transforms the values of the hash fields of each record into an address of a disk block into which the record is placed. To search for records on hash fields, the hash function is applied to the search value, and the address of the disk block is calculated. Only a single disk block access is required!

Taking a highly simplified example, suppose that we have 1000 blocks for our student file. We could hash on the numeric student identifier field (suppose a six digit integer) by taking the final three digits to be the label for the block. Thus, the record for student 123456 will be placed in block labelled 456.

Hashing is in principle a simple technique for file organization, but complexities appear when we consider what happens when disk blocks overflow, or how to ensure as much as possible an even distribution of records through the available disk blocks. Also, it seems at first sight that we need to know in advance the number of disk blocks to be allocated in order to choose a suitable hash function. All these considerations complicate the topic and are not considered here. Further reading on hashing and file organization in general may be found in the bibliography at the end of the chapter.

### 6.1.5   Indexes

We have seen that purely physical organization of files on disks cannot solve all the performance problems that might arise. For example, hashed (ordered) files will only perform well for searches on their hash (ordering) fields. Indexes provide a way of speeding up performance for other fields, at the expense of the extra space required for the index itself. The concept of an index to a file is similar to the index of a book. Embedded in a book is a sequence of more important pieces of text to be directly referenced. Without an index, the book would need to be searched from cover to cover. An index orders the names of the indexed pieces so that it may be quickly searched (see binary search). To find a piece in the main text, the reader searches for the item in the index, and is then referred to the positions (addresses) in the main text where the piece occurs. We will consider single-level and multi-level static indexes, as well as the dynamic B-tree index. In the following description, the file that is to be supplemented by an index will be called the *data file*. The index acts on a single field of the data file, called the *indexing field*. (It is possible for an index to act on more than one field.)

A *single-level index* acts just as a book index described above. It is an ordered file, each of whose records contain the two fields.

- *Index field*:   Contains the value of the indexing field (in sorted order).
- *Pointer field*:   Contains the addresses of disk blocks that have the index value.

In the case that the indexing field is a key field, then the pointer field of each index record will contain just a single pointer. Otherwise there will be multiple pointers and, therefore, either a variable length field or some other structure. It also makes a difference to the construction of the index if the indexing field is an ordering field for an ordered file. An index whose indexing field is both primary and orders the data file is called a *primary index*. Figure 6.1 shows a schematic for an index to student last names in a student data file physically ordered by student-ID.

To retrieve a record from an indexed file, based upon a search condition on the indexing field, the index is searched. As the index file is ordered, the search is binary. Having located the record in the index, the pointer is followed to the record in the data file. Thus, an index allows a speed-up from linear to logarithmic time.

A factor so far neglected in the discussion is the blocking of the index. For a large data file, the index file will itself be substantial and will need to be placed on the disk in many blocks. The search time on the index will be proportional to the logarithm of the number of these blocks. We can cut down the search time even further by allowing the index to be itself indexed, and so on recursively. This is the principle of the *multi-level index*.

As the structures become more complex, there is more trouble with highly dynamic files. As records come and go from the data file, so the index must also be modified, and if the index is multi-level, then modification may have ramifications throughout its levels. A structure that handles these modifications in a particularly

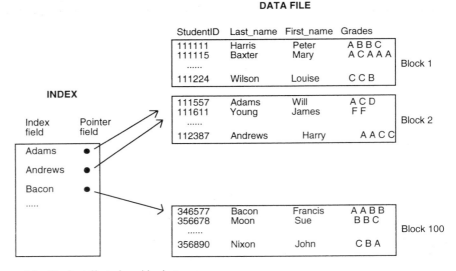

**Figure 6.1**   Student file indexed by last name.

elegant fashion is the *B-tree* (and its variant the *$B^+$-tree*) (Comer, 1979). B-trees remain *balanced* in a dynamic setting, in the sense that the various branches of a balanced tree remain of equal length as the tree grows and shrinks.

Suppose that we have a set of data records, indexed by an integer field (any type that is capable of being linearly ordered will do as an index domain). A B-tree contains two sorts of nodes: internal (non-leaf) nodes and leaf nodes. Internal and leaf nodes have essentially the same structure, except of course that leaf nodes do not contain pointers to descendants. Each index field will contain a pointer to the record that is indexed: for simplicity these pointers are not shown in the description below. Each node of a B-tree contains a list of index fields (with pointers to data) interleaved, if internal, with a list of pointers to immediate descendants. Within each node, the index fields are in strictly increasing order. Figure 6.2 shows an example of an internal node.

In our example, the B-tree is said to be of *fan-out ratio* 4, meaning that each non-leaf node can contain at most three index fields and have at most four immediate descendants. The basic property of a B-tree is that the value of the index field for all descendants is within the range set by the index fields of the ancestor node. In our example, the left-most pointer will point to nodes containing index values less than 3; the second pointer will point to nodes containing index values between 3 and 6; and so on. Figure 6.3 shows a B-tree of order 4.

The B-tree operations of search, insert and delete are now described. Search, based upon a value of the indexing field, is the most straightforward, because no restructuring of the tree is required. The other operations restructure the tree to keep it balanced.

*Search*: The search begins at the root node and depending on the value of the indexing field, a route will be traced to one of the descendants, and so on down the tree until an exact match is found or the bottom of the tree is reached. On exact match, the value of the pointer field retrieves the required records from the data file. If no match is found, then an appropriate message is returned.

*Insertion*: When a record is inserted into the data file, the B-tree is searched to determine where its index record is placed. The insertion of a new record may or

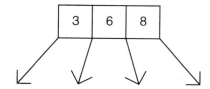

**Figure 6.2**  Internal node of a B-tree.

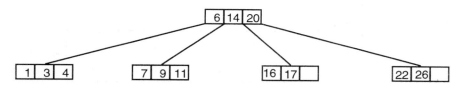

**Figure 6.3**  Two-level B-tree with fan-out ratio 4.

may not require a restructuring of the B-tree. For example, the tree in Figure 6.3
is nearly full. If a record with index value 19 is added to the file, the B-tree search
begins at the root node. As 19 is between 14 and 20, the search takes the third
branch from the left and arrives at a leaf node. This node has room for one more
index, and the index is added to the node. On the other hand, if the record with
index value 10 is added to the database, the B-tree search arrives at a leaf node
already full with values 7, 9 and 11. A restructuring is required, which produces
the new B-tree shown in Figure 6.4. The addition of 10 to node 7–9–11 causes an
overflow and splitting of this node. The middle value 10 is promoted to the next
level. Insertion of 10 into node 6–14–20 also causes an overflow and splitting.
The resulting tree is shown at the bottom of Figure 6.4: note that the tree is still
balanced.

*Deletion*: A similar algorithm ensures a balanced tree on restructuring a tree
following a deletion. If the deletion of the appropriate index value from node $N$
leaves $N$ empty (*underflow*), then a process of merging nodes will take place

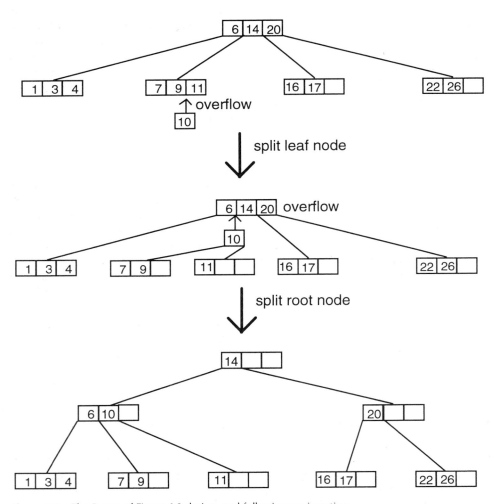

**Figure 6.4**   The B-tree of Figure 6.3 during, and following, an insertion.

propagated from $N$. This process is inverse to the insertion restructuring and will not be described in detail. Of course, if $N$ is still non-empty, then no merging is required.

The B-tree structure has the following properties, which are maintained as the structure is dynamically modified with insertion and deletion of records from the data file. Let $n$ be the number of records in the data file to be indexed.

- A B-tree is completely balanced; at any stage in its evolution, the path from root to leaf is constant regardless of the leaf. Search time is therefore bounded by the length of this path and has complexity $O(log\ n)$.

- Insertion and deletion of a record from the file (and possible restructuring of the index) have time complexity $O(log\ n)$.

- Each node is guaranteed to be at least roughly half-full at all stages of the tree's evolution.

The B-tree provides a balanced multi-level index in a dynamic setting and constitutes one of the first substantive achievements for research into general purpose data structures. B-trees allow indexes to be placed in a dynamic environment, which achieves significant speed-ups. They, or their later modifications, are at the core of most modern database management systems. In fact, the $B^+$-tree is the refinement most used in actual implementations. In a $B^+$-tree, pointers to records are stored only at the leaf nodes. Thus, the leaf nodes must contain every index value and will be more numerous than for the B-tree: on the other hand, the non-leaf nodes will have a simpler structure and will occupy less store.

## 6.2   FROM ONE TO TWO DIMENSIONS

The discussion so far has applied to any general purpose database. A noteworthy characteristic of files in such a database is that they are multi-dimensional, each entity having several independent attributes. The student file used as an example has dimensions corresponding to at least student-ID, first name, last name and grades, and each of these dimensions is usually independent of the others. (There may be dependencies between the attributes in some cases. Dependencies may be of the form studied in database theory and ideally eliminated by normalization or of a more statistical nature, e.g. correlations between job descriptions and salaries in an employee database.)

With spatial data, the dimensions are orthogonal, but there is a dependency between them expressed in Euclidean space by the Euclidean metric. To illustrate the point, consider the data on the positions of key points in the Potteries, as shown in tabular form in Table 6.1 and on a grid in Figure 6.5.

Consider now the effect of retrievals on this data. We distinguish two types of search conditions on the planar points.

- *Point conditions*:   Satisfied by all records whose spatial references are located at a given point.

- *Range conditions*:   Satisfied by all records whose spatial references are located within a given range. The range may be of any shape, although typically a rectangular area will be specified by the coordinates of two opposite vertices or a disc identified by centre and radius.

**Table 6.1**   Places of interest in the Potteries

| Id | Site | East | North |
|----|------|------|-------|
| 1 | Newcastle Museum | 14 | 58 |
| 2 | Waterworld | 30 | 67 |
| 3 | Gladstone Pottery Museum | 73 | 23 |
| 4 | Trentham Gardens | 20 | 00 |
| 5 | New Victoria Theatre | 17 | 55 |
| 6 | Beswick Pottery | 68 | 25 |
| 7 | Coalport Pottery | 54 | 36 |
| 8 | Spode Pottery | 37 | 43 |
| 9 | Minton Pottery | 36 | 39 |
| 10 | Royal Doulton Pottery | 31 | 88 |
| 11 | City Museum | 42 | 63 |
| 12 | Westport Lake | 15 | 93 |
| 13 | Ford Green Hall | 48 | 99 |
| 14 | Park Hall Country Park | 88 | 39 |

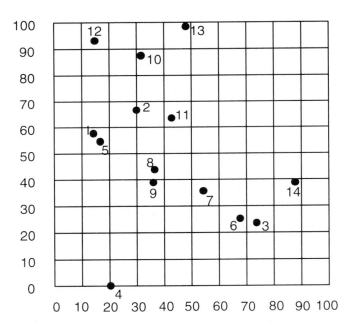

**Figure 6.5**   Location of some places of interest in the Potteries.

Consider the following example queries:

- *Query One (non-spatial query)*:   Retrieve the point location of Trentham Gardens.

- *Query Two (point query)*:   Retrieve any site at location (37, 43).

- *Query Three (range query)*:   Retrieve any site in the rectangle with south-west and north-east vertices (20, 20), (40, 50), respectively.

The first query, although involving spatial location in the result is concerned only with searching the file using a non-spatial field, and so the general purpose access methods discussed above would be appropriate. The second query is a retrieval using a spatial search condition. Without any index or file ordering, a linear time search would be necessary. A typical approach would be:

Step 1.   open Potteries file
Step 2.   while there are records to examine do steps 3.1 to 3.4
　　3.1   get the next record
　　3.2   if the value of the first coordinate field is 37
　　3.3   then   if the value of the second coordinate field in 43
　　3.4          then retrieve site name from record

For the third range query, a similar linear-time search may be adopted.

Step 1.   open Potteries file
Step 2.   while there are records to examine do steps 3.1 to 3.4
　　3.1   get the next record
　　3.2   if the value of the first coordinate field is in the range [20, 40]
　　3.3   then   if the value of the second coordinate field is in the range [20, 50]
　　3.4          then retrieve site name from record

Let us build two indexes for the two spatial coordinate fields, using the general purpose database approach. The indexes are shown (with the site names indicating the pointer fields) in Table 6.2, and the two sequences of points are shown as paths in Figure 6.6. For the second query, we conduct a binary search of the east index to locate records whose first coordinates have value 37. We then go to the full records to check if the second coordinates have value 43, and retrieve those records for which this is the case. For the third query, we could do a range [20, 40] search on the east index, resulting in a list of pointers to the data file. For each pointer in the list, the data file record may be accessed and the north value checked to be in the range [20, 50], in which case the site name is retrieved.

What is striking in this discussion is the observation that only one of the indexes is used in these retrievals. What we would like to do is to construct an index that takes advantage of ordering in two dimensions. Particularly for range searches, we

**Table 6.2**   East and North indexes to the places of interest in the Potteries

| East | Site | North | Site |
|------|------|-------|------|
| 14 | Newcastle Museum | 00 | Trentham Gardens |
| 15 | Westport Lake | 23 | Gladstone Pottery Museum |
| 17 | New Victoria Theatre | 25 | Beswick Pottery |
| 20 | Trentham Gardens | 36 | Coalport Pottery |
| 30 | Waterworld | 39 | Minton Pottery |
| 31 | Royal Doulton Pottery | 39 | Park Hall Country Park |
| 36 | Minton Pottery | 43 | Spode Pottery |
| 37 | Spode Pottery | 55 | New Victoria Theatre |
| 42 | City Museum | 58 | Newcastle Museum |
| 48 | Ford Green Hall | 63 | City Museum |
| 54 | Coalport Pottery | 67 | Waterworld |
| 68 | Beswick Pottery | 88 | Royal Doulton Pottery |
| 73 | Gladstone Pottery Museum | 93 | Westport Lake |
| 88 | Park Hall Country Park | 99 | Ford Green Hall |

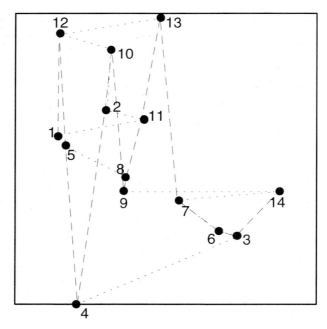

**Figure 6.6**  The two indexes shown as paths through the places of interest in the Potteries.

would like to order the records in such a way that records that are near each other in space are near each other in the index.

### 6.2.1   Two-dimensional orderings

The principal problem confronting designers of data structures for two-dimensional and multi-dimensional Euclidean space is that computer storage is essentially one-dimensional. (Even though a disk of course physically exists in three-dimensional space, the computational model of it is one-dimensional, as can be seen by noting its one-dimensional addressing scheme.) Thus, we have to unravel the higher-dimensional space in such a way as to minimize the distortions thereby caused, in a similar way to unravelling a piece of two-dimensional knitting into one-dimensional yarn. This section examines some of the ingenious techniques that have been proposed for this unravelling process in the case of two dimensions.

More formally, let us take a finite portion of the Euclidean plane $P$: without loss of generality, let $P$ be the unit square $[0, 1] \times [0, 1]$. Let $S$ be the real interval $[0, 1]$ where the ordering is the usual ordering of real numbers. We are looking for a bijective function: $f: P \rightarrow S$ that has the property that, for all $x, y \in P$, $x$ is near $y$ in the region if and only if $f(x)$ is near $f(y)$ in the interval. In the discrete case, assume that $S$ is regularly tessellated into squares. Then the function $f$ can be represented as a path through the grid. Suitable functions, called *tile indexes*, will have the property that points close in $P$ will be close on the path, and vice versa. In practice, it is not possible to find tile indexes that have this nearness property for all pairs of points, but functions have been found that work well in most cases.

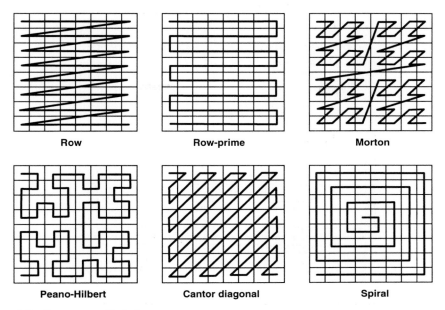

**Figure 6.7**   Six common tile indexes.

Figure 6.7 shows six common tile indexes. The *row ordering* shown in the top-left of the figure is the first and most obvious possibility. It corresponds to the scan of a raster image across the rows of pixels. Although simple, it has the unfortunate property that close points at row ends are often quite distant on the path. The *row-prime* order provides a simple modification that allows some improvement, but there is still a problem with half of the end-points. The *Morton* (top-right) and *Peano-Hilbert* (bottom-left) orderings, are rather more ingenious. The Morton ordering can be seen as a recursive 'Z', showing the letter 'Z' at various scales, or if rotated through a right-angle, recursive 'N'. It may be arrived at by interleaving the bits of the row and column numbers. The Peano-Hilbert ordering relates to the Morton ordering as the row-prime to the row ordering, and is a discretization of the space-filling Peano-Hilbert curve. Both these orderings are much better at preserving nearness. The last two curves are simpler to see as a pattern, namely the *Cantor-diagonal* order (row-prime rotated) and the *spiral* order. In general terms, the indexes that best preserve nearness are the Peano-Hilbert and Morton orderings. The bibliography contains some further references on the properties of these orderings.

### 6.3   RASTER STRUCTURES

This section reviews some of the basic methods for storing and compressing raster data. A *raster* is of course an array or grid of cells, often referred to as *pixels*. The three-dimensional equivalent is a three-dimensional array of cubic cells, called *voxels*. Each cell in a raster is addressed by its position in the array (row and column number). For example, in Figure 6.8, the raster may be represented as a $32 \times 32$ array, each cell occupied by 0 or 1. Rasters are able to represent a large

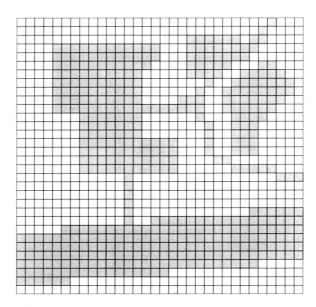

**Figure 6.8**  An example raster structure.

range of computable spatial objects: a point may be represented by a single cell, a strand by a sequence of neighbouring cells and a connected area by a collection of contiguous cells. Rasters are managed naturally in computers, for all commonly used programming languages well support array handling. However, a raster when stored in a raw state with no compression can be extremely inefficient in terms of computer storage. For example, a large uniform area with no special characteristics must be stored as a large collection of cells, each holding the same value. In this section, we examine ways of improving raster space efficiency.

### 6.3.1   Chain codes, run-length codes and block codes

*Chain codes* represent the raster boundary of a region by giving a starting point and the cardinal direction (north, south, east, west) to follow as we progress around the boundary. Consider as an example the boundary of a region that is part of the raster in Figure 6.8, shown in Figure 6.9. Starting at the south-west corner and proceeding anti-clockwise: go north 7, west 3, north 7, and so on. This can be chain-coded as:

[N7, W3, N7, E11, S1, W1, S1, W1, S9, E4, S3, W10].

*Run-length encoding* was discussed in the context of image compression in Chapter 1. *Block coding* is a generalization of run-length encoding to two dimensions. Instead of sequences (runs) of 0s or 1s, square blocks are counted. Each block is defined, using the medial axis transform (Rosenfeld and Pfaltz, 1966; Rosenfeld, 1980), in terms of a distinguished point (centre or south-west vertex, for example) and the length of a side. The previous example is represented using block coding as shown in Figure 6.9.

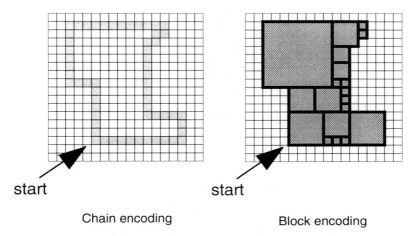

start                             start

Chain encoding                    Block encoding

**Figure 6.9**    Chain code and block code for the raster in Figure 6.8.

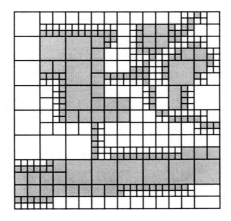

**Figure 6.10**    Quadtree subdivision for the raster in Figure 6.8.

### 6.3.2    Region quadtrees

A very widely used structure for holding planar areal raster data is the *region quadtree*. The term *quadtree* arises because the underlying data structure is a levelled tree where all non-leaf nodes have exactly four descendants. The principle of the region quadtree is a recursive subdivision of a non-homogeneous square array of pixels into four equal-sized quadrants. The decomposition is applied to each sub-array until there is homogeneity. This process has been applied to the image in Figure 6.8, and the result shown in Figure 6.10. To see the process of subdivision in detail, Figure 6.11 shows successive subdivisions of the north-west quadrant of the image.

Quadtrees are stored in a levelled-tree data structure, with the root at the top level (level 0). For each non-leaf node, its four constituent quadrants are represented by its four descendent nodes. A homogeneous quadrant, where no further subdivision is required is stored as a leaf node. Leaf nodes may have attributes associated

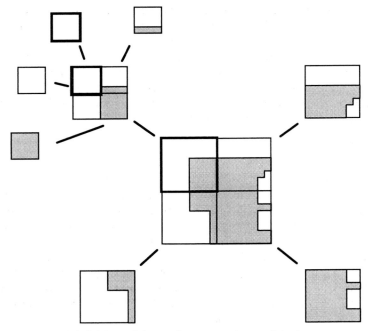

**Figure 6.11**   Successive subdivisions of the north-west quadrants of the image.

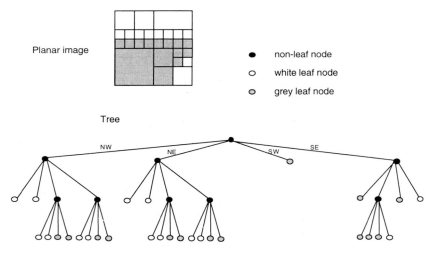

**Figure 6.12**   The quadtree for part of the raster in Figure 6.8.

with them, such as a colour code, or they may point to records of more detailed information in a data file. We need a convention for the order of the descendants, say NW, NE, SW, SE in each level of the tree. Figure 6.12, shows the tree corresponding to the subdivision shown in Figure 6.10, where for reasons of space we have omitted the top levels of the image and begun with one of the third-level quadrants as the root.

The great advantages of the region quadtree structure are that it takes full advantage of the two-dimensional nature of the data (clearly not the case for run-length encoding) and it has variable resolution, devoting more depth to the tree only in places where necessary detail is to be captured. For a given image size, the pattern that makes the most storage demands is the alternating chequer pattern, because there are no homogeneous regions to be amalgamated. There is a body of results on space complexity of the quadtree structure. An example is the following proposition of Rosenfeld *et al.*, quoted by Samet (1989b).

> The quadtree representing a polygon with perimeter $m$ embedded in an image of size $2^n \times 2^n$ contains $O(m + n)$ nodes in the worst case.

Despite its clear strengths, the region quadtree structure has some associated problems. The structure is highly dependent on the embedding of the spatial objects in the image space: a small translation or rotation of the object will usually result in a quite different quadtree structure. Related to this is the problem of major reorganization required by changes to the image: this will result in inefficiencies for dynamic images. A region quadtree will not provide as great benefits for a highly irregular image (often the kind found in geographic applications). If the image is largely composed of points and lines, once again, there may be difficulties with using the region quadtree approach. Indeed, an entirely different approach based upon the vector model may be more appropriate.

Algorithms for Boolean operations such as union, intersection and difference of images in a raster may be elegantly coded and perform well, taking advantage of the

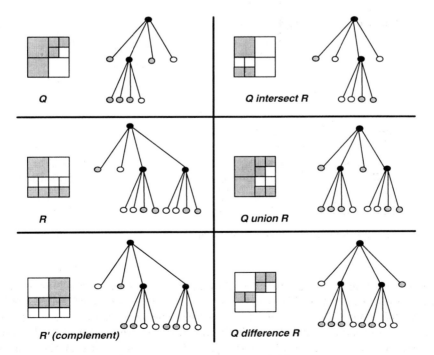

**Figure 6.13**   Quadtree complement, intersection, union and difference.

recursion and variable resolution inherent in the quadtree structure. Algorithms for other operations, such as labelling and counting the number of connected components in an areal spatial object, also perform well. Samet (1989b) provides much detail on these and other algorithms. To give a feel for the approach, we consider the designs of recursive algorithms to compute the quadtrees of the complement, intersection, union and difference of two planar spatial objects represented as quadtrees, assuming that the images are binary (see Figure 6.13). The algorithm for complement is a simple recursion.

### Quadtree complement algorithm
**Input:** A binary quadtree $Q$
**Output:** A binary quadtree that represents the inverse image (black becomes white and white becomes black)
**Procedure:**
Step 1.    $n \leftarrow$ root of $Q$
Step 2.    if $n$ is a leaf node
    2.1    then invert the value of $n$ (0 to 1 or 1 to 0)
    2.2    else apply quadtree complement algorithm to each of the descendants of $Q$
Step 3.    end

The intersection algorithm is given in a less formal way: formal descriptions may be found in Samet (1989b) and Sedgewick (1983). Suppose that we are given two binary quadtrees $Q$ and $R$, and require to construct a quadtree $S$ representing the image formed by the Boolean intersection of the images represented by $Q$ and $R$. We construct an output tree as we go along. Begin by making the roots of $Q$ and $R$ the active nodes $q$ and $r$, respectively. There are now several possible cases:

- *Case 1:    Either q or r (or both) is a leaf node representing a white area.* The resultant tree is constructed with a node in the position of $q$ or $r$ representing a white area.

- *Case 2:    Node q is a leaf node representing a black area.* The resultant tree is formed as a copy of tree $R$.

- *Case 3:    Node r is a leaf node representing a black area.* The resultant tree is formed as a copy of tree $Q$.

- *Case 4:    Both nodes q and r are non-leaf nodes.* Apply the algorithm recursively to the four corresponding descendant subtrees of $q$ and $r$. The root of the resultant tree is a non-leaf node and the subtrees formed by recursive application of the algorithm are joined as subtrees to this root.

Other Boolean operations can be composed from the complement and intersection operations, using equational relationships such as De Morgan's laws. For example, $Q \cup R = (Q' \cap R')$ and $Q \backslash R = Q \cap R'$.

A link with the previous section on two-dimensional orderings becomes clear from Figure 6.14. If the chequer pattern is structured as a quadtree and the leaf nodes of the quadtree are read from left to right, then the pattern that the trail follows in the plane corresponds to the Morton ordering of the plane.

The quadtree is simply generalized to higher dimensions. Thus, in three dimensions we have the octree, where each non-leaf node has eight ($2^3$) descendants.

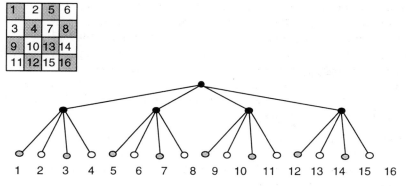

**Figure 6.14**  Region quadtree is a Morton order of its leaf nodes.

## 6.4  POINT OBJECT STRUCTURES

This section considers data structures that are designed specifically with point objects in mind. The three structures that are considered are grid files, point quadtrees and 2D-trees. As usual, assume that the point objects are located in the plane by means of two coordinates, and may have other attributes.

### 6.4.1  Grid file structures

As an introduction to this approach, and to the bucket methods that underlie it, we begin with the fixed-grid or fixed-cell structure, that allows binary search methods to be applied in more than one dimension. It will be seen that this has some disadvantages that are overcome with the variable grid structure of Nievergelt *et al.* (1984). We will use the example dataset from the Potteries given in Table 6.1 and Figure 6.5.

*Fixed grid structure*

The *fixed grid* structure is a partition of the planar region into equal sized cells: squares in our example. Each packet of data whose point locations share the same cell of the grid is stored in a contiguous area of secondary storage. The usual term to describe a cell of this structure is a *bucket*. Each bucket represents a physical location in storage where the complete records are held. Thus, the grid cell structure is a way of partitioning objects that have planar point references so that neighbouring objects are more likely to have their attribute data stored in the same or nearby areas of storage. The performance of range queries is thus improved. An example of a fixed grid structure for our Potteries point dataset is given in Figure 6.15. The region has been partitioned into 16 squares, each containing between zero and two points. Boundary ambiguities are resolved by making the convention that each square owns the points on its south and west boundaries, but not points on the other two boundaries. Thus point labelled '6' belongs to the square containing the point labelled '7'.

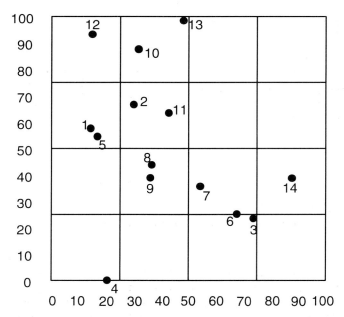

**Figure 6.15**   Fixed grid partition of Potteries point data from Table 6.1.

The decision regarding the size of the partition is dependent upon at least two considerations. First, the larger the total number of points the more cells will be required, because the buckets correspond to blocks of secondary storage that can hold only a limited number of records. Second, the size of a cell is dependent on the magnitude of the range in the average range query supported by the system. There is no purpose to be served by making the cell size much smaller than the average range.

Looking at Figure 6.15 immediately shows a major disadvantage with the fixed grid method. The population of cells with points varies, and in some cases there are no points in cells. This problem becomes more acute for less uniformly distributed points, and is analogous to problems with hashing for one-dimensional files. An ideal hash function distributes the records uniformly over the storage area. If the distribution is too far from uniform, there will be many empty cells, while other cells will be full to overflowing, and special overflow bucket structures will have to be introduced. Returning to two dimensions, we would like a partition that took account of and adapted to the density of points in particular local areas. This leads us to the next structure and the later point quadtree. However, if the points are uniformly distributed, then the fixed grid structure provides a simple and easily implementable solution. The fixed grid is applicable to an arbitrary number of dimensions. It may be appropriate to use a tile planar ordering (e.g. Morton order) for the cells of the grid.

### Grid file

The grid-file devised by Nievergelt *et al.* (1984) is an extension of the fixed grid structure that allows the vertical and horizontal subdivision lines to be at arbitrary positions, taking into account the distribution of the points. This structure is

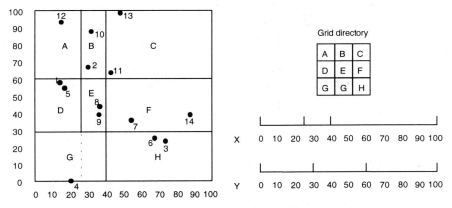

**Figure 6.16** Grid file structure for Potteries point data from Table 6.1.

designed for dynamic data, entering and leaving the system as time passes. A new level of indirection, called the *grid directory*, is introduced. Grid cells may have their data sharing the same bucket only if their union is a rectangle. Figure 6.16 shows an example, where a bucket size of two records is assumed (much smaller than a real application). The total area is divided as a three-by-three grid: the two grid cells in the south-west region are amalgamated into a single bucket, because there is spare bucket capacity at that location and the amalgamation results in a rectangle. The grid directory shows the relationships between cells and buckets, while the two linear scales, dynamically updated, show the positions of the partitions.

The grid cell is designed to expand and contract as new data are inserted and deleted. A rectangle may be divided if it becomes too full and cells may be amalgamated if the space becomes too empty. Note that in our example, we have reduced from 16 buckets in the fixed grid structure to eight buckets with this structure. The grid file is designed to be applicable to an arbitrary number of dimensions.

### 6.4.2 Point quadtree

The point quadtree (Finkel and Bentley, 1974) combines the grid approach with a multi-dimensional generalization of the binary search tree. This quadtree has many characteristics in common with the region quadtree, described earlier in the chapter. Assuming as usual that the data are planar, each non-leaf node is associated with a data record for a point location and has four descendants (NW, NE, SW, SE). Thus, the data record itself will have four fields pointing to the four descendants, two fields to hold the coordinates and further fields to hold the data associated with the point, such as the name of the city at that point location (see Figure 6.17).

The point quadtree has the property that the position of each partition into quadrants is centred on a data point, unlike the region quadtree where a quadrant is always partitioned into four equal-sized subquadrants. A data structure where the positions of subdivisions are independent of the data points is called a *trie* structure, in opposition to a *tree* structure where the positions of subdivisions are dependent. Thus, a more strict term for the region quadtree is the *region quadtrie*. The point quadtree is a true tree structure, because the centre of a subdivision into four is always a data point.

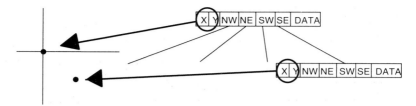

**Figure 6.17**    Quadtree records and the points to which they refer.

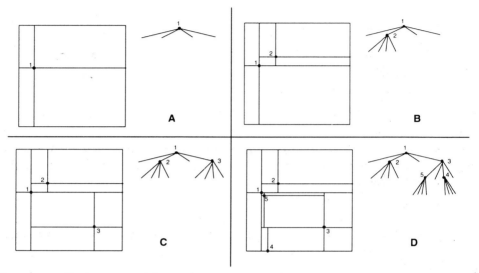

**Figure 6.18**    The first stages of the quadtree construction for the Potteries point data.

The point quadtree is described with reference to an example. Consider once again the point data from the Potteries given in Table 6.1 and Figure 6.5. Assume that the points are to be entered into a quadtree index in alphabetical order (the numerical order is also strict, from '1' to '14'). The first point (Newcastle Museum) is placed at the root of the tree, and the plane is divided into four quadrants as shown in Figure 6.18(a). The second point (Waterworld) is compared with the root. We assume the usual ordering of the cardinal directions, NW, NE, SW, SE. Point '2' is north-east of point '1', and so is placed as the NE (second) descendant of the root. The north-east quadrant of the plane is divided into four quadrants, as shown in Figure 6.18(b). Point '3' (Gladstone Pottery Museum) must now be placed in the tree. It is compared first with the root (SE), and as there are no further nodes along the SE branch, it is placed as the SE descendant of point '1'. The south-east quadrant of the plane is divided into four quadrants, as shown in Figure 6.18(c). Point '4' is processed next: it is compared first with the root (SE), and then with point '3', the SE descendant of the root (SW). As there are no further nodes for comparison, point '4' is placed as the SW descendant of point '3'. The appropriate quadrant of the plane is divided into four quadrants. Figure 6.18(d) shows the insertion of points '4' and '5' into the tree and the resulting planar subdivisions. Figure 6.19 shows the full quadtree and planar subdivisions for this dataset.

The algorithm for insertion into a quadtree is now presented in a general setting.

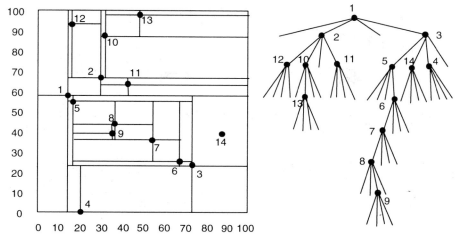

**Figure 6.19**  Point quadtree for the Potteries point data.

The quadtree is a tree in which each non-leaf node has four descendants. A preliminary procedure is presented that determines the appropriate branch for a new point coming into the quadtree.

**Quadtree branch algorithm**
**Input:**  Given a point $p$ and a non-null quadtree $Q$
**Output:**  The label (NW, NE, SW, SE) of the quadrant in which point $p$ belongs with respect to the root of $Q$
**Procedure:**
Step 1.  if $x$-coordinate of $p$ less than $x$-coordinate of root of $Q$
Step 2.  then if $y$-coordinate of $p$ less than $y$-coordinate of root of $Q$
Step 3.    then return quadrant SW
Step 4.    else return quadrant NW
Step 5.  else if $y$-coordinate of $p$ less than $y$-coordinate of root of $Q$
Step 6.    then return quadrant SE
Step 7.    else return quadrant NE

We are now ready to give the recursive algorithm that takes a new point and inserts it into the quadtree.

**Quadtree insert algorithm**
**Input:**  Given a point $p$ and a quadtree $Q$
**Output:**  The quadtree $Q$ updated by the insertion of the point $p$
**Procedure:**
Step 1.  if $Q$ is a null tree with no nodes
Step 2.  then make a quadtree with root $p$ and null trees as its four descendants
Step 3.  else use the quadtree branch algorithm to evaluate the appropriate subtree
Step 4.  recursively apply the algorithm to insert $p$ into that subtree

The shape of a quadtree is highly dependent upon the order in which the points are inserted into it. Figure 6.20 shows the quadtree associated with an insertion of

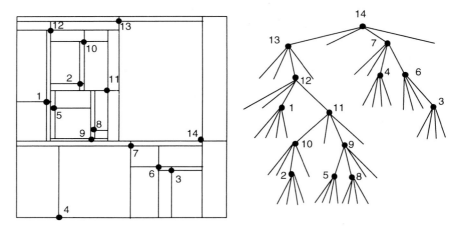

**Figure 6.20**   Point quadtree for a reversed insertion of the Potteries point data.

the Potteries data in inverted order: 14–1. This dependence has implications for dynamic data. If a point near the top of the tree is deleted, the tree may be substantially different (unless a dummy marker is retained). Therefore, point quadtrees are not immediately suited to dynamic information systems: point quadtrees are possible, however.

Some worst-case performance measures may be given for the point quadtree. Let $n$ be the number of points structured by the quadtree. Then, the quadtree build time is proportional to the total path length $O(n \log n)$, and the point query search time is $O(\log n)$. We have not presented the algorithms for point and range query for this structure: these operations will be given in detail for the next structure and may be modified for the point quadtree structure.

### 6.4.3   2D-tree

The point quadtree is excellent in that it takes full advantage of the embedding of the points in a Euclidean plane. However, there remain problems with dangling nodes and the exponential increase in the number of descendants of non-leaf nodes as the dimension of the embedding space increases: for $k$ dimensions, each node has $2^k$ descendants. The 2D-tree (in general, $k$D-tree solves these problems at the expense of a deeper tree structure. The $k$D-tree is a binary tree (each non-leaf node has two descendants) regardless of the dimension $k$ of the embedding space.

For planar embeddings, the 2D-tree does not compare points with respect to both dimensions at all depths, but compares $x$-coordinates at even depths and $y$-coordinates at odd depths (assume the root is at depth 0). Each record with a point location has two fields pointing to the descendant records, two fields storing the coordinates of the point and additional data fields. For the binary tree structure, we make the convention that the left descendant is less than the right descendant, when compared with respect to the appropriate coordinate (see Figure 6.21).

The 2D-tree and planar decomposition for our Potteries point data inserted in numerical order is given in Figure 6.22. Point '1' is inserted at the root; point '2' has a greater $x$-coordinate than point '1' and so is inserted as the right descendant of the

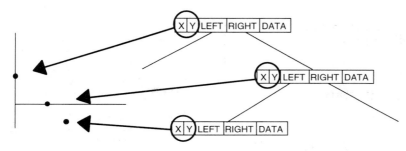

**Figure 6.21**    2D-tree records and the points to which they refer.

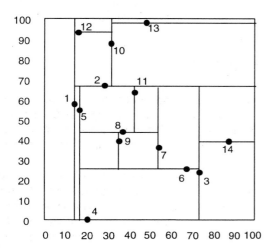
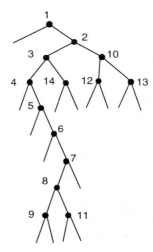

**Figure 6.22**    2D-tree for the Potteries point data.

root; point '3' has a greater $x$-coordinate than point '1' and a lesser $y$-coordinate than point '2', so is inserted as the left descendant of point '2'. The process is continued with the 14 points. As before, we have problems with restructuring for dynamic data. However, the tree is a simpler structure and there are less (about half as many) dangling nodes.

Algorithms for point insertion and range queries are now given. Point query is a special case of the range query algorithm. For the insertion algorithm, to handle the alternating $x$-coordinate and $y$-coordinate levels, we define two separate algorithms that call each other.

### 2D-tree even level insert algorithm
**Input:**   Given a point $p$ and a 2D-tree $T$
**Output:**   The quadtree $T$ updated by the insertion at an even level of the point $p$
**Procedure:**
Step 1:   if $T$ is a null tree with no nodes
Step 2.   then make a binary with root $p$ and two descendant null trees
Step 3.   else   if $x$-coordinate of $p$ less than $x$-coordinate of root of $T$
Step 4.            then recursively apply the odd level insert algorithm
                          to insert $p$ into the left descendant subtree of $T$

Step 5.          else recursively apply the odd level insert algorithm
                 to insert $p$ into the right descendant subtree of $T$

### 2D-tree odd level insert algorithm
**Input:**   Given a point $p$ and a 2D-tree $T$
**Output:**   The quadtree $T$ updated by the insertion at an odd level of the point $p$
**Procedure:**
Step 1.   if $T$ is a null tree with no nodes
Step 2.   then make a binary with root $p$ and two descendant null trees
Step 3.   else   if $y$-coordinate of $p$ less than $y$-coordinate of root of $T$
Step 4.          then recursively apply the even level insert algorithm
                 to insert $p$ into the left descendant subtree of $T$
Step 5.          else recursively apply the even level insert algorithm
                 to insert $p$ into the right descendant subtree of $T$

We are now ready to present the 2D-tree insertion algorithm, which calls the preceding algorithms.

### 2D-tree insert algorithm
**Input:**   Given a point $p$ and a 2D-tree $T$
**Output:**   The quadtree $T$ updated by the insertion of the point $p$
**Procedure:**
Step 1.   apply the 2D-tree even level insert algorithm to $p$ and $T$

For range searches, assume a rectangular search range specified by diagonal points $r_1$ (south-west extreme) and $r_2$ (north-east extreme). We will test a point at a given position in the 2D-tree against the range using the appropriate dimension ($x$ or $y$) for that level of the tree. The global level referred to in the algorithm is the level of the point with respect to the entire 2D-tree.

### 2D-tree range query algorithm
**Input:**   Given a rectangular range specified by diagonal points $r_1$, $r_2$ and a 2D-tree $T$
**Output:**   The points of $T$ in the given range
**Procedure:**
Step 1.     if $T$ is not a null tree then do steps 1.1 to 1.12
Step 1.1     if global level is even
Step 1.2     then if $x$-coordinate of root of $T$ > $x$-coordinate of $r_1$
Step 1.3        then recursively apply the algorithm to the left subtree of $T$
Step 1.4     else if $y$-coordinate of root of $T$ > $y$-coordinate of $r_1$
Step 1.5        then recursively apply the algorithm to the left subtree of $T$
Step 1.6     if the root of $T$ is within the rectangular range
Step 1.7     then output the record at the root of $T$
Step 1.8     if global level is even
Step 1.9     then if $x$-coordinate of root of $T$ < $x$-coordinate of $r_2$
Step 1.10       then recursively apply the algorithm to the right subtree of $T$
Step 1.11    else if $y$-coordinate of root of $T$ < $y$-coordinate of $r_2$
Step 1.12    then recursively apply the algorithm to the right subtree of $T$

Figure 6.23 shows an example of the algorithm in action with the usual Potteries point data in a 2D-tree. The rectangular range is shown with a thick boundary. The search begins with point 1 at the root of the tree (even level). This point is to the left

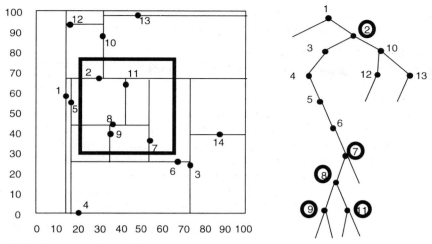

**Figure 6.23** Range search in a 2D-tree.

of the left boundary of the search rectangle, so the left subtree is not examined (Steps 1.2–1.3): in any case it is null. Point 1 is not in the rectangle, so it is not retrieved (Steps 1.6–1.7). Point 1 is to the left of the right boundary of the search rectangle, so the right subtree is examined (Steps 1.9–1.10). The algorithm is applied to the root, point 2, of this subtree (odd level). As point 2 is within the search rectangle, its record is retrieved and its left and right subtrees are searched according to the algorithm, which continues recursively to the bottom of the tree. The figure shows only the part of the tree directly examined by the algorithm. Points 2, 7, 8, 9 and 11, shown circled in the figure, are retrieved.

As with the point quadtree, the 2D-tree suffers from the problem of its structure being dependent upon the order in which points are inserted. In the worst case, the insertion of each point will result in a further level of the tree and a finished tree with the same number of levels as points. Attempts have been made to alleviate this difficulty. For example, Bentley and Friedman (1979) proposed the adaptive kD-tree, where the hyperplanes (lines in the 2D-case) dividing the space do not pass through the data points. However, problems with dynamic data, being inserted into and deleted from the database, remain. Robinson (1981) has combined the kD-tree structure with the B-tree idea, resulting in the kDB-tree, which is appropriate for dynamic situations where the points are stored in secondary storage.

## 6.5  LINEAR OBJECTS

The spatial data structures described so far have handled points and raster areas. The next classes of spatial objects to consider are those that are represented using vector structures and that are more than just points. Examples would be polylines, linear networks, or the boundary of an areal object. This section considers structures suitable for handling single items belonging to these classes of objects: in the next section we look at structures for collections of such objects.

For linear objects, a first thought might be to use the region quadtree structure, considering linear objects as very thin regions. Unfortunately, the unmodified region quadtree is not usually appropriate, using too much space and resulting in unnecessarily deep tree structures. However, specialized quadtrees have been constructed for handling linear objects, and it is to one class of examples of these that we now turn our attention. A longer list of such structures, with greater detail given, is contained in Chapter 4 of Samet (1989a).

### 6.5.1  PM quadtrees

The class of PM quadtrees is a variant of the region quadtree specially designed for the structuring of polygonal objects. There are several variants within the class, including $PM_1$, $PM_2$ and $PM_3$ quadtrees. We will describe the $PM_1$ quadtree, which can be formulated in the general setting of a planar network of vertices and edges, all of which are straight line segments. Assume such a network of vertices and edges enclosed in a square region of the plane. The region is divided into quadrants as for the region quadtree (i.e. the trie subdivision). The subdivision is such that vertices and edges are separated into distinct leaf nodes. To be precise, the region is divided into the minimum number of quadrants, subquadrants, etc., such that the quadtree satisfies the following constraints:

1. Each leaf node of the quadtree represents a region that contains at most one vertex of the network.

2. If the leaf node of the quadtree represents a region that contains one vertex of the network, then it can contain no part of an edge of the network unless the edge is incident with that vertex.

3. If the leaf node of the quadtree represents a region that contains no vertex of the network, then it can contain only one part of an edge of the network.

Figure 6.24 shows the partial construction of a $PM_1$ quadtree after the first two subdivisions have been made. All quadrants except those labelled 'A', 'B' and 'C' satisfy the three constraints above, therefore require no further subdivision. Quadrants 'A' and 'C' violate constraint 2 and quadrant 'B' violates constraint 1, and so must be further subdivided. The subdivisions result in regions 'D' and 'F', which violate constraint 3, and 'E' violating constraint 1, therefore these must be further subdivided.

The structuring of records in the $PM_1$ quadtree allows different field structures for vertices and edges. In fact, entire edges are not stored in a quadtree node, but only parts of the edges clipped by the appropriate subquadrants – called *q-edges* by

**Figure 6.24**  Staged construction of the $PM_1$ quadtree for the boundary of the raster region in Figure 6.8.

Samet (1989a). When an edge is inserted into the $PM_1$ quadtree, the tree is searched for the appropriate places to insert parts of the edge, and as we move down the levels of the tree, the edge is successively clipped by the boundaries of the sub-quadrants. Deletion is a similar but inverse process.

The $PM_2$ and $PM_3$ quadtrees are variants of the $PM_1$ quadtree, where the constraints are altered to provide advantages under some conditions. The $PM_2$ quadtree has a structure that is more stable under translations and rotations of the planar graph. Both $PM_2$ and $PM_3$ quadtrees have a smaller tree structure but correspondingly more complex associated records.

## 6.6 STRUCTURES FOR COLLECTIONS OF INTERVALS, RECTANGLES, POLYGONS AND COMPLEX SPATIAL OBJECTS

A very common requirement is to structure a large collection of objects so as to facilitate good performance on point and range queries: retrieving all those objects located at a given point or within a given range. We have already seen examples of such structures when the constituent objects in the collection are all points (see point quadtree and 2D-tree). Now the collection is allowed to have more general classes of constituent objects. We begin with the *segment tree*, which structures a collection of linear intervals so that it becomes efficient to retrieve all intervals enclosing a given point. The spatial objects here are one-dimensional and so at first sight are not in the mainstream of considerations for this section. However, they are useful, both for later work with rectangles and also they are relevant to temporal indexing.

After segment trees, structures designed specifically for collections of rectangles will be considered. At first sight, this class also seems rather restricted, but as we will see, structures based upon collections of rectangles can be used to facilitate retrievals from collections of any type of bounded planar object. Later sections deal with more complex spatial objects.

### 6.6.1 Segment trees

The segment tree (Bentley, 1977) is a binary tree structure for a finite set of bounded intervals on the real line that makes efficient queries of type:

Given a point $p$ on the line, retrieve all intervals in the set that enclose $p$.

It may be noticed that such a structure will allow efficient processing of point-in-rectangle queries, provided that the rectangles have sides parallel to the two axes of the coordinate frame. All that is required is to take two collections of intervals, one being the vertical boundaries of the rectangles and the other the horizontal boundaries. Then, to test whether a point is in a rectangle, find all the vertical intervals and all the horizontal intervals that respectively contain the $y$-coordinate and $x$-coordinate of the point. If a pair of retrieved horizontal and vertical intervals corresponds to the same rectangle, then that rectangle will enclose the point.

The segment tree structure is described, mainly by reference to an example. Suppose that we have a finite set of intervals, each given in terms of its boundary points. The first step is to sequence the boundary points in ascending order. For example, if the intervals are:

A[1, 10], B[3, 15], C[4, 7], D[4, 9], E[6, 15], F[8, 12], G[10, 14], H[16, 22], I[18, 20], J[10, 19]

where, to simplify the example, we assume that the domain from which the points are taken is the set of integers, then the sequence is 1, 3, 4, 6, 7, 8, 9, 10, 12, 14, 15, 16, 18, 19, 20, 22. These points become labels for the leaves of a binary tree, as shown in Figure 6.25. Each segment is now split into maximal chunks and distributed as labels of nodes in the tree. For example, segment A labels a single-level node [1, 10], whereas B labels leaf-nodes [3] and [15] and non-leaf nodes [4, 6], [7, 10] and [12, 14]. Each node therefore represents an interval (or point) and is labelled by zero or more intervals.

Given a point $p$, the process of searching the tree for intervals containing that point is straightforward. We begin at the root. If the root is labelled with any intervals, insert them into the result list. Examine the immediate descendants of the root. If $p$ does not belong to the interval represented by either of the nodes, then halt the procedure. Otherwise go to the node representing the interval containing $p$, add any intervals labelling the node to the result list, and continue this procedure until the leaf is reached or the procedure halts. The result list will hold only the intervals containing $p$. For example, to search for intervals containing the point 5, start at the root. Proceed to the node [1, 10] and add interval $A$ to the result list. Proceed to node [1, 6], then to node [4, 6] and add intervals $B$, $C$ and $D$ to the result list. Halt the procedure. The intervals found are $A$, $B$, $C$ and $D$.

The segment tree is a static structure, since the set of interval end-points is given in advance. However, new intervals whose end-points are in the given set may be dynamically inserted into the tree structure by labelling appropriate nodes. Also, intervals may be deleted from the structure.

Blankenagel and Güting (1994) have extended the scope of the segment tree in a non-trivial way to the *external segment tree*, which works for data held in secondary storage. It is shown that for a point query, the retrieved pages are each at least half-full of records containing the given point. Also, as the tree develops dynamically, page filling of at least 50 per cent is guaranteed. The tree is balanced, with efficient update algorithms provided in the cited paper.

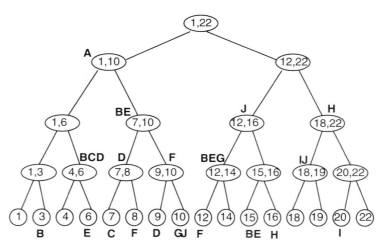

**Figure 6.25**   Example of a segment tree.

### 6.6.2   Rectangles and minimum bounding boxes

Efficient indexes for rectangles are important because rectangles can be used to approximate bounded planar spatial objects. Each geometric object is enclosed in its smallest bounding rectangle with sides parallel to the axes of the Cartesian frame (see Figure 6.26). Each such rectangle is known as a *minimum bounding box (mbb)*. An efficient indexing of the minimum bounding boxes facilitates queries on the objects themselves.

Normally, the detailed geometry of the bounded object will be stored separately from its mbb, for example in the form of an appropriate form of quadtree, and this geometry is referenced by the record associated with its bounding box. Thus, detailed investigation of the geometry of the object is possible: however, it is not always necessary. For example, consider the range query 'Find all the objects which lie in their entirety in a specified disc', applied to the objects shown in Figure 6.27. In order to answer this, the following steps are required:

1. Identify all the mbbs that lie wholly inside the given circular range. The objects inside these bounding boxes must lie in their entirety in the disc and therefore are to be retrieved. Figure 6.27 shows this case applying to object 'C'.

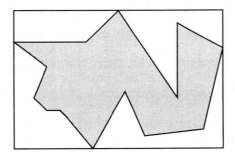

**Figure 6.26**   A minimum bounding box for a complex planar object.

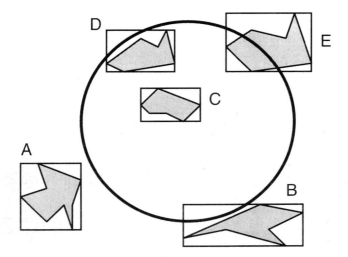

**Figure 6.27**   Using minimum bounding boxes for a range query.

2. Identify all the mbbs that intersect with, but do not lie wholly inside, the given circular range. The objects inside these mbbs may or may not lie in their entirety in the disc. For these cases, the detailed geometry of each object must be referenced and further computation performed to test whether it lies in its entirety within the given range. Figure 6.27 shows object 'D' lying entirely in the range and to be retrieved; object 'E' lying partially in the range and not to be retrieved; and object 'F' lying wholly outside the range and not to be retrieved.

It is therefore important to find satisfactory ways of indexing rectangles, and this is our next concern.

### 6.6.3   R-trees and R⁺-trees

The R-tree (Guttman, 1984) is a multi-dimensional extension of the B-tree (see section 6.1.5), and may be applied to index large collections of points in multi-dimensional space. R-trees guarantee avoidance of index page overflow and underflow. We look at the special case of point-pairs in 2-D space that define axes-parallel rectangles.

The R-tree is a rooted tree where each node represents a rectangle and usually corresponds to a page of secondary storage. The leaf nodes represent containers for the actual rectangles required to be indexed. Each higher level node represents the smallest axes-parallel rectangle that contains the rectangles represented by its descendants. It should be noted that the set of containing rectangles at any given level may well overlap each other. The R-tree is a dynamic structure that is constructed in a similar way to a B-tree. As rectangles are inserted into the structure, and leaf nodes become full, the effect of this is propagated to higher level nodes and

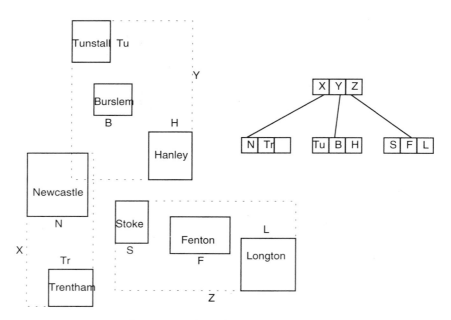

**Figure 6.28**    R-tree structure for minimum bounding rectangles of the Potteries towns.

the tree grows. Similarly, the tree contracts as rectangles are dropped from the collection. The overall tree structure remains balanced, and Guttman's Insert and Delete algorithms ensure that each node is always at least half-full.

The primary question is how to divide the space of rectangles into groups (represented by higher level nodes) as the rectangle collection expands: conversely, how to coalesce groups as the collection contracts. There are several reasonable alternative ways of dividing the space, for example, minimize the total area of the containing rectangles, or minimize the area of the overlap of the containing rectangles. A good subdivision will have both low total area and low overlap. The Pack algorithm of Roussopoulos and Leifker (1985) creates an initial tree that has a good subdivision structure.

An example of a one-level subdivision and corresponding R-tree for a set of mbbs for the towns in the Potteries is shown in Figure 6.28. We assume a branching factor of three, such that each node may contain at most three rectangles and thus have at most three descendants.

A problem with the R-tree structure, particularly for rectangles that are large compared with the total space, is caused by overlap of containing rectangles. Point and range searches may be inefficient, because the search for an object may have to take place in many different subtrees, even though the object is stored in only one of them. The R$^+$-tree structure (Stonebraker *et al.*, 1986) is a refinement of the R-tree that does not permit overlapping in the containing rectangles associated with non-leaf nodes, although it of course cannot prevent overlapping of leaf rectangles. It achieves this by partitioning rectangles and storing parts in different nodes of the tree. Studies have shown that it admits more efficient searches for large objects. However, unless complex insertion and deletion algorithms are used, the filling of each node to at least half is not guaranteed. The Potteries example is structured using an R$^+$-tree in Figure 6.29.

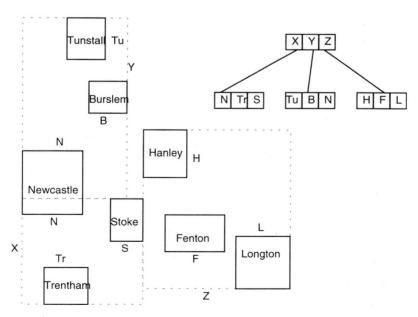

**Figure 6.29**   R$^+$-tree structure for minimum bounding rectangles of the Potteries towns.

### 6.6.4   Field tree

The field tree (Frank, 1983; Frank and Barrera, 1989), with features in common with the grid file and quadtree, was one of the earliest structures specifically designed to structure complex vector-based objects (polylines and polygons). The basis of the structure is a collection of regular rectilinear grids overlaying the plane. The grids have different resolutions and displacements. Each cell of a grid acts as a container for objects. An object may be inserted into a grid-cell if it does not overlap the grid-cell boundaries and there is no finer-meshed grid that will hold the object. There are no cell overflow and handling facilities apart from the use of overflow pages.

### 6.6.5   Cell tree

The cell tree (Günther, 1987) is another structure designed to improve the efficiency of retrieval of polygonal objects. It has characteristics in common with the BSP-tree (section 6.6.6) and the $R^+$-tree (above). As with the $R^+$-tree, the space is hierarchically partitioned into non-overlapping regions. However, while the R- and $R^+$-trees structure objects as contained in their minimum bounding rectangles, with the cell-tree, the leaves of the tree contain convex polygons (polyhedra in the higher dimensional cases). Arbitrary polygons may be handled by decomposing them as the union of convex polygons. As with the $R^+$-tree, the difficulty comes when deciding how to suitably split the plane into cells to accommodate in each cell the appropriate number of sub-cells (convex polygons at the leaf level). As with the $R^+$-tree, polygons may be partitioned so that different parts are stored in different nodes of the tree. For the cell tree, the problem is more complicated than with $R^+$-trees, due to the increased complexity of the objects – general convex polygons rather than rectangles. The splitting may be done in a way similar to that for BSP-trees, now to be described.

### 6.6.6   BSP-tree

The Binary Space Partitioning (BSP) tree (Fuchs et al., 1980) is a binary tree that hierarchically decomposes the plane into polygonal regions. Its original application was in computer graphics, where it may be used to structure a static 3D polyhedral configuration in such a way that 2D images from arbitrary view angles may be efficiently retrieved, taking care of hidden surface removal. The two-dimensional BSP tree may be used to hierarchically structure sets of directed line segments. Given a sequence of directed line segments $s_1, s_2, \ldots, s_n$ in the plane, the construction of the two-dimensional BSP-tree takes place as follows.

1. Place segment $s_1$ at the root of the tree.
2. Extend segment $s_1$ in both directions to form the infinite directed line $l_1$. If any other segment in the sequence is cut by $l_1$, then replace that segment with two segments, taking account of the cut.
3. Examine the next segment, say $s_2'$ in the (possibly revised) sequence. Determine whether $s_2'$ is to the left or right of $l_1$ and place $s_2'$ in the tree as the left or right descendant of $s_1$.

4. Extend segment $s_2'$ in both directions either indefinitely or up to its intersection with $l_1$ to form the directed line $l_2$. If any other segment in the sequence is cut by $l_2$, then replace that segment with two segments taking account of the cut.

5. Continue in this way until every line segment is added to the tree.

An example of this construction is given in Figure 6.30. The segment sequence is initially $a, b, c, d, e, f, g$, as shown in Figure 6.30(a). Segment $a$ becomes the root of the BSP-tree: its associated line (shown as a dotted line) cuts segment $e$ into parts $e_1$ and $e_2$. Segment $b$ is processed next: it falls to the left of $a$ and thus becomes the left descendant of $a$ in the tree. The process continues to give the planar subdivision shown as Figure 6.30(b) and tree shown as Figure 6.30(c).

A balanced tree structure implies minimal depth for the leaves and efficient tree searching. As the figure shows, the BSP-tree may be unbalanced. The structure of the tree is dependent upon the order of insertion of the segments. It is possible to find cases, such as edges of a convex polygon with all segments oriented in the same direction, where no ordering of the nodes leads to a balanced tree. The other factor influencing efficiency of searching and retrieval is the number of edge cuts that are made. Of course, the more edge cuts, the more nodes in the tree.

Van Oosterom (1993) has a discussion of static and dynamic BSP-tree balancing and the trade-off between a balanced tree and the number of cuts. In the same reference, Van Oosterom presents extensions to the BSP-tree that can handle more complex objects than directed line segments, for example polylines and polygons.

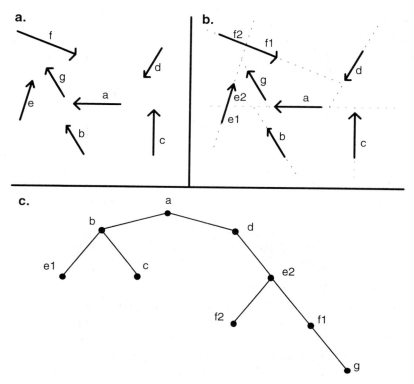

**Figure 6.30** Construction of BSP-tree from a collection of line segments.

### 6.6.7  Bang file

The BANG file (Freeston, 1987) may be extended to provide a further way of structuring planar object data. As with other structures in this section, it has much in common with a B-tree. Also, like an $R^+$-tree, there are no repetitions of the same object in different places in the structure. The structure maintains some of the spatial relationships between its constituent objects. Freeston proposes a dual structure containing trees that represent two distinct plane partitioning strategies:

- *Point partitioning*: handling queries of type 'retrieve the points completely enclosed in a given region'.
- *Object partitioning*: handling queries of type 'retrieve the objects completely enclosed in or overlapping a given region'.

Amongst the considerations in the work by Freeston (1989) is the problem of highly nested objects, as in Figure 6.31. In this case, it is impossible to place the dividing line so that it completely separates at least one object from the rest of the pack. In Freeston (1993), this problem is considered in greater generality. Allowed is the possibility that one partitioning region may fall completely inside another at the same level in the tree. The only condition imposed is that boundaries of regions may not intersect. It may be shown that for any such collection of areal objects, it is always possible, by placing a boundary around some of them that does not intersect with any existing boundary, to divide the collection into two groups having the

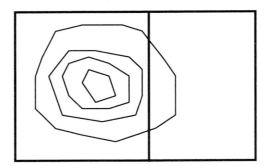

**Figure 6.31**  Dividing a collection of highly nested objects.

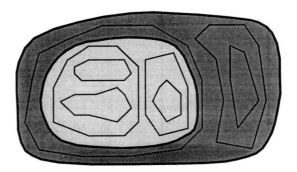

**Figure 6.32**  A separation of nested polygons into two regions.

property that the distribution of objects between the two groups is never more unbalanced than one-third to two-thirds. Figure 6.32 shows an example where a judicious separation divides the collection in the ratio 5 : 3. Freeston uses this partitioning scheme as the basis of a multi-dimensional indexing structure that may have wide applicability.

## 6.7 SPHERICAL DATA STRUCTURES

Geodata are referenced to the earth, and so planar models can be only an approximation to the true surface in which the spatial references are embedded. For many applications, this first approximation is sufficient, but more and more applications are global and in these cases a need arises for a structure that truly reflects the topology and geometry of the Earth. A second approximation is to assume that the Earth is a sphere. At present, such structures remain research topics and to the author's knowledge have not been used in proprietary systems.

It is possible to tessellate a sphere into spherical rectangles using lines of latitude and longitude. Such tessellations have been used as the basis for extensions of the quadtree structure to spherical surfaces (see, for example, Tobler and Chen, 1986). However, there is the difficulty that the shape and size of the quadtree cells vary as we move from the equator to the pole. Also, at a pole, the cells will be triangular. As discussed in section 5.4.5, recent approaches have used central projections of the Platonic solids onto the surface of the globe. The five Platonic solids are tetrahedron, octahedron, icosahedron, cube and dodecahedron. The cube has the advantage that its faces are squares and so quadtree structures can be used on them with little modification. However, it is not possible to orient the cube so that poles and equator respectively correspond to two of its vertices and a set of its edges. In fact, the octahedron is the only Platonic solid that has this property and it is this object that is the basis of a spherical representation now described.

The *Quaternary Triangular Mesh (QTM)*, originated by Dutton (1984), recursively approximates locations on the surface of a sphere by a nested collection of equilateral triangles. A particular case of a QTM arises from the central projection of the surface of the globe onto the surface of an octahedron (see Figure 6.33). The

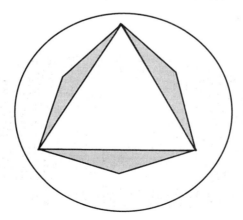

**Figure 6.33**   An octahedron inscribed in a sphere.

surface of an octahedron consists of eight triangular faces and each face may be recursively decomposed using the QTM, as follows.

Figure 6.34 shows the first three levels of decomposition of an equilateral triangle. At the *first level*, the nodes of the triangle are labelled 1, 2 and 3. The triangle is partitioned as shown into four equilateral triangles. Each sub-triangle is labelled according to the label of the node nearest to it, except that the inner sub-triangle is labelled 0.

At the *second level*, the unlabelled nodes of the three sub-triangles are labelled so that each edge contains the labels 1, 2 and 3 in some order. Each sub-triangle is now further divided into four. Each sub-sub-triangle is labelled with two digits: the left digit is the label of the sub-triangle to which it belongs, and the right digit is the label of the node nearest to it, except that the inner sub-sub-triangle has right digit 0. This process continues as shown for the *third level* and for subsequent levels, as far as desired with the recursion.

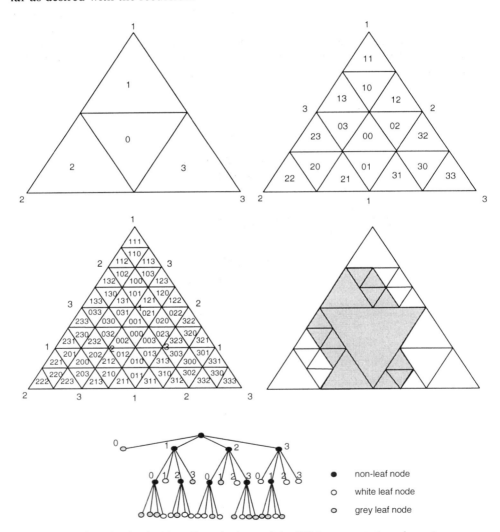

**Figure 6.34**   Three levels of QTM cell numbering and the QTM representation of a region.

The QTM may be used as a triangular version of the region quadtree. An example of this use is shown in Figure 6.34, where a region is decomposed into its maximal triangular pieces. In the traditional rectangular region quadtree, each node has four descendants, corresponding to the north-west, north-east, south-west and south-east subregions represented by its descendent nodes. Similarly with this 'QTM region quadtree', each node has four subnodes, corresponding to the sub-triangles labelled 0, 1, 2 and 3. As with the traditional region quadtree, the QTM region quadtree is a trie structure.

This construction then leads us to find triangular versions of other standard quadtree structures. For example, it is possible to imagine a triangular version of the point-quadtree as shown in Figure 6.35. A triangular frame abc surrounds the data points. These points are inserted into the triangular frame in the order 1, 2, 3, . . . as labelled. The first insertion induces a partition of the triangle into three sub-triangles and a consistent labelling of these sub-triangles as shown; further inser-tions induce further subdivisions. The corresponding tree is shown in Figure 6.35. This structure has the advantage over the standard point quadtree that each node has only three descendants rather than four. However, determining which triangle contains a point is much more computationally expensive and so it is unlikely that this approach would have any more than curiosity-value.

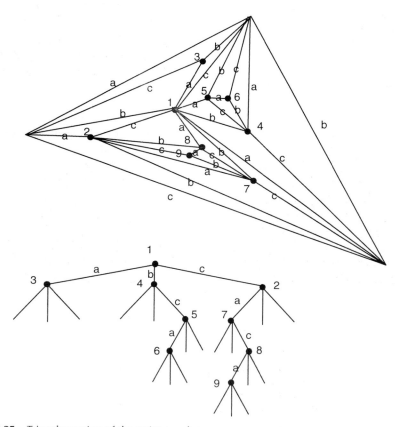

**Figure 6.35**  Triangle version of the point quadtree.

## BIBLIOGRAPHIC NOTES

Physical file organization is covered in detail in Smith and Barnes (1987) and more generally in Elmasri and Navathe (1994). For specific material on hashing, a basic reference is Morris (1968) with details of hashing for dynamic files in Fagin *et al.* (1979). Indexes, including B- and B$^+$-trees, are also introduced in Elmasri and Navathe (1994) and Smith and Barnes (1987). A comprehensive survey of B-trees, B$^+$-trees and other variations is Comer (1979).

Further information and plenty of references on planar orderings is given in Samet (1989a) and van Oosterom (1993). Abel and Mark (1990) consider the suitability of the orderings with regard to spatial query types, such as point and range queries. Another ordering not discussed in the main text is based on bit-interleaving Gray codes: for details, see van Oosterom (1993) and Faloutsos (1988). Some further analysis of the properties of tiling indexes is given by Goodchild (1989).

This chapter contains only a few of the many spatial data structures described in a large research literature. General surveys of spatial data structures may be found in Nievergelt (1989), Samet (1989a, 1989b) and van Oosterom (1993). The point quadtree is due to Finkel and Bentley (1974). The three-dimensional generalization of the quadtree to the *octree* is found in Hunter (1978) and Reddy and Rubin (1978). An excellent history of the development of the quadtree and related structures is to be found in Samet (1989a, 1989b). Many of the algorithms for operations on quadtrees and *k*D-trees may be found in Samet (1989b) and Sedgewick (1983). The class of PM quadtrees for structuring polygonal maps was originally given in Webber (1984), Samet and Webber (1985) and Nelson and Samet (1986): they are described in detail in Samet (1989a). Measures of the complexity of spatial objects, necessary for the development of metrics on spatial access methods, are discussed in Brinkhoff *et al.* (1995).

The fixed grid file is introduced in Knuth (1973) and Bentley and Friedman (1979). The segment tree was introduced by Bentley (1977). This reference is not easily accessible, but summaries of work on segment trees and their application to the rectangle intersection problem may be found in Samet (1989a). The segment tree is an internal memory structure: work on extending the concept to secondary storage is discussed by Blankenagel and Güting (1994). The segment tree was extended to the range tree by Bentley (1979).

The R-tree structure for rectangles was originally given by Guttman (1984). The non-overlapping R$^+$-tree structure is described in papers by Stonebraker *et al.* (1986), Sellis *et al.* (1987) and Faloutsos *et al.* (1987). A variant of the R-tree, the R*-tree, is used as the basis for a query processor in Kriegel *et al.* (1993). Full details of the field tree may be found in Frank (1983) and Frank and Barrera (1989). The cell tree is described by Günther (1987) and Günther and Bilmes (1988, 1991). The Binary Space Partitioning (BSP)-tree is discussed in Fuchs *et al.* (1980) and Fuchs *et al.* (1983). Extensions to the BSP-tree, including the object BSP-tree, are considered in Van Oosterom (1993). The BANG file is introduced by Freeston (1987) and used by him in spatial applications in Freeston (1989). Later work that takes a new approach to partitioning and indexing collections of complex (not necessarily spatial) objects is described in Freeston (1993). (The reader is warned that the author of this paper has proposed some corrections and additions, distributed at the conference but not generally available.)

Regarding spherical structures, Dutton (1994) proposed the Quaternary Triangular Mesh (QTM) to consistently reference any location on a planet. The work is followed up by several papers of Dutton (including Dutton, 1989, 1990). Tobler and Chen (1986) propose a tessellation that still allows the usual quadtree structures. Fekete and Davis (1984) introduced a structure similar to that of Dutton but based upon the central projection of the icosahedron. Other recent work on this topic includes Otoo and Zhu (1993) and Goodchild and Shiren (1992), both papers describing work closely related to that of Dutton.

# CHAPTER SEVEN

# Architectures and interfaces

The end is to build well. Well building hath three Conditions.
Commodity, Firmness and Delight.

Sir Henry Wotton, *Elements of Architecture*, 1624

The overall objective of a GIS is to allow the efficient analysis and sharing of geo-information. This objective is supported by the data models, structures, management systems and access methods discussed in earlier chapters. However, these are not the only considerations: the overall architecture of the system is also critical. Relational database technology has proved its capability to manage the information resources for many kinds of applications, but has not yet been successfully applied in a direct way to areas where data are of complex structure, such as CAD/CAM, office information and geo-information. Geodata are typically voluminous with a naturally imposed hierarchical structure (e.g. regions bounded by polylines, composed of line segments, composed of points). Geodata processing is characterized by transactions that are much longer than a typical standard relational database transaction (e.g. the design of a pipeline or implementation of an electrical network may require transaction times of several months compared with the adjustment of a bank balance or flight booking). These and other features impact the overall architecture of GIS.

Computer systems should exist to serve human needs, but computing practitioners have sometimes been guilty of viewing the system as being devoid of people. Successful computer support for human activity can only be effectively achieved if people are viewed as an integral, indeed as the most important part of any computer-based system. Recent history can provide salutary lessons by showing examples of what can happen if human requirements and capabilities are not taken fully into account in system design. A humorous instance is quoted in Preece *et al.* (1994, p. 14), where the vendors of a proprietary police system received a call from one of their users reporting 'your terminal is dead – come and get it'. On arrival, they found the terminal with two bullet holes in it: a police user had received the on-screen message 'Do not understand' one time too many. There are many cases where a lack of account of human factors in design has caused systems to be unusable at critical times, for example in the design of aircraft and air-traffic control systems. Poor design of control panels was officially reported to be a major factor in the near disaster at the Three Mile Island nuclear power station. It is crucially important to consider the human aspect of any system during its development and use.

The emphasis of this book as a whole is on models, representations, structures and management of spatial information. However, no text that purports to give a computing perspective on GIS would be complete without a consideration of architectural and human factors in the context of GIS. This chapter therefore summarizes some of the main issues in these areas and provides pointers to further reading in the bibliography.

## 7.1   DATABASE ARCHITECTURES AND IMPLEMENTATIONS

A general feature of spatial databases is the size of their data volumes. Healey (1991) gives some examples of sizes of the early global and continental large spatial databases. Current large projects such as EOS (Earth Observation System) and Sequoia make huge demands upon handling data volume. Problems must be solved at the hardware level (large fast storage devices), data structure level (efficient storage and retrieval methods) and metadata level (giving the user some appreciation of the kinds of data in the database). Also, the nature of graphics data is such that in order to present an image on the display unit, it may be necessary to retrieve many objects from the database in a fraction of a second. Architectures for GIS conveniently may be divided into two categories: hybrid and integrated (Bracken and Webster, 1989), both being special cases of distributed architectures.

### 7.1.1   Hybrid and integrated architectures

The hybrid approach to GIS architecture manages the spatial data independently and in a different module from the non-spatial data. Figure 7.1 shows schematically a hybrid architecture. The left-hand box represents the geometric and topological engines, while the right-hand box holds the non-spatial data. Thus, for a land parcel, the geometry of the parcel area along with topological relationships (e.g. adjacencies to other areas) are held in the left unit, while the name, address, owner and other

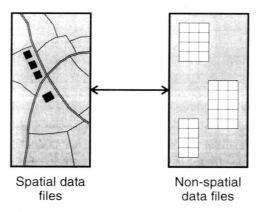

Spatial data
files

Non-spatial
data files

**Figure 7.1**   Hybrid GIS architecture.

information about the parcel are held in the right. Typically, the left unit is a collection of files at the operating system level while the right box may be a relational database. Linkage between filestore and database is provided by pointers from records in files to unique identifiers of tuples in the database.

The primary motivation for the hybrid architecture is the heterogeneity of management strategies for spatial and non-spatial data, caused by the special nature of spatially referenced data and the specialized models and structures that have arisen in GIS to handle them. Often, such models and structures are not immediately compatible with other systems, for example with relational database systems (Waugh and Healey, 1987; Worboys *et al.*, 1993). If spatial and non-spatial data handling are separated in the system architecture, then data management can be optimized for the two components separately.

Often in current hybrid systems, spatial data are stored in a set of system files and non-spatial data stored in a relational database, the generic form being the *geo-relational model* (Waugh and Healey, 1987). The prime proprietary example is ARC/INFO, the very name of which suggests the hybrid architecture of the system, with ARC being the graphics and spatial data engine and the non-spatial data handled in the INFO database. In practice it is possible to connect the graphics engine with any of the major proprietary relational databases, such as ORACLE (see Figure 7.2).

Although a hybrid architecture has the very important practical benefit of optimizing performance depending on data type, it suffers from the disadvantage that the spatial data are handled outside the database and therefore cannot gain from standard database functions such as concurrent access management, integrity, security and reliability. It would be better if both spatial and non-spatial data could be stored in the same database, and this leads to the development of integrated architectures.

*Integrated architectures*

For an integrated architecture, the basic idea is to manage the data in an integrated fashion in a single database. The spatial data are handled in the same way as the

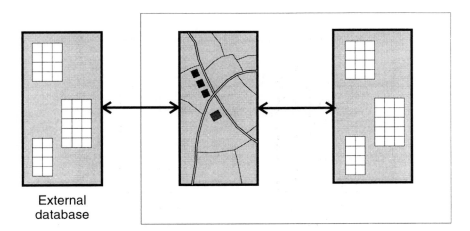

External
database

**Figure 7.2**   Hybrid GIS linked to an external RDB.

non-spatial (attribute) data. Earlier models of integrated architectures are based upon relational databases (Healey, 1991). Figure 7.3 shows schematically this approach. All data are held in tabular form within a relational database.

Earlier chapters have considered E-R modelling and logical relational database design for point, line and polygon spatial data. The problem with such an integrated solution, at least if conducted in a unmodified manner, is that performance on retrieval of spatial data is poor, due to the large number of relational accesses and joins that are required to reconstruct the spatial objects (points to chains to polygons), also to connect spatial objects to their other attributes. Another difficulty is that standard proprietary databases do not provide the wide range of indexing mechanisms required for spatial data, as discussed in Chapter 6: typically, only

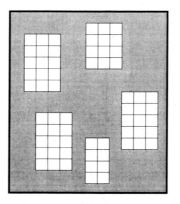

### Spatial and non-spatial data in the database together

**Figure 7.3**   Integrated relational architecture.

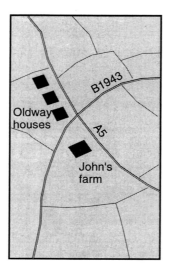

**Figure 7.4**   Integrated object-oriented architecture.

B-tree type indexes are provided. Such problems have led to limited take-up of the integrated relational approach. The bibliography at the end of this chapter refers to research on extensible relational systems that overcomes some of these problems.

Another direction that integrated architectures have taken is to integrate within database paradigms other than relational. Object-oriented DBMS hold most promise in this direction, and the architecture is shown schematically in Figure 7.4. Objects in an object-oriented database may have both spatial and non-spatial references, and so the object-oriented paradigm leads most naturally to an integrated architecture. Object-oriented databases and object-oriented database management systems have been discussed in Chapter 2. Object-oriented GIS (OOGIS) are relatively new, and the bibliography at the end of this chapter describes some research projects (not proprietary systems) in this fruitful area.

Even though technical problems remain to be solved with integrated architectures, there are great benefits arising from the uniform treatment of all the data held in the system. Future systems that rely heavily upon DBMS functionality, such as concurrency, distribution, security and integrity, may find the integrated approach appealing.

### 7.1.2 Distributed architectures

Future systems will take advantage of the dramatic growth in communications technology evinced by information highways, the Internet and World Wide Web. GIS, whether of hybrid or integrated architecture will play its role in distributed systems (Goodchild, 1994). Figure 7.5 shows a case where an integrated system (OOGIS) is linked with two relational databases, from which it draws non-spatial data about its spatial objects. For example, the GIS may hold land parcels along with their basic attributes, but link up to a land registration system to get more detailed ownership information. The need to use several databases as data sources for GIS functionality can be found in many existing real-world applications, where alongside the GIS and its dedicated database there usually exists at least one large proprietary database. The GIS should be able to access data from the corporate database and any other external databases for which connections are established.

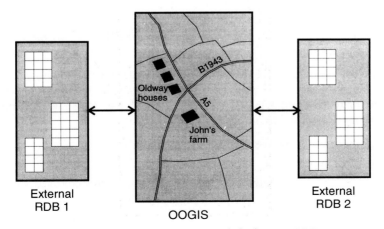

External
RDB 1

External
RDB 2

OOGIS

**Figure 7.5** Distributed GIS architecture with an OOGIS linked to two RDBs.

The architectures described above and the problem of connecting GIS to other databases (and other types of system) in a distributed network brings us to an important beneficial property of a spatial database network. A spatial database is *seamless* if users can navigate through it without meeting any artificial boundaries. These boundaries might manifest as the edges of printed map-sheets from which the data were captured or from the 'boundaries of knowledge' in a particular database connected to the network. It is now widely accepted that GIS applications should be built over a seamless database or at least a virtual database which behaves seamlessly. The *virtual* database, which may consist of many different databases, should from the user's perspective represent a transparent interface to all its heterogeneous constituents. It should enable the total functionality of the GIS to apply to the entire distributed collection of data.

### 7.1.3   Version management

Two key problems for GIS implementations that attempt a coupling of GIS functionality with traditional, relational database technology are the provision for the simultaneous existence of different versions and the support for long transactions in this dual environment. The principles involved in the management of long-term and short-term transactions are fundamentally different. The assumption implicit in standard general purpose databases is that transactions are short-term events, so that it does not matter if all users (except for the user involved in the transaction) are locked out during that short time. This provides a mechanism for preserving database integrity. However, transactions in GIS are usually of a much longer duration. For example, the design of a pipeline or implementation of an electrical network may require transaction times of several months.

Several solutions have been proposed in order to overcome these problems, with varying degrees of success. These include check-out, multiple check-out, and versioning.

- *Check-out* extracts appropriate data physically from the database to a file for update, but other users are locked out of write-access while the update takes place.
- *Multiple check-out* allows multiple extraction of the same data from the database to different files, allowing several users to continue to work with the data while updates are in progress, but problems arise with conflicts when the updated data are checked back in.

All forms of check-out are slow. The most elegant principle to accommodate long-term transactions is the principle of *versioning*, initially developed within the CAD industry (Chou and Kim, 1986; Katz and Chang, 1988). Versioning may be summarized by the following points:

- Users may have their own versions of the database.
- Each version is kept entirely within the database.
- Upon update, a copy is made of the master record and this changed copy becomes part of the user's version.
- Different versions share most of the data and only changed records are kept separately.

- At the end of the transaction, versions are integrated with the master and conflicts resolved within the database.

Although versioning is a solution for handling long transactions in a multi-user environment, problems arise when external databases (not the database dedicated to the GIS) are involved. Typically, the version-managed GIS will have interfaces to major proprietary databases but these interfaces only provide data management at the level of the proprietary database. Functionality, such as long transaction management, is provided solely within the GIS version-managed datastore. The extension of this functionality to interfaces with external databases is essential because the versioning approach does not cater for short transactions and there may be the need to keep some GIS data within the short transaction environment.

Despite all the benefits of the version-managed database, from the GIS perspective it is not possible to expect transfer of the vast amount of data from current commercial database applications just because that data could be useful to the GIS. So it is very important to maintain a well integrated interface with existing DBMS. Full integration would be the ideal case, where it is possible to have long and short transactions over all the data within version-managed and non-version-managed environments.

## 7.2   HCI FOR GIS

Computer systems do not exist in isolation but are connected to other systems, *and with human users.* Ultimately, the success of any computer system rests on whether it can be effectively used by people. The term 'human-computer interaction' (HCI) was first used during the 1980s to define the study of computer systems with the emphasis on the human users. People rather than machines are placed at centre stage. Although the system interface between computer and human is very important, HCI is a wider study, which includes an understanding of the psychology of humans when interacting with machines, training, organization and health issues. This section looks at some of the main aspects of HCI from the GIS perspective.

### 7.2.1   HCI and system design

Preece *et al.* (1994, p. 46) identify the following three key components for successful design, taking account of HCI issues.

- *User-centred*:  As with any technology, there is a temptation to leave it to the technologists: this should be resisted. Users of the system should be involved at all stages of the design process. The users need to be identified: they include those who will be actually using the end product at the user interface (the *end-users*), but also anybody who has an interest in the success of the system.

- *Integrated*:  The design process should not be too narrow but should integrate expertise from several disciplines: these include computer science, artificial intelligence, linguistics, philosophy, sociology, anthropology, ergonomics and human factors, engineering, design, graphics, social psychology and cognitive psychology.

- *Iterative*: When the design process is iterative, users and the integrating disciplines in the previous item can get involved at each stage as needed.

### 7.2.2   Interface types

The interface is a very important component of any user-centred system. Interfaces may be described according to the types of interaction that they allow between the human user and computer system, and classified as *command entry, menus, forms, natural language dialogue*, or *direct manipulation*.

#### Command entry

This is the earliest form of interaction style between people and machines. The user is presented with a *system prompt*, indicating that the system is awaiting the user's command. Most readers will be familiar with the command entry interaction style as exemplified by the MS-DOS operating system. The user is prompted to issue a command to the system by means of a terse prompt, such as 'C: > '. The only on-screen information thus provided by the system is that the current drive is labelled C (probably the local resident hard-disk). The disadvantage of this form of interaction is that the user must know in advance the type of legal responses that the system expects: no hints are given by the system. If the user does not conform precisely to the format expected by the system, then the user's command will not be recognized. However, commands do enable users to express instructions clearly and concisely, and may be an effective style for the experienced user.

#### Menus

A menu is a list of options displayed on the screen, from which the user selects one or more items, thus determining the next state of the system. The user does not need to remember commands nor complex command syntax. As with all types of interface there is a trade-off between the amount of menu functionality provided and the ease of use of the menu system. Menus can clog up the space on a screen, particularly when the problem domain is complex. One solution is to place a *menu bar* at a suitable place on the screen (usually horizontally near the top). When the user selects an item from the bar, a *pull-down* menu expands from the item. When the item is selected from the menu, the menu usually disappears. For example, the Microsoft Windows File Manager has a menu bar containing options File, Disk, Tree, etc. If the File option is selected, a pull-down menu for handling files is expanded and contains options such as Open, Move, Copy and Delete. Quick access to commonly used menu items is often provided on a *ribbon* near the top of the screen. Another possibility is to use a *pop-up* menu, that results from the selection of an item from the main part of the screen. For example, the X-Windows protocol sometimes has a pop-up menu in the main window that allows the user to define new clients, change settings, etc. A *roll-up* menu may be minimized by the user to an icon when not in use, while at the same time being always accessible for selection and maximization.

## Forms

A form facilitates the interaction between human and machine, where the interaction involves the user in entering several heterogeneous pieces of information into a form. In a form-based front-end to a GIS it may be possible to select an object on the screen by means of a mouse, resulting in a pop-up form that allows various types of editing upon the object, depending upon the nature of the object selected. Thus, a house object may have a pop-up form that allows the user to specify owner and address, and also to edit the geometry by changing the outline of the land in which the house is situated. As with menus, there is a danger that the screen will become crowded with too many forms: the same mechanisms may be used to partially alleviate this problem, such as *pull-down* and *roll-up* forms.

Some fields in a form may be optional and others mandatory, requiring an appropriate response from the user before the system accepts the data entered from the form. A data entry form for a relational database may insist that the user provides a unique entry into record key fields. Spreadsheets have the feel of a form-based interaction medium.

A more general type of form is the *dialogue box*. For instance, when the user wants to save a file, the system may display a dialogue box that provides options such as file name, type, path, etc. These options may be presented to the user in a variety of ways:

- *Radio button*:   To be turned on or off.
- *Check-box*:   To be checked if a particular option presented is desired.
- *Command button*:   To be pressed for further options to be presented.
- *Scroll arrows*:   To be pressed to alter values of a parameter, often numeric with up and down arrows to increase or decrease the value.
- *OK/Cancel buttons*:   To be pressed when all options are set and agreed/cancelled.

## Natural language dialogue

The ideal would be for humans to communicate with computers in their naturally spoken, heard and written language. The language (called *natural language*) could be spoken or entered via a keyboard. However, progress in computer science towards this goal has been slow: there are some very difficult problems encountered along the route. The problems are not just concerned with grammatical rules (syntax). For a traditional example, consider the following two sentences:

1. Time flies like an arrow.
2. Fruit flies like a banana.

At first glance, these two sentences have the same grammatical structure. However, when the meaning of each sentence is grasped, it becomes clear that the two structures are quite different. In the first sentence, the word 'flies' acts as a verb and the word 'like' acts as an adverb. In the second, the word 'flies' acts as a noun and the word 'like' acts as a verb. The two sentences vividly demonstrate that some semantic knowledge is required in order to determine the structure of a sentence of natural language: syntax cannot be divorced from semantics. Thus, for a computer to converse in natural language, some domain knowledge will be needed.

Even though the general problem is unsolved, it is quite feasible to expect natural language interaction with computers in restricted domains. The advantage that natural language has is flexibility and ease of use for the completely novice user. However, natural language can often be verbose and ambiguous, and for many tasks a menu or forms interface is more efficient.

*Direct manipulation of icons in a Graphical User Interface (GUI)*

A *Graphical User Interface (GUI)* enables the user to interact with the computer system by pointing to pictorial representations (*icons*) and lists of menu items on the screen. Pointing is usually accomplished with mouse or keyboard. *Icons* are visual mnemonics and allow the user to interact with the computer without having to remember commands or type them at the keyboard. The archetypal direct manipulation using an icon is the act of deleting a file on the Apple Macintosh by moving an icon representing the file to the waste basket. Icons exist in a *metaphoric framework*. For example, the Apple Macintosh metaphor is the desktop and office space. This metaphoric space is populated with files, filing cabinets, waste baskets and clipboards. Operations on the icons include moving them about the space, pasting them to the clipboard, and dropping them into the filing cabinet or the waste basket. Such interactive methods are often called object-oriented, for the reason that they are based upon classes of objects and operations that they support. In fact, all the interactive methods described in this section are based upon metaphors, whether the command, menu or form: it is a matter of degree. Direct manipulation assumes that the objects of interest are visible and capable of manipulation by direct action that has a visible effect. Cartographic data are rich in metaphors and a very interesting branch of GIS research has explored this question.

GUIs are components of menu and forms interfaces as well as supporting interaction by direct manipulation. A particular GUI is the window-icon-mouse-pop-up-menu (WIMP) interface. GUIs allow system designers and developers to construct systems without having to worry about details of how things are displayed on the screen or the precise code for input and output. Common user interactions, such as deleting or saving a file are built into the GUI and so do not need to be coded from scratch for each new application. They are independent of the particular piece of hardware (the device) and so contribute to the open systems philosophy.

An *iconic language* provides a structure for the systematic manipulation of icons to express and convey meaning. We may view an icon as having *syntax*, being the image of the icon, and *semantics*, being the meaning expressed by the icon. Thus an image of an exclamation mark '!' in a triangle (syntax) may convey to a motorist the meaning of 'danger ahead' (semantics). Icons may be *composite*, and therefore capable of decomposition into primitive elements. Icons support *icon operations*, being the legal actions that may be applied to them. For example, a waste basket icon may accept the movement of a file icon to the basket's position on the screen, but may not accept a rotation of the basket. Lee and Chin (1995) develop in detail the theory and application of iconic languages in the context of GIS.

### 7.2.3 Metaphors

In human language, a metaphor is the use of a word or phrase denoting one kind of idea or object in place of another for the purpose of suggesting a likeness between the two. More generally, metaphor has been defined by Johnson (1987) as 'a pervasive mode of understanding by which we project patterns from one domain of experience in order to structure another domain of a different kind'. The *source domain* is projected by the metaphor into the *target domain*. (Note the equivalence of notation between metaphor and model (section 4.4.1). This equivalence may itself constitute a metaphor!) Members of the source and object domains in a metaphor are usually taken from different linguistic categories. Madsen (1994) sets out some of the major linguistic categories as *animate versus inanimate*, and *human versus animal versus physical*. It is the ingrained nature of metaphoric understanding in humans that makes it a powerful vehicle for human-computer interaction.

Kuhn and Frank (1991) discuss a formal approach to these metaphoric functions as well as considering spatial applications. In fact, the desktop metaphor itself *spatializes* an essentially unspatial activity of deleting, saving, copying and moving(!) files. They note some key spatial metaphors: we may 'navigate' through a database, 'zoom' in on a detail or 'scan' a file. All these metaphors are part of a more general metaphor, transforming operations in a visual field to operations on a computer system. It is revealing that the widely pervasive term for the most common type of GUI object is the 'window', through which the user views the interaction with the system as if looking out into a visual field. This field itself is part of an even more general metaphor based upon the notion of 'container' from which objects can come and go (into and out of sight).

The challenge for GIS research is to find appropriate metaphors for spatial data handling. The discipline of cartography gives some pointers, but a hard-copy map is an essentially static object, while computer systems offer us a dynamic domain. Johnson (1987) uses the term *image schema* to describe entire metaphoric structures. Mark (1989) analyses geographic concepts in the context of image schemata and discusses the relation between users' views of space and GIS operations.

In the final chapter, current research on spatio-temporal systems is discussed. Temporal metaphors are almost always spatial, thus, we 'go into the future' (moving through a field), 'time passes quickly', 'tomorrow is just round the corner', etc. Maybe these spatial metaphors can be used to facilitate effective interaction with the next generation of temporal GIS.

### 7.2.4   Dialogues

We may summarize the discussion up to now in the context of an interaction between human and machine, termed a *dialogue*. This interaction takes place in a metaphoric space, for example with the human using *views*, *tools* and *agents*. The metaphors for the content of the dialogue include *paper document*, *desktop*, *office space* and *map*. Transitions in the dialogue are effected by *fade*, *zoom* and *pan*, for example. The types of interaction are *commands, forms, menus, direct manipulation of icons, natural language*. The dialogue has input from humans using, for example:

- *Keyboard*:   By means of commands, forms and menus.
- *Mouse*:   By means of selection, collection (lassooing), translation, etc.
- *Pen*:   By means of character-recognition.
- *Voice*:   By means of voice-recognition.
- *Screen*:   By means of touch-sensitive screen.

A constituent of the dialogue is display on the machine, using, for example, a screen with windows, scrolling displays and animations. Output issues include visual symbolism, timely response, crowding in field-of-view, focus of output. Current window systems include X, Windows, Motif, Open Look and Macintosh.

### 7.2.5  GIS interfaces

A GIS integrates database functionality for heterogeneous types of data, with special emphasis on spatial types. An interface to a GIS should at least allow users to effectively select and retrieve relevant spatial data, manipulate the data, and present the results of analysis. Relational database query languages, such as SQL (section 2.2.2), are designed mainly to handle database retrieval. Traditional SQL statements are often embedded in a standard programming language program, for the purpose of data manipulation and presentation. Standard corporate databases do have some manipulation and presentational functions (report writers, for example). For a GIS, all three aspects, retrieval, manipulation and presentation, are traditionally handled by the system. This section looks at some features that a GIS interface requires.

*Data selection*

Selection of data is made by providing filters that may contain both spatial and non-spatial attributes, or combinations. For example:

- Retrieve all potteries whose name contains the word 'Spode'.
- Retrieve all potteries that are within three miles of the Wedgwood Pottery.
- Retrieve all potteries that are within three miles of the Wedgwood Pottery and whose names do not contain the word 'Spode'.

The user may select data by providing commands, selecting from a menu, filling in a form (all standard database approaches) or by direct manipulation – pointing to the items or providing a window, using a mouse. Selections based upon commands require knowledge of a database query language, such as SQL. Any such database language will need to be extended to allow spatial filters. The appropriateness of SQL for GIS has been the subject of much debate and will shortly be discussed. In the case of direct manipulation, the object data must be clearly visible on the screen. Pure direct manipulation does not naturally allow logical combinations of filters, as given by the third of the three queries above. However, a mixture of menu (form) and direct manipulation may be appropriate. For example, we might select a road on the screen and request its name or the average annual traffic that it supports. Conversely, we might use a form to highlight all roads that are class 'A' and then use a spatial circular window indicated with a mouse (assisted by menu to choose window shape) to select all A-roads within three miles of Stoke city centre.

Selection by direct manipulation is accomplished by a pointing tool, usually an arrow on the screen whose position is controlled by a mouse. Pointing can be ambiguous if several objects occupy the same position. This may be overcome by a pop-up menu, allowing the user to choose, or the layering of objects as with a graphics system. It would be unusual for all database objects to be clearly visible on

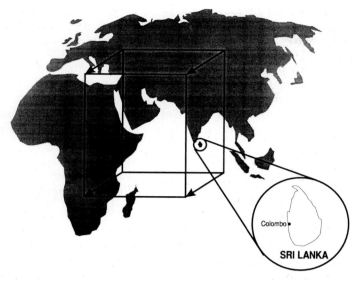

Colombo

**SRI LANKA**

**Figure 7.6** Panning (shown by arrows) and zooming in part of a global GIS.

a single screen. Extra operations required are *pan*, moving the viewing window to a different part of the database, and *zoom*, providing greater or lesser detail (see Figure 7.6).

*Data presentation*

Direct manipulation allows the integration and delivery of results from several queries. A good presentation provides an effective medium in which the user can work by direct manipulation to produce further analyses and results. For example, with the previous query it may be useful to integrate the selected roads with the point sites of cinemas in the Potteries region and then perform a proximity analysis to determine catchment areas for each cinema, based upon the road network.

Data presentation is an area where GIS allows considerably more functionality than a standard database. A relational database will allow some incorporation of the results of a database query into a report, maybe using a report writer. GIS can use its graphics functionality to present data to the user in a multiplicity of ways.

It is important to present the spatial objects with appropriate rendering and in an appropriate context. Graphical rendering allows many possibilities for conveying large amounts of information and provision of different views of the data to a heterogeneous user population. Similarities and differences of rendering may convey to the viewer that the rendered objects have similarities, maybe belonging to the same object class. Some of the variables that we have at our disposal during graphical display are:

- *Areal outline*: Colour, grey-scale or black and white; pen characteristics (e.g. thickness, nib shape).

- *Areal fill*: Colour, grey-scale or black and white; pattern (e.g. hatched, dotted).

- *Line*: Colour, grey-scale or black and white; pen characteristics (e.g. thickness, nib shape).

- *Symbol*:   Type; shape; size; positioning in the field of view, in relation to other objects; amount of detail.

The *context* is important for spatial information and might include a larger area in which the results are positioned, orientation and scale details. As with most information, its power lies in seeing how it relates to other information in the system. For an example of context, see Figure 1.2 in Chapter 1 where the inset in the map shows the Potteries region in the context of England. Even though orientation is not explicitly shown, it is implicitly given by the context, relative to the orientation of England.

For graphics presentations that have any complexity, the meaning of symbolism provided by the graphical rendering and context will not be obvious to the viewer. In that case a mapping, or key, from the symbol set to the meaning (*semantics*) must be explicitly given. This key is often called the *legend* (a term coming from cartography).

### Data manipulation

Data manipulation requires sequenced application of the spatial and non-spatial operations discussed earlier in the book. Some of these operations can be facilitated by the interface. For instance, direct manipulation will allow natural geometrical transformation and overlay operations to be performed. There is a strong object-oriented feel in the direct manipulation of spatial data. Traditional non-spatial databases, used for straightforward applications, demand only simple user models of the information contained within them. The objects in a bank database can be modelled using such simple types as positive and negative integers, character strings and Booleans. However, spatial information is considerably more complex and problems arise when models that are too low-level are forced upon the user. Ideally, the user of spatial information should be helped to apply whatever high-level metaphor is appropriate to the points, lines and areas on the screen. It has been found that an object-oriented paradigm facilitates effective interaction, encouraging the user to manipulate classes of objects with well-defined operations. Even if the GIS is not based upon an object-oriented database or does not support an object-oriented programming environment, significant improvements in user effectiveness can be gained by at least providing an object-based user interface.

Egenhofer (1990) gives a useful analysis of the direct manipulations to graphical renderings that can effect operations upon the GIS. Database updates and screen manipulations are two different concepts, but direct manipulation of objects on the screen can easily be linked metaphorically to database updates, maybe with the provision of **commit** and **rollback** functions as buttons on the screen. There is a direct parallel with the notion of WYSIWYG (What You See Is What You Get) environments, for example with modern word processors. A set of such operations, with no pretension to completeness, is given below.

- **Create**:   Creates a new spatial object in the database based on a created graphic object (maybe with non-spatial attributes provided by filling a form).
- **Delete**:   Deletes a selected spatial object from the database.
- **Collect**:   Forms a collection of objects from those selected on the screen.

- **Refresh**: Gives the results of the last operation only (all other material is removed from the screen).

- **Overlay**: Combines two collections of objects, placing them together in the field of view; equivalent to set union; for example overlaying cinemas and a major road network.

- **Intersect**: Combines two collections of objects, placing those that occur in both collections in the field of view; equivalent to set intersection; for example, intersecting public houses with buildings of special historic interest.

- **Remove**: Combines two collections of objects, removing those that occur in the second collection from the first and displaying the new first collection in the field of view; equivalent to set difference; for example, removing public houses from the collection of buildings of special historic interest.

- **Transform**: Effects a geometrical transformation (scaling, translation, reflection, affine transformation, etc.) upon a group of selected objects.

- **Checkpoint**: Sets a checkpoint to which we can return if unhappy with the updates after the checkpoint.

- **Rollback**: Allows a return to the state at the last checkpoint.

- **Commit**: Transfers all direct manipulations to the database as updates; irreversible.

The map-overlay underlies many conceptualizations and interactions with geographic information. Egenhofer and Richards (1993) make a detailed examination of the map-overlay metaphor as a medium for interaction with a geographic database using direct manipulation. The authors apply their analysis to suggest a direct manipulation interface for overlay-based operations.

*SQL debate*

There is an ongoing debate as to whether the relational database language SQL can be successfully extended to handle spatial queries. Arguments for SQL include its near universal use as a database query language and the emerging standard SQL/MM, a part of SQL3 for multi-media use that will include a comprehensive set of GIS operations. Arguments against SQL for GIS have been that SQL queries are difficult to formulate for complex retrievals where several tables may need to be joined, and that the presentational aspects of current SQL are poor. The weaknesses of SQL are linked with the weakness of relational technology to handle spatial information and the outcome of the debate probably depends on the architectures that become common for managing spatial data.

The new standards for SQL offer a conceptual change and much greater possibilities. In traditional SQL, a basic database retrieval functionality was provided by the query language. Any further functionality could only be achieved by embedding SQL in a computationally complete programming language, such as C. In newly emerging standards, the converse holds (see Figure 7.7). The functionality of SQL may itself be extended by allowing the user to define his or her own operations, and embed them in SQL commands. Thus, a spatial filter, such as 'within $x$ miles of' may be coded and the operation used in an SQL statement. Also the range

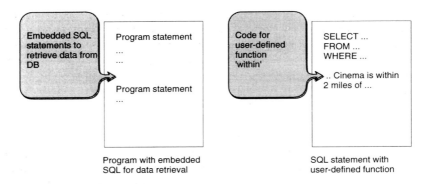

**Figure 7.7**   Alternative types of old and new SQLs.

of types to be allowed is to be greatly extended. The new standard is giving SQL a much more object-oriented emphasis, with a wide range of data types and operations (see Appendix 2).

As GIS increasingly becomes just one component of a distributed system in which individual units have specialized functionality, it may be that the SQL debate will become less relevant. It is not necessarily important that the user is able to directly manipulate spatial data in a relational database environment. What is more important is that the combined manipulation of spatial data within a GIS and non-spatial data in a relational or object-oriented DBMS should be compatible and transparent to the user.

*Visualization*

The discussion on GIS interaction has covered data retrieval, manipulation and presentation. Important missing links are *analysis* and *interpretation*. As most of the information in a GIS is graphics-based, its analysis and interpretation is often effected by *visualization*. Visualization has been defined by Buttenfield (1991) as follows:

> Visualization is the process of representing information synoptically for the purpose of recognizing, communicating and interpreting pattern and structure. Its domain encompasses the computational, cognitive and mechanical acts of generating, organizing, manipulating and comprehending such representations.

As the definition suggests, the application of visualization is considerably broader than spatial information analysis and interpretation. It has been used in a wide range of situations, where complex multi-variate processes are involved. As a non-spatial example, visualization has been used in software support to provide a means of understanding programs and system development. A *program visualization tool* may take a traditional program and animate its execution. The same principle applies to system design, assisting designers to simulate their incomplete designs.

Spatial analysis concerns the perception of patterns and special features of the information, extraction of parameters and discrimination of special classes of objects. For GIS, it could be described as part of the process of transforming spatial information into knowledge. Interpretation relates the knowledge gained from analyzing the information with contextual knowledge so as to make deductions, theorize or gain understanding. Visualization is primarily a human cognitive activity, and the types of visualization processes that may be used should be taken into account in the design of the system. Thus, the chapter ends by emphasizing the human interaction component of the system as critical.

## BIBLIOGRAPHIC NOTES

An early paper on extensible relational databases is by Haskin and Lorie (1982). There are at least two major projects, both in the USA, that are exploring the suitability of extensible relational databases as integrated management systems for geographic information. The Starburst project, from the IBM Research Laboratory at Almaden, California, has been examining extensions to standard relational database systems (see, for example, Haas *et al.* (1990), Lohman *et al.* (1991). The Sequoia project (Stonebraker and Dozier, 1991; Guptill and Stonebraker, 1992) has used updated versions of the extensible relational system Postgres to manage spatial data captured from satellites. Goodchild (1994) provides a readable overview of the current and future effects of communications networks on GIS.

There is now a moderately sized literature on the applications of the object-oriented approach to GIS. Survey papers are by Worboys *et al.* (1990), Egenhofer and Frank (1992) and Worboys (1994b). Choi and Luk (1992), extend a generic object-oriented database system with GIS functionality. Milne, Milton and Smith (1993), discuss the construction of an object-oriented GIS based on the general purpose object-oriented DBMS, ONTOS. David, Raynal, Schorter and Mansart (1993) describe the implementation of an object-oriented GIS using the OODBMS $O_2$ from $O_2$ Technology (Bancilhon *et al.*, 1992). Other applications of object-oriented technology to GIS are described by Worboys (1992), Van Oosterom and Van den Bos (1989), Scholl and Voisard (1992a, 1992b). Research papers describing proprietary GIS with object-oriented features are by Chance *et al.* (1990), Herring (1991a) and Newell (1992). Version management is described in the context of a proprietary system in Easterfield, Newell and Theriault (1990). Kämpke (1994) considers some of the issues of versioning in an abstracted realm of polygon objects.

A good recent text on human computer interaction is by Preece *et al.* (1994). A collection of papers on cognitive and formal aspects of GIS research is found in Mark and Frank (1991). Another text concentrating on HCI for GIS is by Medyckyj-Scott and Hearnshaw (1994). Lee and Chin (1995) develop in detail the theory and application of iconic languages in the context of GIS. Raper and Bundock (1991) present a complete layer-based model of a user interface for a GIS, called UGIX. Useful general material on spatial data interface functionality may be found in Frank and Mark (1991). Egenhofer (1990, 1994) provides discussions of the main issues of database query interaction applied to GIS. In Egenhofer and Richards (1993), the authors move on to make a thorough examination of the map-overlay metaphor as a medium for interaction with a geographic database using direct manipulation. A further overview paper is by Egenhofer and Herring (1994). The appropriateness of SQL as a GIS interaction language is discussed by Raper and Bundock (1991) and Egenhofer (1989a, 1991a, 1992). Visualization is discussed with regard to GIS by Buttenfield (1991) and in the system development context by Shu (1988).

# CHAPTER EIGHT

# Next-generation systems

I have seen the future, and it works.

Lincoln Steffens

GIS is not just the name of a system for specialized applications but is also an interdisciplinary research activity with its own conferences, journals, and centres of excellence. This chapter will describe three areas of current GIS research where computing science has a major role, preceded in section 8.1 by a list of some of the items of the current GIS research agenda for which it is felt that computing science is a principle player. Section 8.2 will describe some research to introduce temporality into GIS; section 8.3 is devoted to three-dimensional systems; and section 8.4 concerns the application of logic-based approaches and knowledge-based systems to GIS. This sample is the result of the author's own taste and interests, and therefore necessarily idiosyncratic.

## 8.1 INTRODUCTION: THE RESEARCH AGENDA

The GIS research agenda is set by researchers and consumers, individually and collectively. Individually, research in GIS is undertaken by a global and interdisciplinary community, with input from geographers, environmental scientists, computer scientists, and people from many other areas. Collectively, the community has its own conferences and journals (see Appendix 3) as well as some major grant-funded initiatives, including the US National Center for Geographic Information and Analysis (NCGIA, 1989, 1992) and the European Science Foundation's GISDATA scientific programme (Arnaud et al., 1993).

Goodchild (1992) identified a need to do science in this field, by attempting to answer the scientific questions that geographic information handling raises and by pursuing scientific goals using the technology that the systems provide. He identified key issues in this scientific endeavour: data collection and measurement, data capture, spatial statistics, data modelling and theories of spatial data, data structures, algorithms and processes, display, analytical tools, and institutional, managerial and ethical issues. This section considers the subset of the overall GIS research agenda that is important in the context of computing science. Günther and

Buchmann (1990) presented a list of research areas for technical GIS, and the following is a modified and updated list based upon that work and the other sources above.

- *Models and user requirements*: Sound and complete specifications that capture the semantics of generic and particular geo-domains are needed. These models should facilitate requirements capture from domain experts. They should allow full definitions of data quality. Hierarchical models will allow degrees of abstraction/generalization on the data.

- *Representations*: Multiple representations are needed that allow data exchange and integration of distributed heterogeneous systems. Cross-representation specifications of spatial objects and operators are important to define multiple representations. This issue is very important in view of the dramatic advances in communications and distributed systems.

- *Spatial access methods*: There is currently a debate about the usefulness of further work on spatial access methods. A view propounded by Stonebraker in his keynote address to the International Symposium on Large Spatial Databases (Singapore, 1993) is that new methods make only minute differences and that hardware advances dominate. However, it is unlikely that major advances are not still to be made.

- *Computational paradigms, architectures and data management*: Research is needed on the applications of newer computational paradigms, such as object-orientation, logic-based approaches, parallel architectures and neural networks, to the handling of spatial information. Dynamic data and spatial database update require further research, as does the legacy issue (updating existing systems and their data).

- *Interfaces and languages*: An interface to a spatial information system should help users to take advantage of the data within. Research includes database languages, metaphors, visualization methods and approaches to metadata handling. Data mining and exploratory uses of information systems hold much promise.

- *Finite resolution, data quality and multiple resolution systems*: A major problem with spatial data is the control of error propagation under spatial operations. Further research is needed on finite precision geometry and multiple resolution techniques.

- *Extensions to the two-dimensional spatial model of geo-information*: Research on three-dimensional and spatio-temporal information systems is still required at all levels, including high-level models, structures, representations, interfaces, architectures and performance issues.

## 8.2  SPATIO-TEMPORAL SYSTEMS

Temporality is an inherent aspect of geo-information, but until recently has been relatively neglected by GIS research, the emphasis being on the spatial dimensions. Indeed, the term 'geographic information system' has been synonymous with 'geo-spatial information system'. GIS technology has, quite naturally, concerned itself with the problems of capture, storage, manipulation and presentation of information that is referenced to space. This history of neglect is probably partly due to the inadequacy of the technological support, both in terms of hardware and software. However, the situation is changing: the speed and capacity of hardware, along with the software that is now becoming available, make temporal information systems possible.

Section 1.2.6 provided a general motivation of spatio-temporal information systems. The maps in Figure 1.10 show the changing scene within one small area of the Potteries region over a period of just over 30 years, with snapshots recorded at three times: 1937, 1959 and 1969. Geographic phenomena are observed and recorded in a multi-dimensional setting that includes spatial and temporal dimensions. For example, a storm exists in space and time, and its intensity is dependent upon both of these. The spatial dimensions include three that determine its spatial configuration with respect to the Earth and at least two more that determine how it is presented graphically in a particular system. Its temporal dimensions include at least one dimension with which to measure its occurrence as it happens in the real world and one with which to account for its existence in the computer system (beginning with its birth upon insertion into the database and concluding with its deletion). Other dimensions will be required to measure rainfall levels, wind velocity vectors, temperatures, and so on.

### 8.2.1   Temporal extension to the field versus object dichotomy

Within the relational model, measurable geographic phenomena may be recorded as collections of tuples, each tuple containing values for each of the dimensions. Each value is measured according to some scale, which may be nominal, ordinal, interval or ratio, depending on the nature of the dimension. Thus, a storm may be recorded as a collection of tuples, each containing values measuring location, periods of time for which it exists both in the real world and in the computer system, rainfall level, wind velocity vector, etc.

We have seen in Chapter 4 that spatial models may be broadly divided into two groups: field-based and object-based. One way in which we conceptualized this dichotomy was as division of relations by columns and rows, respectively. Thus, geographic fields were seen as variations over a spatial framework, while geographic objects had references to spatial objects. This approach may be extended by the addition of temporal dimensions.

Figure 8.1 shows the field-based approach extended so that the spatial framework becomes the *spatio-temporal (ST) framework*. Thus, the field-based approach to modelling of spatio-temporal phenomena admits the definition of *ST-field functions*, each of which models the variation of a measurable attribute though space and time. For example, the temperature attribute of a storm will vary with respect to both space and time. A term that sometimes describes these kinds of ST-field functions is *process*, which may be defined as a phenomenon with an extension and variation in spatial and temporal dimensions.

On the other hand, Figure 8.2 shows the object-based approach extended so that spatio-temporally referenced objects in the application domain are modelled to reference not just pure spatial but spatio-temporal references. For example, areas representing incidence of vegetation types may have time invariant attributes, such as species' names and properties, but reference areas that vary with time.

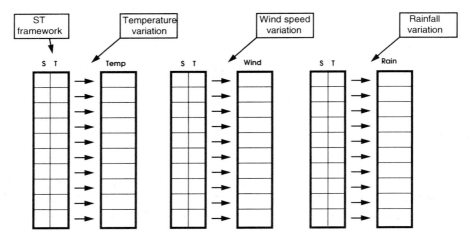

**Figure 8.1**   ST-field model of geographic phenomena.

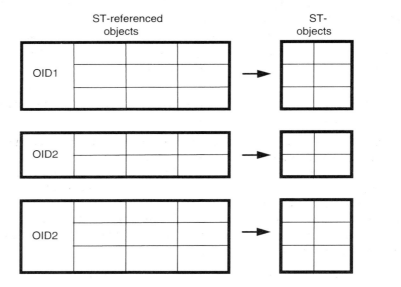

**Figure 8.2**   ST-object model of geographic phenomena.

### 8.2.2   Spatio-temporal examples

In order to motivate some of the ideas introduced in this section and developed further below, we describe some examples.

*Example 1: Road planning*

This example was first constructed in Worboys (1994a). Figure 8.3 shows three stages in the development of a road bypass around a town. Continuous and dotted lines indicate built and projected road, respectively. There are two temporal dimensions operative here: *transaction time*, the time that the system is updated

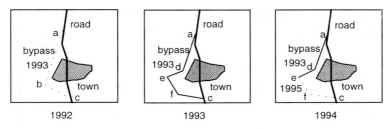

**Figure 8.3**    Spatial and temporal information on a bypass development.

and *valid time*, the time that an event occurred in the application domain. The three boxes indicate the state of the database in the years (transaction times) 1992, 1993 and 1994, respectively. Within each box events are represented that are referenced by valid time. Details are:

- *State of the database, 1 January 1992 (transaction time)*:   The bypass is projected to be constructed during the year 1992 (valid time), and completed by 1 January 1993 (valid time), as shown by the chain *abc*.

- *State of the database, 1 January 1993 (transaction time)*:   The bypass has been completed, but follows the revised route given by the chain *adefc*. (Maybe the road had to be re-routed around a conservation area at point *b*).

- *State of the database, 1 January 1994 (transaction time)*:   The bypass has not yet been completed. The stretch of road indicated by chain *efc* still remains to be built. It is projected to be completed by 1 January 1995.

In the first box (1992), we see a linking of a temporal element with a spatial object to indicate a projected event. The second box (1993) shows a modification, not due to error but due to a change of plan. The third box (1994) indicates that the state of the database in transaction time 1993 was in error, since one stretch that was indicated built in 1993 had yet to be built in 1994.

We would like to be able to model and design a system that will hold the subtle shades of temporal information indicated in this example. Instances of the kinds of queries that the system should support are:

1. What was the state of information about the bypass in 1992?

2. How did the state of information about the bypass differ in 1992 from 1993?

3. When is road segment *ef* due for construction?

Note that the answer to the third query will depend upon when the query is put to the system. This example clearly shows the need for at least two-dimensional time that is linked to spatial objects.

*Example 2: Hydrographic information*

This example is discussed by Langran (1992) and is motivated by the requirement to have navigational information that is both as accurate as possible in real time but also provides accurate historical records of previous states of the information system. To illustrate the need for two temporal dimensions, an investigation into a shipping accident would need to know not only what at that time were the locations of ships, lights and hazards in neighbouring waters, *but also what information was available to the ships' navigators about these locations*. Note that there is a critical distinction between the state of the information in a system at a previous time and the actual configuration of hazards, etc., as they really existed at that time.

Langran's example deals with point data concerning beacons, shoals, hazards, bells and lights. Information is available on the positions of such features at different times. Also known and recorded by the system are the times at which transactions with the information system occur. Transactions include the following, all of which may be made pro-actively or post-actively, and with each transaction, the transaction time is recorded:

- Insertion of an object into the system. The object's attributes include position (given by two coordinates) and valid time.
- Deletion of an object from the database.
- Modification of an object in the database.
- Restoration of an object to the database.

For concreteness, we give Langran's sequence of transactions. A fictitious initial state (as of 30 August) of the information system is shown in Table 8.1. Further transactions are shown below:

*1st September*

A buoy (45, 46) was added pro-actively (valid from 8 Sep) as there were plans to place it in the harbour.
A light (12, 18) was added (valid from 18 Aug).
The shoal's coordinates changed to (15, 21) (valid from 18 Aug) following further survey information.
The hazard and bell deleted.

*5th September*

The hazard is restored without change.
A correction is made to the light's position (18, 18). The alteration was made post-actively, to apply from 18 Aug, following the discovery of a data entry error.

Close examination of these transactions reveals some ambiguities, for example, the hazard's position was defined in the initial state, deleted and then reinstated without change. Are we to assume that in the period when the hazard was deleted from the database it was not present in the ocean, or was the reinstatement due to an incorrect deletion? Such imprecisions are a source of difficulty for temporal information systems. Once more, we see the events being modelled referenced to at least two temporal dimensions: transaction time and valid time.

**Table 8.1** Initial state of Langran's hydrographic example system

| Object | Position | Valid from |
|--------|----------|------------|
| Beacon | (38, 38) | 8 Mar |
| Shoal | (23, 32) | 2 May |
| Hazard | (56, 34) | 3 Mar |
| Bell | (44, 28) | 8 Aug |

*Example 3: Administrative areas*

Administrative regions reference spatial objects that define the areal extent of the regions. Assume, as in the UK, that every 10 years when the new census is taken, the boundaries may have been changed. Census and other statistics relate to these regions. In order to conduct longitudinal studies, it is necessary to relate statistics to the correct historical versions of the regions to which the data relate. Thus, what is required is an information system containing historical spatial data on the regions. Let us assume that data referenced to transaction time, while useful, are not essential here.

Figure 8.4 shows a highly simplified picture of change of regional structure over time. In the census year 1971, the district $D$ is divided into two regions $R$ and $R'$. In 1981, the spatial extent of $R$ is reduced and that of $R'$ is correspondingly enlarged. In 1991, the spatial extent of the entire district is increased and the spatial extent of $R'$ is reduced to accommodate a new region $R''$.

Suppose that there exist various non-spatial datasets referenced to these administrative areas. For instance, suppose that data exist on population totals referenced to regions and mortality figures referenced to districts. Furthermore, imagine a road running through the district, with a spatial extent that may vary with time. Then, there will be spatio-temporal relationships between the regions and the road. The following questions are examples of those that may be posed:

- What variation has there been in the population density of $R'$ between 1971 and 1991?
- What variation has there been in the morbidity ratio (ratio of deaths to population) of $R'$ between 1971 and 1991?
- Has the road ever passed through region $R$?
- Does the road currently pass through land that has ever belonged to region $R'$?
- Does the road currently pass through land that has always belonged to region $R'$?

The characteristics exemplified here are of information referenced to complex spatial objects that suffer discrete change with respect to time. The maintenance of object identity under change of attributes can cause difficulties. When is it appropriate to say that an object has changed so much that it has a new identity? This is a question for those working and specializing in the particular application domain, as well as for system experts.

A simplifying feature is that the temporal component is essentially one-dimensional, in the sense that we are concerned mainly here with discrete changes in the real world being modelled in the information system, rather than the changes to

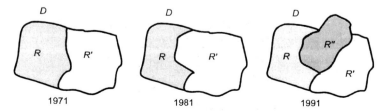

**Figure 8.4**   Change of regional structure through time.

the information system itself: we are dealing with an *historical* information system. The temporal dimension is also linear, in that neither future nor past branching nor any cyclicity is admitted. These simplifications are too stringent for some of the systems under current consideration. For example, we may wish to hold information on projections of future boundaries under consideration by the UK Boundary Commission. Also, we may want to be able to post-actively change a boundary without deleting the earlier version.

*Example 4: Land information system*

This example shows some of the temporal requirements that land information systems may have. Information in the system relates to land ownership and comprises spatial, temporal, legal and other components. Ownership is affected by contracts, death and inheritance, legal proceedings, fire, etc. The example we use is a modified version of one given by Al-Taha (1992). Figure 8.5 shows the spatial and ownership variation in a fictitious land area through some decades of this century. The chronology of events is as follows.

> *1908 (Original records):* Information is held on a street, a land parcel owned by Jeff (parcel 1), a land parcel owned by Jane (parcel 2), further parcels and buildings.
>
> *1920:* Jane has incorporated parcel number 3 into her ownership, now named parcel number 5.
>
> *1938:* Parcel 4 has been enlarged and a school has been built on it.
>
> *1958:* Jane's house has been destroyed by fire and Jane has died. Jack now owns the building and land in parcel 1.

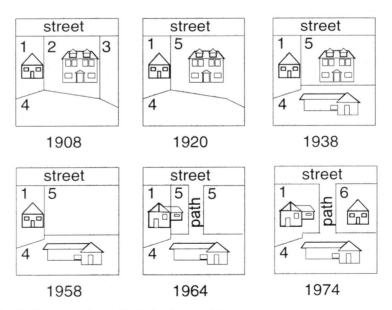

**Figure 8.5**   Spatio-temporal variation in land ownership.

*1960*:   The council gives notice of its intention to build a path through parcel 5 in 1962 so as to give better access to the school.

*1962*:   The building of the path is postponed until 1964.

*1964*:   Jack has built an extension that intrudes partly into parcel 5. The council has built the path through parcel 5.

*1974*:   Jack has incorporated part of parcel 5 (by adverse possession) into his ownership of parcel 1. Jill has taken possession of land parcel 6 and built a house upon it.

There are references in this example to transaction and valid time dimensions. For example, in 1962 (transaction time) information is received that the path, originally forecast in 1960 (transaction time) to be built in 1962 (valid time) is postponed until 1964 (valid time). Many of the valid times are unclear from the information provided, for example, the exact year in which Jane's house was burned down is unknown.

### 8.2.3   Temporal dimensions and structure

The four examples above provide an indication of the concepts that are important for temporal analysis. It is clear that time in information systems is measured along at least two separate dimensions.

- *Transaction time dimension*:   also called *database time* or *system time*. This is the time when transactions take place within an information system.

- *Valid time dimension*:   also called *event time* or *real-world time*, when the events occur in the application domain.

It is also possible (see Figure 8.6) to distinguish different structural types for each of the temporal dimensions. Time may be measured as a *discrete* or *continuous*

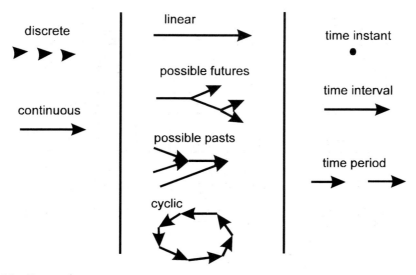

**Figure 8.6**   Temporal structure.

variable. A continuous temporal variable is used where processes require the measurement of time to be possible at arbitrary levels of precision. For example, the
position of a marker on a glacier may be measured at various time points, and the
underlying model might assume a theory of interpolation for other time points. A
discrete temporal variable is used where time is measured at certain points or intervals and the variation is discontinuous between these points. An administrative
boundary may occupy one position at time $t$ and another position at time $t'$, but it
probably does not make sense to say that the boundary occupies some intermediate
position between times $t$ and $t'$. It does not creep continuously but jumps discretely.

Differences in structure arise from the differing temporal orderings. Time may be
*linearly ordered*, such that for any two different temporal events, one will be earlier
than the other. Time may be *partially ordered*, allowing the possibility of alternative
futures, pasts, or both. Events may form a temporal cycle, exhibiting periodicities.

Temporal events are usually associated with time points, intervals (durations) or
disjoint unions of time intervals. The object class hierarchy for purely temporal
objects is in general simpler than the spatial case, at least for the special case where
the temporal dimensions are assumed to be linear. Just as we have made the distinction between spatial and graphical object classes, so should we do the same for time.
There is a distinction to be made between time in the real world, whether the valid
time dimension of events or the transaction time dimension of interactions with the
information system, and time as presented to the user of the system, for example in
animations.

With regard to any single temporal dimension, we may record the occurrence of
an event at a *temporal instant* (analogous to points in one-dimensional space) or the
duration of an event throughout a connected *temporal interval* (analogous to a
segment in 1-D space). In practice, time instants may be approximated by short
intervals that are the smallest indivisible temporal intervals, often called *chronons*.
Disjoint temporal intervals may be collected together to form a *temporal period* or
*temporal element*.

A *bitemporal interval* is the two-dimensional version of a time interval. It indicates the extent of an interval event in valid and transaction time and is the

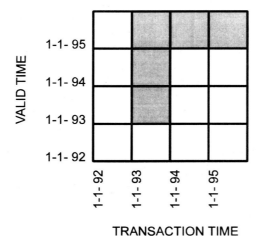

**Figure 8.7**  Bitemporal period for road segment *ef*.

Cartesian product of a *transaction temporal interval* (*T-interval*) and a (*valid temporal interval* (*V-interval*). Thus, the T-interval gives the insertion and deletion times and the *V-interval* describes the interval of temporal existence of the event in the real world. A *bitemporal period* generalizes a temporal period to two dimensions, being the union of a set of bitemporal intervals. A bitemporal period is constrained so that each pair of distinct constituent bitemporal intervals do not overlap (i.e. do not impinge into each other's interior).

To illustrate the concept of bitemporal period, Figure 8.7 shows the bitemporal period associated with the spatial chain *ef* of the bypass in Example 1 and Figure 8.3. Grey areas indicate existence in the bi-times shown on the valid time and transaction time axes. Thus, to the system in 1992, the section *ef* did not exist at any valid time, but in 1993, information had been entered that showed its existence from 1993 onwards. In 1994, this information has been revised to show that segment *ef* exists only from 1995 onwards.

Temporal operations, as with purely spatial operations, may be classified as set-theoretic, topological or Euclidean. An example of a topological temporal relationship is 'temporally overlaps' and of a Euclidean relationship is 'one year after'. Such operations have been discussed by Allen in a series of papers (for example, Allen, 1984).

Clifford and Isakowitz (1994) note some interesting points about temporal models, in particular the use of 'variables' such as **now**, $\infty$ (denoting infinite future) and $-\infty$ (denoting infinite past). They classify databases that can handle such concepts as *variable databases* and posit the need for a third temporal dimension called *reference time*, being the time that a database is observed.

### 8.2.4 Temporal information systems and query languages

Temporal information systems can be subdivided into four types: *static, historical, rollback* and *bitemporal*. Each type supports zero to two temporal dimensions. Table 8.2 shows the temporal dimension(s) that are supported in each case. *Static* systems support neither transaction nor valid time, thus providing only a single snapshot, probably the current state, of information in the application domain. Almost all proprietary information systems are essentially static systems. *Historical* systems support only valid time, and are appropriate in cases where the history of activity in the application domain is important. *Rollback* systems support only transaction time and are required for applications where the history of the information system is needed. Proprietary systems have some rollback capabilities, often in the form of a transaction log which records previous transactions in the case of system failure. However, the ability to query past states in the same way as current states is not

**Table 8.2**  System types supporting combinations of time dimensions

|            | Valid time | Transaction time |
|------------|------------|------------------|
| Static     | x          | x                |
| Historical | ✓          | x                |
| Rollback   | x          | ✓                |
| Bitemporal | ✓          | ✓                |

usually supported, as it would be in a true rollback system. *Bitemporal* systems (sometimes just termed *temporal*) support both database and event times and are the goal of much temporal information system research.

There has been an attempt to standardize the terminology associated with temporal information systems. The paper by Jensen *et al.* (1994) is the result of a consensus effort from 25 researchers and contains definitions and terminology for many of the most widely used temporal database concepts. Work is also in progress on temporal extensions to database languages. A promising temporal language is TQuel (Snodgrass, 1987), a temporal extension to Quel, the query language for the proprietary database INGRES. TQuel supports both valid and transaction time dimensions. However, the research that may eventually have the most visibility is on temporal extensions to SQL-92 and SQL3. Snodgrass *et al.* (1994) present a specification of the language TSQL2, designed to extend SQL-92: Pissinou *et al.* (1994) discusses progress on a temporal extension to SQL3.

### 8.2.5  Spatio-temporal models, representation and implementation

There are sound reasons to construct a conceptual model that brings together the purely spatial with the purely temporal by forming an aggregated object consisting of a member of the spatial hierarchy and bitemporal reference. The class of such aggregates we term the class of spatio-temporal objects. Such a class is sufficient to model many of the situations that are required in a TGIS. An approach taken by

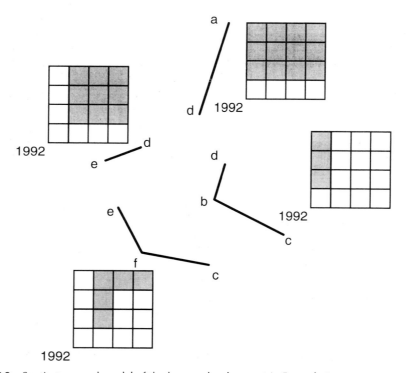

**Figure 8.8**  Spatio-temporal model of the bypass development in Example 1.

the author (Worboys, 1994a) is to form classes of primitive spatio-temporal objects by associating two-dimensional temporal elements with pure spatial objects. Composite spatio-temporal classes are created, where the composition is between primitive spatial object classes and primitive temporal object classes. In the bypass of Example 1, the spatio-temporal object class referenced by the bypass is decomposed into five spatio-temporal polylines (see Figure 8.8). Each composite object shows the bitemporal extent of a portion of the spatial layout of the bypass. For example, spatial polyline *dbc* is shown to exist in the information system in 1992 with a projected valid time life-span from 1993 onwards.

Operations may be defined that act upon spatio-temporal object types. First, there will be *projection operators* that map spatio-temporal objects to purely spatial and temporal component objects. Some of the projection operators are listed below:

**Lifetime:** Projects an ST-object onto the total temporal (or bitemporal) period for which it exists.

**Max-S-project:** Project an ST-object onto its total spatial extent, i.e. those parts of the spatial extent of the object that have existed *at some* period of the object's lifetime.

**Min-S-project:** Projects an ST-object onto its largest common extent, i.e. those parts of the spatial extent of the object that have existed *at all* periods of the object's lifetime.

Figure 8.9 shows these operators acting on the bypass of Example 1. The leftmost section gives the bitemporal period for which some part of the bypass has existed (**lifetime**). The central panel gives the total spatial extent that has existed at some time during the lifetime of the bypass (**max-S-project**). The right-hand panel shows those part of the spatial extent that have always existed during the lifetime of the bypass (**min-S-project**). Having decomposed into purely spatial and temporal objects by means of the projection operators, it is then possible to use the spatial and temporal operators already discussed. There may also be the need for operators that cannot be decomposed into actions upon purely spatial and temporal projections.

Langran (1993) discusses some of the practical questions that arise with the representation of spatio-temporal data and implementation of spatio-temporal information systems. A major issue is what to timestamp. For composite data, as spatial data usually are, there are decisions to be made on which level to associate the

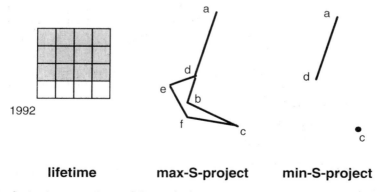

lifetime          max-S-project          min-S-project

**Figure 8.9** Projection operators applying to the bypass.

temporal references. If too high a level is chosen, then the detail of temporal variation will be missed: if the level is too low, then storage will be wasted and the system may perform at unacceptable levels.

Just as generalization is an issue for spatial GIS, so it may be applied to both the spatial and temporal dimensions. Imagine a feature that was coming and going in a quasi-periodic fashion. If a snapshot is taken, then the feature may or may not be present. It would be nice to present the nature of the temporal fluctuation of this feature in some way.

Spatio-temporal systems have been implemented: indeed, Langran (1993) discusses an automated hydrographic charting system for the US National Ocean Service. Relational databases have also been used: for example, Raafat et al. (1994) describe the logical model of a temporal relational system that will handle spatial data referenced by valid time.

### 8.3    THREE-DIMENSIONAL GIS

The Earth is usually thought of as existing in three spatial dimensions. Therefore, the current technology applying to two dimensions (plus a notional extra half dimension for surfaces) restricts the terrestrial phenomena that we can model. Examples of application areas that require a full treatment of the third spatial dimension are climatology, geology, computer-aided design and manufacturing (CAD/CAM), medical imaging and robotics.

A three-dimensional spatial information system should be able to model, represent, manage, manipulate, analyse and support decisions based upon information associated with three-dimensional phenomena. Chapter 5 described the increase in complexity of object types and operations that was necessarily incurred in moving from one to two dimensions. The transition to a third dimension results in an even greater diversity of object types and spatial relationships, not to say very large data volumes. This section surveys the kinds of issues that arise in three-dimensional GIS and points the reader to the relevant research literature.

#### 8.3.1    Euclidean three-space

Metrics and topologies may be defined for three-dimensional space, just as they can for two dimensions. We shall focus on Euclidean 3-space, $\mathscr{R}^3$, although more general or exotic spaces may also be defined. Assume a coordinatization with respect to a fixed *origin* and three mutually orthogonal axes. A *point* in $\mathscr{R}^3$ has associated with it a unique triple of real numbers $(x, y, z)$ measuring its distance from the origin in the direction of each axis. These points $(x, y, z)$ may be treated as *vectors* (see Figure 8.10). Thus, they may be added together and multiplied by scalars according to the rules:

$$(x_1, y_1, z_1) + (x_2, y_2, z_2) = (x_1 + x_2, y_1 + y_2, z_1 + z_2)$$

$$k(x, y, z) = (kx, ky, kz)$$

Given a point vector, $\mathbf{x} = (x, y, z)$, we may form its *norm*, defined as follows:

$$\|\mathbf{x}\| = \sqrt{(x^2 + y^2 + z^2)}$$

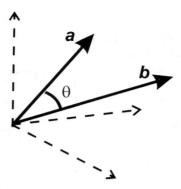

**Figure 8.10**  Vectors **a**, **b** in $\mathcal{R}^3$.

*Euclidean 3-space* has the usual measures of distance and angle. Suppose given points (vectors) **a**, **b** in $\mathcal{R}^3$, having coordinates $(a_1, a_2, a_3)$ and $(b_1, b_2, b_3)$, respectively. Then, the distance $|\mathbf{ab}|$ is given by:

$$|\mathbf{ab}| = \|\mathbf{a} - \mathbf{b}\| = \sqrt{(b_1 - a_1)^2 + (b_2 - a_2)^2 + (b_3 - a_3)^2}.$$

The *angle*, $\theta$, between vectors **a** and **b** is given as the solution of the equation:

$$\cos \theta = (a_1 b_1 + a_2 b_2 + a_3 b_3)/(\|\mathbf{a}\| \cdot \|\mathbf{b}\|)$$

Other constructions in Euclidean 3-space may be made by extending the constructions of Chapter 3. Although the definitions are usually straightforward, the extra degree of freedom allows a very large increase of possibilities for the embeddings. For example, the variety of possibilities for embedding an arc in 3-space leads to the complexities of the mathematical theory of *knots*.

### 8.3.2    Field and object models in three dimensions

As with two dimensions, three-dimensional models of geo-phenomena may be divided into field-based and object-based approaches. In the field-based approach, the spatial framework will be a portion of three-dimensional, probably Euclidean, space. Fields will then be functions from locations in the three-dimensional spatial framework to attribute domains. Thus, an atmospheric field may be a variation of temperature in a portion of 3-space.

To characterize object-based models for 3-D GIS, we need to describe the object types and operations that can act upon them. As in Chapter 4, a three-dimensional GIS model is populated by spatially referenced objects, each of which references one or more spatial objects. The definition of all possible forms of spatial object that may exist in three dimensions is non-trivial. We present a much simplified discussion.

Assume an underlying three-dimensional Euclidean space with the usual topology, in which spatial objects are embedded. Objects in three dimensions can be divided according to dimension as **point**, **arc**, **surface** and **body** objects (see Figure 8.11). The type **body** can have a high degree of topological complexity, including tunnels and internal cavities.

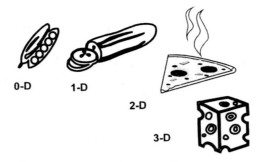

**Figure 8.11**    Objects in $\mathscr{R}^3$ of differing spatial dimensionality.

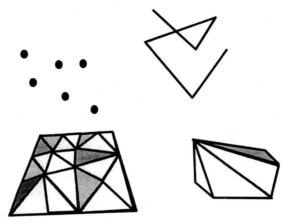

**Figure 8.12**    Discrete object types **point, polyline, discrete surface, polyhedron**, embedded in $\mathscr{R}^3$.

In discretized forms, these objects may be represented as **discrete point, discrete 1-extent, discrete surface** and **discrete body** objects (see Figure 8.12). All these objects are thought to be embedded in 3-space.

- **Discrete point:**   Defined as in Chapter 4 except that the three-dimensional nature of the embedding is expressed in the number of coordinates $(x, y, z)$.

- **Discrete 1-extent:**   Defined as in Chapter 4 as line segments, polylines and B-splines. Their embedding in 3-space may be a complex knot.

- **Discrete surface:**   A discrete surface may be defined as the topological equivalent of a disc that may be subdivided into segments, each of which is a discrete 2-extent. The discrete 2-extent most usual in the modelling of surfaces is the planar polygonal facet. Thus, a surface is modelled as a union of such facets. However, it is possible to allow a wider definition of a discrete 2-extent, when embedded in 3-space, just as a discrete 1-extent embedded in two dimensions can be a spline, for example. The corresponding surficial object is the *parametric bicubic surface*, examples of which are Hermite, Bézier and B-spline surfaces.

- **Discrete body:**   A discrete body is a topological equivalent of the unit open ball, whose boundary is a discrete surface. In the case that the surface is a union of planar polygonal facets, the body will be polyhedron. However, there are more general discrete forms.

Typologies of operations on three-dimensional objects have been studied in the literature, but have not greatly extended the two-dimensional typologies discussed in Chapter 4 (see the bibliography at the end of this chapter).

### 8.3.3 Three-dimensional representations, structures and display

The reader may recall from Chapters 5 and 6 that for 2-D objects embedded in 2-space, there are two alternative approaches to representation: either as areas (e.g. grid cells, possibly structured as region quadtrees) or using their boundaries (e.g. polygons). In a similar way, three-dimensional objects can be represented either as volumes or by means of their boundary surfaces (so-called *B-reps*).

An example of volume representation is the voxel set. A *voxel* is a simple volume, such as a cube, that may form part of a tessellation approximating to the body to be represented. This is the three-dimensional equivalent of the two-dimensional pixel. The disadvantage with a voxel representation is that no topological information about the volume is expressed explicitly, thus performance can be poor. General retrievals can be improved by the use of three-dimensional run encoding using 3-space filling curves. All the plane filling curves discussed in Chapter 6 (e.g. Morton, Peano-Hilbert and Cantor) can be extended to three dimensions. The region quadtree structure (Chapter 6) may be extended to three dimensions as the *region octree*. As with the region quadtree, the region octree is based upon the recursive regular subdivision of space until homogeneity is achieved (or we have reached the pre-defined bottom level). With an octree, the initial cube containing the volume is cut in half in three orthogonal directions at each level, resulting in eight pieces (see Figure 8.13). A very simple example of the use of the octree structure is shown in Figure 8.14, where the black nodes indicate internal nodes that need further decomposition; grey nodes indicate presence of homogeneous volume and white nodes indicate absence of homogeneous volume. Constructing a region octree can be very time-consuming, due to the large number of data items that must be processed.

Another approach to representation in terms of volumes is *constructive solid geometry (CSG)*, in which occurrences of primitive spatial three-dimensional object

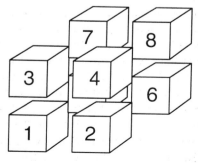

**Figure 8.13** The eight pieces of an octree subdivision.

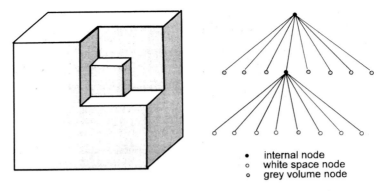

- ● internal node
- ○ white space node
- ○ grey volume node

**Figure 8.14**   A simple example of an octree structure.

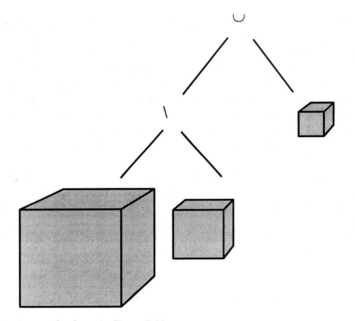

**Figure 8.15**   CSG-tree of volume in Figure 8.14.

types, such as **cube**, **cuboid**, **sphere** and **cylinder**, are combined using geometrical transformations and regularized set operations. Figure 8.15 shows an example, where the volume in Figure 8.14 is structured as a tree, providing the composition in terms of primitive spatial objects (in this case, cubes). First, the top corner cube is removed from the largest cube (indicated by set difference, \). Then the small cube is inserted (indicated by the set union operation ∪. The composition tree is not unique and may indicate the way that the object is to be constructed in engineering applications, for which CSG is a common approach. Set difference can also show parts that have to be removed, for example to make holes.

In boundary representations (B-reps) for three-dimensional spatial objects, each volume is represented by its bounding surface. There are several possible three-dimensional B-reps, which are extensions of representations of planar configurations given in Chapter 5. The surface of a volume is divided into sets of faces, edges and

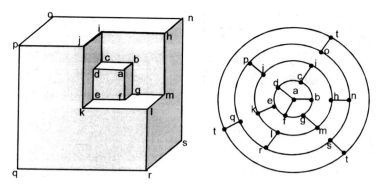

**Figure 8.16** Boundary representation of the body in Figure 8.14.

nodes. The topology of the surface is given in terms of the relationships between the components of the surface. Each face, edge and node is modelled as a discrete 2-extent, discrete 1-extent and point, respectively (see above). Figure 8.16 shows an example of a B-rep using the body from Figure 8.14, which is topologically equivalent to a ball and has no holes. Note that the three occurrences in the planar configuration of node *t*, the unseen corner of the cube, are to be identified. Only the topological relationships are shown in this B-rep: the adjacencies of the unseen faces are also not given in full. However, it would be possible to use one of the tabulations of Chapter 5, for example a DCEL structure, to fully represent all the topological information and relationships.

Regarding the display of three-dimensional information, research by the computer graphics community has led to the efficient handling of the more obvious areas, such as body illumination, texture, perspective and hidden-surface removal. Visualization techniques are harder: Eddy and Looney (1993) describe interesting applications of three-dimensional imaging to the integration of observations and analyses into interpretative models of environmental data.

## 8.4 LOGIC-BASED APPROACHES TO GIS

This section discusses logic-based approaches to GIS. Work in this area is heavily influenced by developments in formal logic and formalisms for reasoning. The modern school of formal logic is generally recognized to have been founded by Gottlob Frege (1848–1925), and followed by major contributions from the logicians Bertrand Russell, Alfred Tarski and Kurt Gödel. We introduce some of the issues and expose the reader to some of the formalism. The aim is not to make the reader an expert logician but to demonstrate the kinds of concepts and formalism required.

### 8.4.1 Knowledge and logic

Knowledge is not easy to define. Like 'object-oriented', 'knowledge-based' has been used in the computing industry to sell systems, and therefore its meaning has been weakened. Knowledge is generally taken to be one further step along the path starting with data and continuing with information. An old definition is 'justifiable, true belief': the word 'justifiable' is important, because it signals that knowledge should be supportable by argument. Examples of pieces of knowledge are:

1. London is the capital of England.
2. All cinemas in the Potteries are currently showing the film *Brief Encounter*.
3. Any road passing through Hanley will go within walking distance of the City Museum.
4. At some time in the past, the footballer Sir Stanley Matthews lived in Hanley.

A distinction that is often made is between a *fact* and a *rule*. In formal logic, the concept of a *fact* has a precise technical meaning, in that a fact is an atomic statement that cannot be divided into a logical combination of simpler statements. Facts also have the property that they are specific: 'Socrates is mortal' is a fact, but 'All people are mortal' is not. Statement 1 in the list above is a fact. Facts are the type of knowledge usually handled in traditional information systems, such as relational databases. Thus, statement 1 may well be inserted into a relation called CAPITALS, with attributes CITY and COUNTRY. Statements 2, 3 and 4 are composite statements, called *rules*. Statement 2 is also a *universal statement*, in that it makes a proposition that is true *for all* cinemas in the Potteries. To make the structure of statement 2 more explicit, we might say something like:

2′. For all *x*, if *x* is a Potteries cinema, then *x* is showing the film *Brief Encounter*.

Statement 2′ introduces the notion of a *variable*, over which quantification may take place. In this case, the quantification is *universal quantification*. The other common kind of quantification is *existential quantification*, stating the existence of something. For example:

5. There is a Potteries cinema currently showing *Casablanca*.

or

5′. There exists *x*, such that *x* is a Potteries cinema and *x* is showing the film *Casablanca*.

Statements 2 and 3 are *conditional statements*: if something holds, then something else will hold. They both involve universal quantification. Statement 4 is a *modal temporal statement*, asserting that a fact held at some time in the past. We could make statements 3 and 4 a little more precise as:

3′. For all *x*, if *x* is a road passing through Hanley, then *x* will go within walking distance of the City Museum.
4′. There exists a past time *t* such that the footballer Sir Stanley Matthews lived in Hanley at time *t*.

Traditional databases are good at handling simple facts, but apart from integrity constraints, they have no framework for the expression of more complex and possibly quantified statements. Often, a statement about information in the database (a rule) can be a much more compact form of knowledge than a list of facts. For example, in a database of airline routes, a universal statement such as: 'All flights to Aberdeen have a stop-over at Edinburgh' is more compact than ensuring that all relevant tuples in a relational database of routes have the Edinburgh stop-over data included. The following simple example illustrates some of these points.

Imagine that we are setting up a database of routing information for a UK and Ireland airline called NEW-AIR. The rules by which the airline wish to operate the routes are expressed in the following statements.

- The UK and Irish cities that NEW-AIR has flights between are London, Manchester, Dublin, Belfast and Edinburgh.
- All flights are direct flights.
- If there is a flight from city $x$ to city $y$, then there will be a flight from $y$ to $x$.
- There is a flight from London to all cities in the network.
- Belfast, Edinburgh and Dublin will each run flights to all cities in the network except Manchester.
- The only flights out of Manchester are to London.

In order that this knowledge may form part of computational processes, the statements made in natural language must be formalized. There are several formal languages in which the rules could be expressed: for example, the logic-based language PROLOG would be ideal for this purpose. For the purpose of this exposition we give a more general formalization in terms of first-order logic, with equality. Such a formal system is a collection of formulas, each of which is a logical combination of relations. Relations take variables and constants as arguments. In our example, the constants are 'London', 'Manchester', 'Dublin', 'Belfast' and 'Edinburgh'. The single binary relation is 'flight'. The statements may then be formalized as:

1. $\forall x \, \forall y.$   $\text{flight}(x, y) \Rightarrow \text{flight}(y, x)$
2. $\forall x.$   $\neg(x = \text{London}) \Rightarrow \text{flight}(\text{London}, x)$
3. $\forall x \, \forall y.$   $((x = \text{Belfast} \lor x = \text{Edinburgh} \lor x = \text{Dublin})$
   $\land \neg(y = \text{Manchester}) \land$
   $\neg(x = y))$
   $\Rightarrow \text{flight}(x, y)$
4. $\forall x.$   $\text{flight}(\text{Manchester}, x) \Rightarrow x = \text{London}$

The symbol $\forall$ is short-hand for 'for all'. The symbols $\Rightarrow$, $\lor$, $\land$, $\neg$ are logical symbols meaning 'implies', 'or', 'and', 'not', respectively. Thus, statement 1 can be read 'for all $x$ and $y$, a flight from $x$ to $y$ implies a flight from $y$ to $x$'. Similarly, the other statements can be seen to be formalizations of the natural language statements expressing the rules of the system. If this information were to be handled in a traditional relational database, all the instances that satisfy the rules would have to be entered in the form of the relation FLIGHT, shown in Table 8.3, with the routes shown in Figure 8.17.

The process of moving from premises to conclusions in logic-based systems by means of logical transitions is known as *deduction*. A classical example of such a deduction process is:

All men are mortal, Socrates is a man,
   therefore Socrates is mortal.

Logical transitions of this kind are capable of implementation in a computer, and so the deductive process is essentially mechanistic. In our example, suppose that we

**Table 8.3**    The relation FLIGHT

| FLIGHT | |
| --- | --- |
| Origin | Destination |
| London | Manchester |
| London | Dublin |
| London | Belfast |
| London | Edinburgh |
| Manchester | London |
| Dublin | London |
| Dublin | Belfast |
| Dublin | Edinburgh |
| Belfast | London |
| Belfast | Dublin |
| Belfast | Edinburgh |
| Edinburgh | London |
| Edinburgh | Dublin |
| Edinburgh | Belfast |

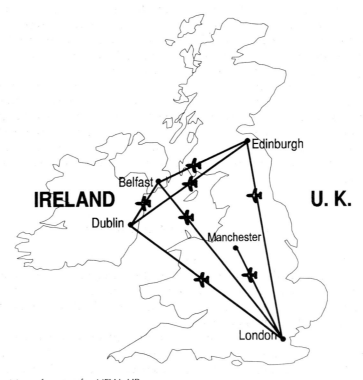

**Figure 8.17**    Map of routes for NEW-AIR.

wished to find out if NEW-AIR operated a flight from Belfast to London. The deduction could take the following form:

Use statement 2 to deduce that there is a flight from London to Belfast.

Then use statement 1 to deduce that there is a flight from Belfast to London.

The logic-based approach expresses knowledge in the form of logical theories. A *theory* comprises a set of formulas, called *axioms*, taken to be the principles upon which the theory is built. In our example, the axioms are statements 1–4. Deduction is used to derive new formulas expressing propositions, or *theorems*. In our example, we derived as a theorem the formula:

flight(Belfast, London)

The set of all formulas that are derivable from the axioms using the principles of deduction is the theory. Logicians would call the mechanistic process of the deduction of a formula from a set of axioms a *proof*.

### 8.4.2  Proof or truth: the notion of model

The reader may have noticed that in the preceding discussion about logic there has been no discussion of truth. Axioms are just formulas and deduction is merely a process for deriving new formulas from old, rather like the legal moves of chess pieces in a game of chess. Clearly, we want to be able to make our deductive systems useful by making them correspond to facts about the world. This correspondence is made by using the modelling process, and is now described.

An *interpretation* for a logical theory is a world in which the expressions of the logic are given meaning, and moreover in which the formulas of the theory are assigned truth-values, **true** or **false**. A *model* of a theory is an interpretation in which all the formulas of the theory are assigned **true**. In our example, we could think of the relation FLIGHTS as providing a model of statements 1–4, because all the statements are true when interpreted in the context of tuples of the relation. Thus, *truth can only be defined in the context of interpretations of the theory*. Interpretations give the theory meaning: the theory itself is but a formal structure without any extrinsic meaning.

### 8.4.3  Deductive database systems

Deductive databases play a key role in current logic-based approaches to information systems. Deductive databases can be thought of as generalizing the concept of a relational database. They comprise an *extension* (similar to a classical relational database) and an *intension*. The intension consists of virtual relations that are defined in terms of other relations using logic.

Suppose that the NEW-AIR airline has expanded to further cities, but has cut down some of its routes, as shown in Table 8.4. Let us define the *extension* of the NEW-AIR deductive database to the basic facts about flights shown in Figure 8.18 and represented by the list of logical formulas in Table 8.4.

For the purposes of the example, the intension of the database will contain a virtual relation '2-accessible', defined in terms of the 'flight' relation. The idea is that two cities are 2-accessible if we may fly from one to the other with exactly one stop-over. The definition of 2-accessible is by means of the formula A below, where the city $z$ in the formula acts as the stop-over.

A:  $\forall x \, \forall y. \ (\text{2-accessible}(x, y) \Leftrightarrow \exists z. \ (\text{flight}(x, z) \land \text{flight}(z, x)))$

**Table 8.4** The extension of the NEW-AIR deductive database

flight(London, Manchester)
flight(London, Dublin)
flight(London, Cardiff)
flight(London, Edinburgh)
flight(Manchester, London)
flight(Cardiff, London)
flight(Dublin, London)
flight(Dublin, Belfast)
flight(Belfast, Dublin)
flight(Belfast, Edinburgh)
flight(Newcastle, Edinburgh)
flight(Aberdeen, Edinburgh)
flight(Edinburgh, London)
flight(Edinburgh, Aberdeen)
flight(Edinburgh, Newcastle)
flight(Edinburgh, Belfast)

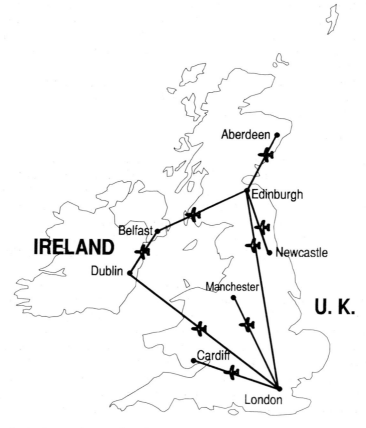

**Figure 8.18** Revised map of routes for NEW-AIR.

The important point is that we have a new relation that is defined, not by specifying tuples, but as a logical combination of other relations: hence the name 'virtual relation'. Thus, deductive databases allow the manipulation of data that are implied by the explicitly stored facts, allowing the meaning of the data to be considered in ways that cannot be accommodated by conventional databases. This ability is potentially relevant to GIS, in which much information is stored in implicit form. The use of logical formalism in a database allows the representation of data with complex structure and inter-relationships: it also allows the representation of knowledge, perhaps of some domain expert.

Computing a query in a deductive database is carried out by using deduction. A conventional database consists of a body of data in some highly structured format (tabular in the case of relational databases), defined and accessed by means of a database language. Most GIS store their information in this way, either in tabular form in a relational database or in some more specialized database structure. In contrast, a deductive database (DDB) aims to use logic as the basis for database definition, query and manipulation. It handles a body of expressions in a logic-based language that allows the expression of propositions, description of their inter-relationships, and management of their storage. It both defines the structure of the data and acts as query language and control mechanism. The database operates by applying deduction to the logical expressions to evaluate their truth or to find values that satisfy them. The significance of DDB technology is the way in which it does not distinguish between the simple statement of a fact and the expression of a relationship or rule. Thus, it greatly reduces the distinction between the data and the programs used to manipulate that data. Deductive databases can enhance simple data with a few rules, perform deductive operations using many rules, or anything in between.

With the NEW-AIR example, suppose that the database receives the query: 'Is it possible to travel from London to Newcastle with exactly one stop-over?' This query $Q$ can be expressed formally as:

$Q$:   2-accessible(London, Newcastle)?

A deductive database attempts to prove this proposition from the axioms, which in this case are the facts about relation 'flight' and the single rule A about '2-accessible'. Using rule A and working backwards (*backward chaining*) from the proposition (formula $Q$) to be proved, it becomes clear that a city $z$ is needed such that flight(London, $z$) and flight($z$, Newcastle). Retrieval from the extension of the database reveals the presence of two facts: flight(London, Edinburgh) and flight(Edinburgh, Newcastle). Thus, the formula $Q$ is proved, and the query is answered positively.

Logic and deduction allow representation of the behaviour of a system, and information about how the database may process it. This could provide sufficient expressive power in the database to allow most or all of an application to be developed within one system that would then meet a wide variety of users' needs. This increase in flexibility is due to the use of deduction to generalize the concept of data retrieval. Rules can define new information in terms of old, and both can be processed in the same way. Deductive databases are also applicable to problems of spatial analysis, since they can apply deduction to provide advanced analytical and modelling capabilities for applications that are not well suited to mechanistic, algorithmic approaches.

Deductive databases are not without their own problems. They constitute an active research area and there are several different approaches to their construction. Perhaps the simplest is to use a traditional (say, relational) database to provide large-scale permanent storage for the extension, linked to deductive capability for the intension. This is known as a *loosely coupled* deductive database. Its advantage is that it is comparatively easy to construct, but achieves little more than persistence of the initial rule base or of final results. Integration with the database is not close enough to enable the data in use to exceed the capacity of the computer's main memory. A more closely coupled approach is required, enabling the processing of data and their storage to be carried out as part of a single operation. However, building a logical language with persistent store much larger than main memory is a non-trivial task.

### 8.4.4  Non-classical logics for GIS

Designers of deductive databases must choose which of the wide variety of logics to implement. Different forms of logic vary in expressive power and in the ease with which they can be implemented. For spatial applications representation of uncertain information is useful, and logics exist that address this issue. Geographic information has spatial and temporal components, and so appropriately extended logics may be applied. For GIS, we really want not just deductive, rule-based tabular data but rule-based complex objects: a system that takes the best of object orientation, databases and deduction and adds support for powerful graphical interactive user interfaces. The system of logic used must be sufficiently expressive, yet not slow the database down with its computational demands, and it should allow a wide variety of possible data models. Data should be stored in a format that is both flexible and compact. Deductive database technology holds out hopes of being able to address and solve these complex problems. Object orientation and logic are not mutually exclusive, and there have been some attempts (Abdelmoty *et al.*, 1993) to unify them in the context of GIS.

With respect to the fundamental support that logic and deduction can provide for GIS, it has long been recognized that classical logics are insufficient to express our observations of the richness of geographic phenomena. In particular, many geographic processes are dynamic and uncertain. Some of the logic-based approaches that have been researched in the context of GIS are now briefly discussed.

*Many-sorted logics*

The classical logic described in section 8.4.1 implicitly assumes a single object type that may be used as an argument in relations. In our example, all arguments of the relations 'flight' and '2-accessible' are cities. Information systems typically contain a variety of types, and it is useful for the logic to be able to distinguish them. This extension of the formalism leads to *many-sorted logics*, where a collection of *sorts* (*types*) is defined. Each relation is assigned a *signature*, that defines how the sorts are associated with its arguments: relations are only syntactically acceptable if the sorts are correctly assigned. For example, suppose that the NEW-AIR system has **city** and **flight** amongst its sorts. Define the signature of relation 'fly' as:

fly(**flight**, **city**, **city**).

Then a particular syntactically correct formula could be:

fly(NA134, London, Manchester).

Many-sorted logics have so far had a limited use in GIS. Clarke's (1981, 1985) calculus of individuals is formulated in a many-sorted logic by Cohn (1987).

*Fuzzy logics, fuzzy set theory and fuzzy spatial objects*

Propositions in classical logic are assumed to take on truth-values **true** or **false** in interpretations. These Boolean values are hard-edged, with no room for any intermediate values. However, of course, many phenomena of interest are not so clear cut. Consider the following propositions:

1. Location $x$ is in the wooded area.
2. The A500 passes close to the nuclear power station.
3. This person will probably vote Labour at the next general election.

The first two sentences express propositions for which it may be difficult to determine a precise truth-value **true** or **false**. In the case of sentence 1, Figure 8.19 shows a fictitious area of woodland with locations $a$, $b$ and $c$. The proposition 'Location $a$ is in the wooded area' is clearly false, 'Location $c$ is in the wooded area' is true, but 'Location $b$ is in the wooded area' may not be so obvious. The proposition 'Location $x$ is in the wooded area' is an example of a *fuzzy proposition*, since it may take truth-values between **true** and **false**. Answers such as 'maybe' could be more appropriate.

Fuzzy logics have been devised (Zadeh, 1988) to handle reasoning with propositions expressed by such as statements 1 and 2. It may be useful to think of truth-value **true** as represented by the number 1, **false** as represented by the number 0, and

**Figure 8.19**   Area of woodland.

intermediate states represented by numbers between 0 and 1. Fuzziness is concerned with imprecision of measurements and propositions, and indistinctness of phenomena. Zadeh has constructed a logical calculus for determining the truth-value of a composite expression. For example, suppose we have established propositions having the given truth-values:

Location $d$ is in the wooded area.        truth-value 0.7

Location $d$ has soil of type marl.        truth-value 0.6

Then we can use Zadeh's fuzzy logic to determine the truth-values of the propositions:

Location $d$ is in the wooded area and has soil of type marl.

Either location $d$ is in the wooded area or it has soil of type marl.

Zadeh provides rules for calculating the truth-values of such composite expressions. However, these rules are often criticized as lacking proper formal foundation and being little more than 'rules of thumb'.

As well as fuzzy logic, it is also possible to devise the closely related fuzzy set theory (Zadeh, 1965), or indeed a fuzzy theory of spatial objects and operations. In the case of the fuzzy set theory, the proposition that 'element $x$ belongs to set $X$' is allowed to have a greater range of truth values than just **true** or **false**. If set $X$ were the set of tall people, then 'Mike belongs to $X$' is not clearly true or false.

To give an idea of what a fuzzy theory of spatial objects and operations might look like, let us model the area of woodland in Figure 8.20 as a **fuzzy region**. The fuzzy region will have a **fuzzy boundary**, **fuzzy interior** and **fuzzy area**. Indeed fuzziness may be applied to most of the usual spatial operations upon **region**.

A *fuzzy region* has been defined by Altman (1994) as a set of concentrations at locations over a regular grid. A *concentration* at a location is a value between zero

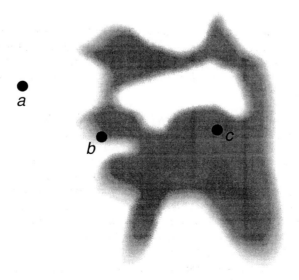

**Figure 8.20**    Fuzzy region object representing the wooded area.

**Table 8.5**  A grid representation of a fuzzy region

| | | | | | | | |
|---|---|---|---|---|---|---|---|
| 0 | 0 | 0.1 | 0.1 | 0.1 | 0 | 0 | 0 |
| 0 | 0.3 | 0.3 | 0.2 | 0.2 | 0.3 | 0 | 0 |
| 0 | 0.4 | 0.7 | 0.3 | 0.3 | 0.4 | 0.2 | 0 |
| 0 | 0.4 | 0.5 | 0 | 0.2 | 0.5 | 0.4 | 0.1 |
| 0.1 | 0.3 | 0.7 | 0.1 | 0.3 | 0.8 | 0.3 | 0 |
| 0 | 0.3 | 0.8 | 0.7 | 0.8 | 0.6 | 0.3 | 0 |
| 0 | 0.1 | 0.4 | 0.5 | 0.6 | 0.5 | 0 | 0 |
| 0 | 0 | 0.1 | 0 | 0.1 | 0.1 | 0 | 0 |

and one that indicates the degree of certainty that the given location belongs to the region. Table 8.5 shows a simple grid of concentrations indicating a central region with a small hole. Unfortunately, the general definition given by Altman provides no way of distinguishing different properties of fuzzy spatial objects, such as the concept that a region is connected (see definition of spatial object type **region** in Chapter 4). Further work in this field would be profitable.

Before leaving fuzziness, we should note that statement 3 above is not a fuzzy statement but concerns probability. The distinction between fuzziness and probability is useful and has been well expressed by Fisher (1994a) in the context of fuzzy viewsheds. In statement 3, when the person votes, she will either vote Labour or not: there is no notion of voting a shade of Labour with a dash of Conservative (assuming a single vote is cast). Even though the outcome of the event is unknown, when the event does take place, it is hard-edged (Boolean). Here, the concern is with probability. On the other hand, the boundary of a wooded area really may be fuzzy, to be distinguished from the probabilistic situation where the boundary is hard-edged but we are uncertain where it is.

*Modal logics*

A limitation of classical logic is that it admits interpretations in a single 'world'. It is not possible to make statements that may have truth values that extend over many worlds of interpretation. Modal logics allow this extension: they can be used for example to express necessity, belief, temporal and spatial propositions. Consider the following examples:

1. It is known that all simply connected regions have no holes.
2. It is believed that a Concorde is within 50 miles of landing at London, Heathrow.

Statement 1 expresses a *necessary* truth that is bound up in the definition of the concept of a simply connected region. It will be true in all possible worlds in which such a definition can make sense. On the other hand, statement 2 (a *possible* proposition) may be true at the Heathrow Air Traffic Control, but not in the cockpit of the aeroplane itself. Modal operators allow these shades of meaning to be expressed in logical formalism. The semantics of possible worlds were constructed in a beautiful piece of work by Saul Kripke (1963).

A particular instance of a modal logic is a temporal logic, where the notion of a necessary proposition translates to a proposition that is true at all times. Possibility translates to the idea that a proposition is true for at least one time. Temporal logic is formed from classical logic by the introduction of some operators that apply to formulas.

$G\phi$: $\phi$ will always be the case

$H\phi$: $\phi$ has always been the case

$P\phi$: $\phi$ has been the case at some time in the past

$F\phi$: $\phi$ will be case at some time in the future

Thus, in the world of the present (as I write this) if $\phi$ is the formula 'Margaret Thatcher is the UK Prime Minister', then $\phi$, $G\phi$ and $H\phi$ are false, $P\phi$ is true and $F\phi$ is unknown. In temporal logic, the worlds represent the state of the world at different times. Time-worlds are connected by an arrow if one immediately follows another. The temporal dimensions for these logics do not have to be structured linearly: it is possible to have parallel simultaneous worlds. There is a correspondence between parallel worlds and versioning in information systems. Figure 8.21 shows a small branching time structure. Worlds are represented by nodes, and connectivity shown by arrows. Each world is labelled by the propositions that are the case in that world (at that time). So, for example, at worlds $a$ and $b$, proposition $\gamma$ is the case. Then two scenarios are considered: world $d$ in which $\gamma$ continues to be the case, and world $c$ in which $\eta$ temporarily becomes the case, followed be the reinstitution of $\gamma$ in world $e$. Examples of modal propositions are the following.

- At world $d$, the proposition $H\gamma$ is the case, since all worlds preceding $d$ in the ordering have $\gamma$.
- At world $b$, the proposition $F\eta$ is the case, since the world $c$ in the future from $b$ has $\eta$.

There has been some research on spatial modal operators. Jeansoulin and Mathieu (1994) have introduced and analysed the spatial modal operators $L_i$, $L_a$, $M_i$, and $M_a$, with the following semantics:

$L_i\phi$: $\phi$ is the case everywhere inside

$L_a\phi$: $\phi$ is the case everywhere in the neighbourhood

$M_i\phi$: $\phi$ is the case somewhere inside

$M_a\phi$: $\phi$ is the case somewhere in the neighbourhood

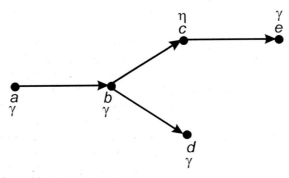

**Figure 8.21**   Possible worlds in branching time.

## BIBLIOGRAPHIC NOTES

A useful general paper that sets a research agenda for technical GIS is given by Günther and Buchmann (1990), from which the list of research topics in section 8.1 was developed.

There has been a history of research amongst the database community into support for the temporal dimensions: Tansel *et al.* (1993) provides a recent collection of papers. Soo (1991) presents a bibliography of temporal database research, updated by Kline (1993). An early paper on historic databases is by Clifford and Tansel (1985). For work in the area of bitemporal systems see Ariav (1986) and Snodgrass (1987); Snodgrass (1992) gives a good overview of these ideas. Clifford and Isokowitz (1994) discuss the need for a third *reference time* dimension for a temporal information system. The paper by Jensen (1994) contains definitions and terminology for many of the most widely used temporal database concepts. The temporal database language TQuel is described by Snodgrass (1987). Snodgrass *et al.* (1994) present a specification of the language TSQL2, designed to extend SQL-92; Pissinou *et al.* (1994) discuss progress so far on a temporal extension to SQL3.

Research into the management of spatially and temporally referenced information in a unified system is relatively novel; Al-Taha *et al.* (1993) give a bibliography. The seminal paper, with a cartographic emphasis, is Langran and Chrisman (1988), and a review of the early work is contained in Langran (1989). The standard and solitary textbook, solely devoted to a discussion of temporal GIS, is Langran (1992). A short general summary from the geographic perspective is Mackaness and Buttenfield (1991). The Cartesian viewpoint of measurable geographic phenomena as collections of tuples has been explored in the literature (Hazelton, 1991; Jensen *et al.*, 1992; O'Conaill *et al.*, 1994). The extension of the field versus object dichotomy to spatio-temporal models is developed in Worboys (1994c). The author's research into the unification of space and time in a GIS can be found in Worboys (1992b) where one-dimensional valid time intervals were associated with spatial objects formalized as simplicial complexes and Worboys (1994a) where two-dimensional temporal classes (so-called *bitemporal elements*) are linked with simplicial complexes. Cheylan and Lardon (1993) develop a generalized conceptual model in the context of polygonal patterns changing through time. Langran (1989, 1991, 1993) takes the argument on to implementation issues for a spatio-temporal information system. Raafat *et al.* (1994) describe the logical model of a temporal relational system that will handle spatial data referenced by valid time. Williams (1995) introduces the notion of a *responsive* GIS, where the data are highly dynamic and decisions must be made quickly.

The most general introductory reference for three-dimensional GIS is Raper and Kelk (1991). Raper (1989) has edited a collection of papers on this theme. There are close links in this field to computational solid modelling, details of which may be found in the texts of Requicha (1980) and Mäntylä (1988). Lienhardt (1991) provides a treatment of boundary representations in *n*-dimensional space. Pigot (1991) gives a formal topological model of 3-space-embedded objects. A more recent paper that is biased towards implementation in a relational database is Rikkers *et al.* (1994). A typology of operations on three-dimensional objects is discussed by Raper (1990). Samet (1989a) (and to a lesser extent Samet (1989b)), discusses three-dimensional data in general and with particular reference to octrees. Jones (1989) gives a summary of data structures for three-dimensional objects. Raper and Kelk (1991) provide a useful summary of architectural and interface considerations for three-dimensional systems. Eddy and Looney (1993) describe applications of three-dimensional imaging to the visualization of environmental data. Mason *et al.* (1994) describe a four-dimensional (three spatial dimensions plus time) system for environmental applications.

Introductory material on deductive databases is contained in Ullman (1988, vol. 2) and Gardarin and Valduriez (1989). Date (1995, Appendix C) has a useful summary of logic-based systems and the associated query language DATALOG. Abdelmoty *et al.* (1993a) and Abdelmoty *et al.* (1993b) indicate some areas where deductive capabilities of information systems can make a specific contribution to GIS. A general article on knowledge bases and GIS is Smith and Yiang (1991).

Fuzzy logics are described by their inventor in Zadeh (1988) and fuzzy sets in Zadeh (1965). A recent discussion of fuzzy sets for spatial analysis is Altman (1994). Fisher (1994a) discusses fuzziness in the context of viewshed analysis and makes a useful distinction between fuzziness and probability in the context of GIS. Leung and Leung (1993) describe an expert system shell based on fuzzy logic for decision support using GIS.

Modal logics are introduced in the clear and not too abstract text of Hughes and Cresswell (1968). The seminal paper on the possible world semantics of modal logics is Kripke (1963). Temporal logic is comprehensively covered in the textbook of Rescher and Urquehart (1971). Applications of modal logics to information systems are discussed in Clifford and Warren (1983) and Thayse (1989): specific applications to GIS are considered in Worboys (1991) and Jeansoulin and Mathieu (1994).

# Cinema relational database example

## CINEMA DATABASE SCHEME

CINEMA (CIN ID, NAME, MANAGER, TEL_NO, TOWN, GRID_REF)

SCREEN (CINEMA ID, SCREEN NO, CAPACITY)

FILM (TITLE, DIRECTOR, CNTRY, YEAR, LENGTH)

SHOW (CINEMA ID, SCREEN NO, FILM NAME, STANDARD, LUXURY)

STAR (NAME, BIRTH_YEAR, GENDER, NTY)

CAST (FILM STAR, FILM TITLE, ROLE)

## SAMPLE TUPLES FROM THE CINEMA RELATIONAL DATABASE

### CINEMA

| INEMA_ID | NAME | MANAGER | TEL_NO | TOWN | GRID_REF |
|---|---|---|---|---|---|
| | Majestic | Julie Jones | 01782764127 | Stoke | SJ 878450 |
| | Regal | Harry Emms | 01782749253 | Hanley | SJ 880471 |
| | Regal | Mary Carr | 01782530152 | Newcastle | SJ 852462 |

**SCREEN**

| CINEMA_ID | SCREEN_NO | CAPACITY |
|-----------|-----------|----------|
| 1 | 1 | 800 |
| 1 | 2 | 750 |
| 1 | 3 | 250 |
| 2 | 1 | 800 |
| 2 | 2 | 800 |
| 3 | 1 | 500 |

**SHOW**

| CINEMA_ID | SCREEN_NO | FILM_NAME | STANDARD | LUXURY |
|-----------|-----------|-----------|----------|--------|
| 1 | 1 | The Godfather | £3.50 | £6.00 |
| 1 | 2 | Superman | £3.50 | £6.00 |
| 1 | 3 | Last Tango in Paris | £3.00 | |
| 2 | 1 | The Godfather | £4.00 | £5.00 |
| 2 | 2 | Jaws | £3.00 | £5.00 |
| 3 | 1 | Last Tango in Paris | £4.00 | |

**FILM**

| TITLE | DIRECTOR | CNTRY | YEAR | LENGTH |
|-------|----------|-------|------|--------|
| Jaws | Spielberg | USA | 1977 | 125 |
| Star Wars | Lucas | USA | 1977 | 121 |
| American Graffiti | Lucas | USA | 1973 | 110 |
| Raiders of the Lost Ark | Spielberg | USA | 1981 | 115 |
| A Bridge Too Far | Attenborough | UK | 1977 | 175 |
| Manhattan | Allen | USA | 1979 | 96 |
| Kramer v Kramer | Benton | USA | 1979 | 105 |
| The Deer Hunter | Cimino | UK | 1978 | 182 |
| The Great Gatsby | Clayton | UK | 1974 | 140 |
| The Godfather | Coppola | USA | 1972 | 175 |
| French Lieut's Woman | Reisz | UK | 1981 | 123 |
| All the President's Men | Pakula | USA | 1976 | 138 |
| Last Tango in Paris | Bertolucci | IT/FR | 1972 | 129 |
| Superman | Donner | GB | 1978 | 143 |
| Marathon Man | Schlessinger | USA | 1976 | 126 |

## STAR

| NAME | BIRTH_YEAR | GENDER | NTY |
|---|---|---|---|
| Marlon Brando | 1924 | M | USA |
| Harrison Ford | 1942 | M | USA |
| Dustin Hoffman | 1937 | M | USA |
| Robert Redford | 1937 | M | USA |
| Meryl Streep | 1949 | F | USA |

## CAST

| FILM_STAR | FILM_TITLE | ROLE |
|---|---|---|
| Marlon Brando | The Godfather | Don Vito Corleone |
| Marlon Brando | Last Tango in Paris | Paul |
| Marlon Brando | Superman | Superman's father |
| Harrison Ford | Star Wars | Han Solo |
| Harrison Ford | Raiders of the Lost Ark | Indiana Jones |
| Harrison Ford | American Graffiti | Bob Flafa |
| Dustin Hoffman | All the President's Men | Carl Bernstein |
| Dustin Hoffman | Marathon Man | Babe Levy |
| Dustin Hoffman | Kramer v Kramer | Ted Kramer |
| Robert Redford | All the President's Men | Bob Woodward |
| Robert Redford | A Bridge Too Far | |
| Robert Redford | Great Gatsby | Gatsby |
| Meryl Streep | Manhattan | Jill |
| Meryl Streep | Kramer v Kramer | Joanna Kramer |
| Meryl Streep | The Deer Hunter | Linda |
| Meryl Streep | French Lieut's Woman | Sarah/Anna |

# Some standards relating to GIS

## SQL STANDARDS

SQL is universally acknowledged as the primary relational database language. The original language was captured in the language SQL1 (American National Standards Institute, 1986) and modified as SQL/89. The current (in 1995) standard for SQL is the International Standard Database Language SQL (1992), usually known as SQL/92 (sometimes SQL2). The large document that describes the language in detail is (International Organization for Standardization (ISO) 1992). A coverage of the features of SQL/92 is given in Date (1995) and Date and Darwen (1993).

SQL/92 has no features that are specially designed for use with GIS. However, the emergent SQL3, due to become an international standard in 1996 or 1997, will have such features. Work so far would indicate that SQL3 will incorporate some object-oriented features, including the capability to support user-defined types, subtypes, operations, polymorphism and object-identity. Management of spatial data will be included within SQL3 as SQL Multimedia and Applications Packages (SQL/MM), which will also handle image, graphics, voice and video. Thus, in SQL3 it should be possible to define spatial object types, such as **polygon**, subtypes such as **rectangle** and supporting operations such as **adjacent**. Further details of the general scope of SQL3 may be found in (American National Standards Institute (ANSI) Database Committee (X3H2), 1994). A discussion of some general difficulties of introducing full object-oriented functionality into SQL3 is contained in Mattos and DeMichiel (1994).

## OBJECT-ORIENTED STANDARDS

The Object Database Standard ODMG-93, defines a standard for object-oriented database management systems that includes both data model and database language. It has been argued that ODMG-93 is to OODBMS as SQL is to RDBMS. ODMG-93 handles:

- object model language (OML): basic object-oriented concepts such as object type and instance definition; operations, inheritance hierarchy, transactions on objects.

- object definition language (ODL): extending the concept of the data definition language of traditional databases; ODL syntax.
- object query language (OQL): syntax, query input and result.
- bindings to object-oriented programming languages: OML, ODL and OQL bindings to C++ and Smalltalk.

A book that discusses ODMG-93 in detail is by Cattell (1994). A wider object-oriented standards group is the Object Management Group (OMG) and its associated Common Object Request Broker Architecture (CORBA).

### SPATIAL DATA EXCHANGE STANDARDS

Spatial data exchange standards must have the capability to transfer vector and raster spatial data with graphical and cartographic characteristics. Purely graphical data have their own standards. These include the Graphics Kernel System (GKS) (two-dimensional data: ISO 7942; three-dimensional data: ISO 8805) and the Programmer's Hierarchical Interactive Graphics Standard (PHIGS) (ISO 9582).

A collection of standard structures for compressing, holding and transmitting images has been developed over the last few years. Most started as proprietary formats, linked to a single company or range of products. Some have developed into general purpose structures, available for transfer between a wide range of applications and products. Common examples follow.

- The CompuServe Graphics Interchange Format (GIF) is a general purpose raster format constructed to allow transmission of images between stations in a network. GIF is the legal copyright of CompuServe. The compression scheme that it uses is a variant of the lossless LZW compression (Section 1.4.3).
- The Aldus Tagged Image File Format (TIFF), widely used in desktop publishing, is a complex and multi-purpose raster structure designed by Aldus and Microsoft. TIFF allows a wide variety of image representations, colour models and compression schemes. A TIFF file has an associated *tag* that describes its dimensions, colour model, bits per cell, compression method, orientation, resolution, and so on. Four types of images are supported: bilevel, grey-scale, colour-mapped (requiring a palette) and RGB-based. Compression type can be uncompressed, RLE or modified Huffman (Section 1.4.3).
- The Encapsulated PostScript Format (EPS) is designed for use within the PostScript language as a means of communicating graphics, as well as more traditional alpha-numeric data forms. It is a development of Adobe Systems. Like any PostScript program, an EPS program may have text, vector graphics and raster images.
- JPEG is our only example of a raster format that uses a lossy compression technique. Unlike the compression of computer programs using for example **zip**, where no information loss may be permitted, images do allow the loss of a certain amount of data before the viewer is able to detect the difference. Compression in JPEG takes account of human perception of images. JPEG is designed especially for continuous tone (grey-scale and colour) images. With such images, compression ratios of 20:1 can be achieved (compared with 2:1 or 3:1 for other approaches).

A standard for digital cartographic data exchange was defined by the US Digital Cartographic Data Standards Task Force (DCDSTF) (1988). The Spatial Data Transfer Specification (SDTS) handles the transfer of raster and vector data as part of the overall DCDSTF proposal. SDTS specifies a set of transfer modules related to global information, data quality, attributes, vector and raster. The United Kingdom standard for the transfer of digital cartographic data is the National Transfer Format (NTF), released in 1987 and updated in 1989. NTF will handle raster and vector data in a tiered approach to transfer. Levels 0–5 allow the transfer of raster, simple vector, complex vector, topology and user-defined formats, respectively. A reference that describes the background to these standards and provides pointers to further reading is Guptill (1991).

A recent Canadian standard in this area is the Spatial Archive and Interchange Format (SAIF). The underlying model of SAIF has evolved from the extended relational model to an object model similar to ODMG-93. SAIF is having a significant input into the spatial requirements of the emerging SQL3 standard. For further details see Surveys and Resource Mapping Branch (1992), Friesen et al. (1993) and Sondheim (1993).

# Sources for further reference

## JOURNALS

*International Journal of Geographical Information Systems*, Taylor & Francis, London, England.

*Communications of the Association for Computing Machinery*, ACM Press, New York, USA.

*Association for Computing Machinery SIGMOD Record*, ACM Press, New York, USA.

*Association for Computing Machinery Transactions on Database Systems*, ACM Press, New York, USA.

## CONFERENCES

Symposium on the Design and Implementation of Large Spatial Databases (SSD): biennial international conference with a computer science emphasis, proceedings published by Springer-Verlag, Berlin, Germany.

International Symposium on Spatial Data Handling (SDH): biennial international conference, alternating with SSD, with a wide disciplinary coverage, latest proceedings published by Taylor & Francis, London, England.

Conference on Spatial Information Theory (COSIT): European conference with a wide disciplinary coverage and emphasis on theory, proceedings published by Springer-Verlag, Berlin, Germany.

GIS Research UK (GISRUK): annual UK research conference with a wide disciplinary coverage, book based on proceedings published as the *Innovations in GIS Series* by Taylor & Francis, London, England.

Association for Computing Machinery SIGMOD Conference (SIGMOD): annual US database conference, proceedings published by ACM Press, New York, USA.

European Geographical Information Systems (EGIS) Annual Conference (EGIS): Proceedings published by the EGIS Foundation, Utrecht, Netherlands.

# Glossary

**Access methods:** The means of getting data items into and out of computer storage. There are four main types: sequential, direct, random and associative. Also, the supporting data structures and indexing techniques.

**Accuracy:** In modelling terms, the degree of match between source and target domains in a model.

**Adjacency:** A topological relationship between areas indicating the sharing of a common arc boundary.

**Affine transformation:** A spatial transformation that preserves the affine properties of embedded objects such as parallelism.

**Aggregation:** The grouping together of several constituent objects to form an aggregate object.

**Algorithm:** A finite sequence of well-defined operations that constitutes a procedure to implement an operation.

**Aliasing:** Unwanted artefacts of sampling a rapidly changing continuous function at discrete intervals. A particular manifestation in GIS is jagged boundaries and staircase effects.

**American National Standards Institute (ANSI):** The US standards body.

**American Standard Code for Information Interchange (ASCII):** A code for information transmission that provides for up to $2^7 = 128$ numeric, alphabetic and control characters.

**Antialiasing:** Reduction of *aliasing*.

**Applications domain:** A specific applications area, such as land-use planning, health care or electricity supply, for which GIS may provide support.

**Applications software:** Software that specifically supports the applications domain for which it has been written. Distinguished from *generic software*, such as a database management system, graphics system, or CAD system.

**Arc:** Principal one-dimensional spatial object, a continuous image of a straight line segment.

**Area:** Principal two-dimensional spatial object, a finite and not necessarily connected collection of *regions*. A regular closed set.

**B-tree:** A balanced multi-level dynamic index to a file. The $B^+$-tree is a refinement where pointers to records are stored only at the leaf nodes.

**Bijection:** A function that is both a surjection and an injection (section 3.3.3).

**Binary Large Object (BLOB):** A data item, usually of large volume, without any internal structure.

**Binary relation:**   A subset of the product of two sets whose ordered pairs show the relationships between members of the first set and members of the second set.

**Binary search:**   A logarithmic time search strategy where a structure is successively divided and subdivided until the item is found or known to be absent.

**Boundary of a set S, ($\partial$S):**   Set-theoretic difference of the closure and the interior of S, in a topological space.

**Boundary representation (B-rep):**   A representation of a geometric object in terms of its boundary, explicitly holding information on structural relationships such as incidence and adjacency between its component parts.

**Buffer:**   Area containing locations within a given range of a given set of features. Buffers are commonly circular or rectangular around points, and corridors of constant width about linear and areal features.

**Bus:**   A communication channel between two or more internal modules of hardware within a computer system. A **hardware buffer** is an area of main memory that holds a block of data.

**Cartesian plane:**   Set of ordered pairs $(x, y)$, where $x, y$ are real numbers. Often written as $\mathcal{R}^2$.

**Cell tree:**   A data structure designed to handle complex planar vector-based objects (polylines and polygons).

**Cell:**   A two-dimensional areal spatial object, topologically equivalent to a disc. Also, the primitive unit of a raster.

**Central Processing Unit (CPU):**   A module of hardware that handles the processing of data in the computer. The CPU is itself made up of several subcomponents: Arithmetic-Logic Unit (ALU), control unit and register file.

**Centroid:**   The centre of gravity of an areal object, sometimes used as its point reference.

**Chain code:**   A raster structure that represents the raster boundary of a region by a starting point and a sequence of cardinal directions (north, south, east, west) to follow around the boundary.

**Classification:**   A construct used in object modelling which allows the formation of an *object class* that contains objects with common structures and behaviours.

**Client-server computing:**   A form of distributed computing in which some stations act as servers and some as clients. A *server* is a station that holds information, usually in a database system, available for transmission to *clients* in the network.

**Closed polyline:**   A polyline with no extreme points.

**Closed set:**   The complement of an open set in a topological space. A closed set contains all its limit points and boundary.

**Closure of a set S (S⁻):**   The intersection of all closed sets containing S, in a topological space.

**Command entry:**   An interaction style where the user is presented with a *system prompt*, indicating that the system is awaiting the user's command (e.g. MS-DOS).

**Complexity of an algorithm:**   The amount of computation time or space the algorithm takes as a function of the input size.

**Composite object:**   An object made up of constituent objects, but treated as a unity. Particular types of object composition are aggregation, association and ordered association.

**Computability:**   A function or problem is computable if it is possible to find some algorithm to compute the function or solve the problem.

**Computational geometry:**   The study of the properties of algorithms for solving geometrical problems.

**Computer network:**   A collection of computers, called *stations*, each attached to a *node*. The nodes are communications devices acting as transmitters and receivers, capable of sending and receiving data. Computer networks are divided into two types, wide-area networks (WAN) and local-area networks (LAN).

**Conceptual data model:**   A data model that is independent of computational paradigm and implementation.

**Connectedness:**   The property of a spatial object $S$ that allows navigation between any two points of $S$ entirely within $S$.

**Connectivity:**   The structure of nodes and edges in a network.

**Constructive solid geometry (CSG):**   Three-dimensional representation in which occurrences of primitive spatial three-dimensional object types, such as **cube, cuboid, sphere** and **cylinder**, are combined using geometrical transformations and regularized set operations.

**Convex polygon:**   A polygon whose interior is a *convex* set.

**Convex set:**   A pointset in Euclidean space which has the property that between any two points of the set may be drawn a straight line segment that lies entirely in the set.

**Coordinatized spatial data model:**   A generic spatial data model consisting of a set of points, each of which has associated a tuple of real numbers which are measurements of distances from a fixed set of lines (called the coordinate frame).

**Data model:**   A model of the structure of the information system, independent of implementation details, and used as a basis for employing algorithms on the data.

**Data quality:**   Lineage, accuracy, completeness, consistency, appropriateness and currency of data.

**Data structure:**   A description of the logical organization of the data in the database. Spatial databases use specialized data structures to provide efficient storage of and access to the spatial references.

**Database management system (DBMS):**   The software system that manages the database. The DBMS handles data definition, manipulation and retrieval, transaction management, performance monitoring, back-up and recovery.

**Database:**   A unified computer-based collection of data, shared by authorized users, with the capability for controlled definition, access, retrieval, manipulation and presentation of data within it.

**Delaunay triangulation:**   A triangulation of a region of the plane such that the circumcircles of the triangles contain no members of the vertex set in their interior.

**Digital elevation model (DEM):**   A special case of the DTM, where the application is to topographical elevation of a surface.

**Digital terrain model (DTM):**   A layer-based model consisting of a single layer, where the underlying spatial framework approximates to the continuum and the layer function is single-valued and represents a continuous variation over the region (see DEMs).

**Digitizer:**   Data capture device that converts an analogue spatial data source to a digital dataset with a vector structure.

**Discretization:**   The process of converting a continuous field or object so that it is computationally tractable. For example, conversion of a smoothly curved arc to a polyline.

**Distributed database:**   A database based upon a distributed system that behaves to the user as if the data are stored in one physical repository even though in reality the data are scattered around the nodes.

**Distributed system:**   A collection of autonomous computers linked in a network, together with software that will support the integration.

**Dithering:**   The clustering of neighbouring pixels of an image cell into a single unit with a wider range of output-values. For example, a bilevel output device can be used to output a multi-level grey-scale image.

**Doubly-connected edge list (DCEL):** A representation that provides a complete representation of the topology of a connected planar graph.

**Encapsulated PostScript Format (EPS):** Designed for use within the PostScript language as a means of communicating graphics.

**Encapsulation:** A feature of objects in an object-oriented system. The internal state of an object is shielded from other objects, which can only communicate with the original object by means of a predefined set of messages.

**Entity occurrence:** A particular instance of *entity type*.

**Entity type:** A basic component of the E-R model. An *entity type* is an abstraction of a collection of similar objects, about which the system holds information.

**Entity-Relationship (E-R) data model:** A conceptual data model, introduced by Chen in 1976, based on the modelling constructs *entity type*, *attribute type* and *relationship type*.

**Equivalence relation:** A binary relation that is reflexive, symmetric and transitive.

**Euclidean plane:** A Cartesian plane upon which the usual Euclidean distance and angle measures are defined. Two-dimensional Euclidean space.

**Euclidean space:** A coordinatized space admitting measurements of distances and bearings between objects.

**Euclidean transformation (congruence):** A spatial transformation that preserves the shape and size of embedded objects.

**Extended Entity Relationship Model (EER model):** A modelling approach which is based upon entities, attributes and relationships, but contains additional modelling constructs, such as the type hierarchy.

**Extensible RDBMS:** An RDBMS enhanced with facilities such as user-defined data types and indexes.

**Field tree:** A data structure designed to handle complex vector-based objects (polylines and polygons), based on a collection of regular rectilinear grids overlaying the plane.

**Field-based models of spatial data:** Geographic phenomena are treated as *spatial fields* over a *spatial framework*.

**File organization:** The logical and the physical organization of the records in a file.

**File:** A sequence of records, usually all of the same type, held in secondary storage.

**Fixed-grid:** Data structure that partitions a planar region into equal sized cells, such that each packet of data whose point locations share the same cell of the grid is stored in a contiguous area of secondary storage.

**Form:** An interaction style with an information system, where the interaction is constrained to consist of entering several heterogeneous pieces of information into a form. Forms may be *pull-down*, *roll-up* and *pop-up*.

**Function:** A binary relation with the property that each member of the first set (*domain*) relates to exactly one member of the second set (*codomain*).

**Fuzzy logic:** An extension of classical logic where the truth value of a proposition may be in the real interval between 0 (false) and 1 (true).

**Generalization (in data modelling):** The process of forming a supertype from one or more entity or object types.

**Geo-relational model:** A form of hybrid GIS architecture where the spatial data are stored in a set of system files and the non-spatial data stored in a relational database.

**Geographic information system (GIS):** A computer-based information system that enables capture, modelling, manipulation, retrieval, analysis and presentation of geographically referenced data.

**Global Positioning System (GPS):** Allows capture of terrestrial position and vehicle tracking, using a network of navigation satellites.

**Graph:** A structure consisting of nodes and edges such that each edge connects two nodes, also called a *network*.

**Graphical User Interface (GUI):** Enables the user to interact with the computer system by pointing to pictorial representations (*icons*) and lists of menu items on the screen. Often associated with a *direct manipulation* interaction style.

**Graphics Interchange Format (GIF):** A general purpose format constructed to allow transmission of images. The legal copyright of CompuServe.

**Graphics Kernel System (GKS):** Graphics standard, 2D in 1985, 3D in 1988.

**Graphics:** Includes 'the creation, storage, and manipulation of models and images of objects' (Foley *et al.* 1990).

**Grid-file:** An extension of the fixed grid structure that allows the vertical and horizontal subdivision lines to be at arbitrary positions, taking into account the distribution of the data points.

**Hashed file:** Organized using a *hash function* that transforms the values of the *hash fields* of each record into an address of a disk block into which the record is placed. To search for records, the hash function is applied to the search value and the address of the disk block is calculated.

**Homeomorphism:** A spatial transformation that preserves topological properties of embedded objects.

**Human-computer interaction (HCI):** The study of computer systems with the emphasis on the human users. HCI includes the user interface, the psychology of humans when interacting with machines, training, organization and health issues.

**Hybrid GIS architecture:** The management of the spatial data independently and in a different module from the non-spatial data.

**Integrated (unified) GIS architecture:** The management of spatial and non-spatial data in a unified fashion in a single database.

**Interior of a set S ($S^{\circ}$):** The union of all the open sets contained in S, in a topological space.

**Internet:** An interconnected collection of networks (called *subnetworks*), each of which retains its own characteristics, but the entire collection appearing as a single network to a user.

**Isoline:** Spatial fields are often visualized as a set of isolines. An *isoline* is the locus of all points in the field with the same attribute value.

**Joint Photographic Experts Group (JPEG):** An image transfer format that uses a lossy compression technique.

**Layer-based data model:** A spatial data model which is a collection of layers.

**Layer:** A computable function from a spatial framework to a set of attributes, indicating the variation of the attribute over the region of space covered by the spatial framework.

**Linear search:** A linear time search strategy where a structure is searched exhaustively, item by item, until the required data are found or known to be absent.

**Logical data model:** A model of the structure of the information system based on a particular database paradigm, e.g. relational. The model is independent of implementation details.

**Lossless compression:** A compression technique where we can reconstruct precisely the original after compression.

**Lossy compression:** A compression technique where we cannot reconstruct precisely the original after compression.

**Many-sorted logic:** An extension to classical logic where constant and variable terms may be chosen from a collection of *sorts*.

**Menu:** A list of options displayed on the screen, from which the user selects one or more, thus determining the next state of the system. Menus may be *pull-down, roll-up* and *pop-up*.

**Message:** A communication between objects in an object-oriented system.

**Method:** The implementation of an operation in an object-oriented system.

**Metric spatial data model:** A generic spatial data model consisting of a set of elements, such that for each pair of elements, it is possible to associate a distance subject to certain mathematical conditions.

**Minimum bounding box:** The smallest rectangle, sides parallel to axes, that contains a bounded planar object.

**Modal logic:** A logic for which interpretations may admit more than one possible 'world'. Examples of modal logics are logics of knowledge, belief and temporality.

**Model:** A model is an artificial construction in which parts of one domain (*source domain*) are represented by means of a structure preserving function (*morphism*) in another domain (*target domain*).

**Monotone chain:** An ordered list of $n$ points in the Euclidean plane such that the projection of the points onto some straight line preserves the ordering of the list.

**Monotone polygon:** A polygon, the boundary of which may be split into two polylines, such that the chain of vertices of each polyline is a monotone chain.

**Multimedia:** The extension of the range of computational data types beyond numbers and basic text to still and animated computer graphics, still and animated images (video), audio and structured text (e.g. hypertext).

**Network:** See graph.

**Node-Arc-Area (NAA) representation:** A topological structure that represents explicitly the adjacency relationships between areas in a subdivision of a surface. Primary constituent entities are **directed arc, node** and **area**.

**Normalization:** In a relational database, normalization concerns the design of an appropriate logical data model. There is a sequence of *normal forms*, imposing conditions upon the structure of the logical model.

**Object:** An independent item (or collection of items) of a data model, which has associated with it an *object type* and a collection of methods.

**Object class:** Implementation of object type in an object-oriented system.

**Object identity:** A fundamental property of objects in an object-oriented system. Each object has a unique identity, independent of the values of any of its instance variables.

**Object interface:** The means by which the visible properties and behaviour of objects in the object model may be observed and changed. All objects in the same class have the same object interface.

**Object type:** A generic name for a set of objects, all of which have the same object interface.

**Object-based model of spatial data:** A data model founded on the concept of an *object*. The geographic space is modelled as populated by a collection of objects that have properties, behaviours and relationships with other objects in the space.

**Object-oriented approach:** Generic name for heterogeneous activities, such as system modelling (OODM), programming (OOPLA), database systems (OODBMS) and GIS (OOGIS). The approach is based upon objects, their properties and behaviours in relation to other objects.

**Open set:** A distinguished set in a topological space that does not contain any of its limit or boundary points.

**Open Systems Interconnection (OSI):** An ISO standard model for distributed computing. OSI is a layered model, each of the seven layers dealing with a separate component or service required, from level 1 (physical) to level 7 (application).

**Operation:** Specifies a functional action on object types that returns other object types.

**Operation polymorphism:** An object-oriented construct, allowing the same operation to be implemented in different ways (by different methods) in different classes.

**Order relation:** A transitive and anti-symmetric relation.

**Overlay:** A Boolean combination of one or more layers by superimposition into a single layer.

**Pixel:** An element of a raster representation of a spatial field.

**Point quadtree:** A type of quadtree for use as a point object data structure.

**Pointset (analytic) topology:** The branch of topology for which the focus is on sets of points and in particular on the concepts of neighbourhood, nearness and open set.

**Polygon (simple):** The area enclosed by a simple closed polyline.

**Polyline:** A finite set of line segments (called *edges*) such that each edge end-point is shared by exactly two edges, except possibly for two points, called the *extremes* of the polyline.

**Precision:** The level of granularity in which measurements can be made in a domain.

**Programmer's Hierarchical Interactive Graphics System (PHIGS):** Full 3D graphics, established in 1988.

**Quad-edge data structure:** Provides representations of subdivisions of not necessarily orientable surfaces without boundaries.

**Quadtree:** Generic term for a variety of data structures that recursively subdivide a region of space, represented by trees in which each non-leaf node has four descendants.

**Quaternary Triangular Mesh (QTM):** Recursively approximates locations on the surface of a sphere by a nested collection of equilateral triangles.

**R-tree ($R^+$-tree):** A multi-dimensional extension of the B-tree that may be applied to index a large collection of points in multi-dimensional space.

**Raster data structure:** The representation of a field as a regular grid of pixels.

**Rasterization:** Conversion to a raster (vector-to-raster conversion).

**Region:** A two-dimensional spatial object, that is a connected regular closed set.

**Region quadtree:** A quadtree (trie) that is used to structure areal image data.

**Regular closed set:** A set $X$ with the property that $X^{\circ -} = X$.

**Relation:** A finite set of tuples associated with a relation scheme in a relational database.

**Relation scheme:** A set of attribute names and a mapping from each attribute name to a domain in a relational database.

**Relational algebra:** The operational structure acting upon relations. Operations include union, intersection, difference, project, restrict, join and divide.

**Relational database management system (RDBMS):** The DBMS that manages a relational database.

**Relational database scheme:** A set of relation schemes in a relational database.

**Relationship:** Connection between entities in the E-R model.

**Remote sensing:** The capture of digital data by means of sensors on satellite or aircraft that provide measurements of reflectances or images of portions of the Earth.

**Resolution:** The *resolution* of a measurable domain is the smallest measurement that it is possible to register in that domain.

**Run-length encoding (RLE):** An image compression technique that counts the runs of equal values in cells of the image and stores the counts.

**Scanner:** Data capture device that converts an analogue data source (usually a printed map, in the case of GIS) into a digital dataset.

**Seamlessness:** A property of an information system allowing users to navigate through it without meeting any artificial boundaries, such as map sheet edges or 'boundaries of knowledge' in a particular database connected to the network.

**Seek time:** When retrieving a block of data from a disk, the time taken for the mechanical movement of the disk heads across the disk to the correct track.

**Segment tree:** A binary tree data structure for a finite set of bounded intervals on the real line that makes efficient retrievals of sets of intervals containing a given point.

**Semi-convex set:** A pointset in Euclidean space having the property that there is a point in the set from which may be drawn a straight line segment to any other point in the set that lies entirely in the set.

**Set-theoretic data model:** A data model founded upon the concepts *element* and *set*. These basic constructs may be connected using the *membership* relation.

**Similarity transformation:** A spatial transformation that preserves the shape but not necessarily the size of embedded objects.

**Simple arc:** A one-dimensional spatial object that is topologically equivalent to a straight line segment.

**Simple loop:** A one-dimensional spatial object that is topologically equivalent to a circle.

**Simple polyline:** A polyline such that no two of its edges intersect at any place other than possibly at their end-points.

**Simple-connectedness:** A property of a connected areal spatial object that has no holes.

**Simplex:** Primitive spatial object types in 0-, 1-, 2-, 3-, . . . dimensional space: point, line segment, triangle, tetrahedron.

**Simplicial complex:** A combination according to prescribed rules of simplices, used as a basis for the formal theory of spatial data models, studied in combinatorial topology.

**Spaghetti data structure:** Representation of a planar configuration of points, arcs and areas as a set of lists of straight line segments. Each such list is the discretization of an arc that might exist independently, or as part of the boundary of an area.

**Spatial analysis:** The application of quantitative techniques to the analysis of spatially referenced information.

**Spatial autocorrelation:** A property of a layer. The degree to which near elements in the underlying spatial framework are near in attribute values.

**Spatial data model:** A data model of information with a spatial component. Spatial variation over a spatial framework is modelled using *field-based spatial data models*. Spatially referenced information may be modelled using an *object-based spatial data model*.

**Spatial field:** A computable function from a spatial framework to a finite attribute domain.

**Spatial framework:** A partition of the given region into a finite tessellation of spatial objects called *locations*.

**Spatio-temporal (ST) information system:** An information system that handles data referenced to both space and time.

**Specialization (in data modelling):** The process of forming one or more subtypes of an entity or object type.

**Structured or Standard Query Language (SQL):** A relational database interaction language that has facilities for data definition, manipulation and control.

**Subtype:** Type $T_1$ is a *subtype* of type $T_2$ if, and only if, each occurrence of $T_1$ is an occurrence of $T_2$.

**Supertype:** Type $T_1$ is a *supertype* of type $T_2$ if, and only if, each occurrence of $T_2$ is an occurrence of $T_1$.

**Tagged Image File Format (TIFF):** A multi-purpose transfer structure designed for raster image formats by Aldus and Microsoft.

**Temporal Information System:** An information system that handles temporally referenced information. Types of temporal system include *historical, rollback* and (*bi*)*temporal*.

**Tessellation:** A partition of the plane or portion of a surface as the union of a set of disjoint areal objects.

**Thresholding:** Converts the intensity of each element of an image to one of two levels, to produce a bilevel image. The converse of dithering.

**Topological spatial data model:** A generic spatial data model consisting of a set of elements, and a collection of sets of these elements, called *open sets*. Topological spaces allow the formulation of general spatial concepts, such as continuity, boundary and connectedness.

**Topological transformation:** A spatial transformation that preserves topological properties of embedded objects.

**Topology:** The study of topological transformations and the properties that are left invariant by them.

**Transaction time dimension** (also called **database time** or **system time**): The time when transactions take place with an information system.

**Transaction:** An atomic unit of interaction between user and database, usually either insertion, modification, deletion or retrieval of data from the database.

**Transitive closure:** An operation on an ordered set that places a direct link between two elements of the set if one is less than the other. In terms of networks, transitive closure augments the edge set of a network by placing an edge between two nodes if they are connected by some path.

**Triangulated Irregular Network (TIN) model:** A layer-based spatial data model, where the underlying spatial framework is a tessellation of irregular triangles. The simplicity of the model is attractive for modelling, for example, topographical altitude, where calculations of slope are easily performed.

**Triangulation:** A set of points in the Euclidean plane that is topologically equivalent to the planar embedding of a simplicial complex.

**Trie:** A tree structure based upon subdivisions of space independent of the position of the objects occupying the space, for example the region quadtree.

**Type:** A modelling abstraction which allows the collection of occurrences of objects or types into a single construct.

**Valid time dimension** (also called **event time** or **valid time**): When the events occur in the application domain.

**Vector data structure:** Representation of an object-based model as a collection of nodes, arcs and polygons.

**Vectorization:** Conversion to the vector data model (raster-to-vector conversion).

**Versioning:** An approach to transaction management in a multi-user environment, developed within the CAD industry. Users have versions of the entire database, stored in the database; different versions share most of the data but changed records are kept separately. Periodically, versions are integrated and conflicts resolved.

**Visualization:** The synoptic representation of information for the purpose of recognizing, communicating and interpreting pattern and structure.

**Voronoi diagram:** A tessellation of a plane into a set of proximal polygons, dual to the Delaunay triangulation.

**Winged-edge:** A boundary representation of a discretized spatial object. See *doubly-connected-edge-list*.

**World-Wide Web:** Runs on the Internet displaying hypermedia objects to members, so that routes through to new documents and related information on different servers on the Internet may be traced.

**X Window System:** A widely used application of the client/server model of distributed computing.

# References

ABDELMOTY, A. I., WILLIAMS, M. H. and PATON, N. W. (1993a). Deduction and deductive databases for geographical information handling. In Abel, D. and Ooi, B. C. (Eds), *Advances in Spatial Databases, Proceedings of SSD'93, Singapore, Lecture Notes in Computer Science 692*, pp. 443–464, Berlin: Springer-Verlag.

ABDELMOTY, A. I., WILLIAMS, M. H. and QUINN, J. M. P. (1993b). A rule-based approach to computerized map reading, *Information and Software Technology*, **35** (10), 587–602.

ABEL, D. J. (1989). SIRO-DBMS: A database toolkit for geographical information systems, *International Journal of Geographical Information Systems*, **3**, 103–115.

ABEL, D. J. and MARK, D. M. (1990). A comparative analysis of some two-dimensional orderings, *International Journal of Geographical Information Systems*, **4** (1), 21–31.

ABITEBOUL, S., HULL, R. and VIANU, V. (1995). *Foundations of Databases*, Reading, MA: Addison-Wesley.

AHO, A. V., HOPCROFT, J. E. and ULLMAN, J. D. (1974). *The Design and Analysis of Computer Algorithms*, Reading, MA: Addison-Wesley.

AKIMA, H. (1978). A method of bivariate interpolation and smooth surface fitting for irregularly distributed data points, *Association for Computing Machinery Transactions on Mathematical Software*, **4** (2), 148–159.

AL-TAHA, K. K. (1992). *Temporal Reasoning in Cadastral Systems*, PhD Thesis, University of Maine, Orono, ME, USA.

AL-TAHA, K. K., SNODGRASS, R. T. and SOO, M. D. (1993). Bibliography on spatiotemporal databases, *Association for Computing Machinery SIGMOD Record*, **22**, 59–67, and *International Journal of Geographical Information Systems*, **8** (1), 95–103.

ALLEN, J. F. (1984). Towards a general theory of action and time, *Artificial Intelligence*, **23**, 123–154.

ALTMAN, D. (1994). Fuzzy set theoretic approaches for handling imprecision in spatial analysis, *International Journal of Geographical Information Systems*, **8** (3), 271–289.

AMERICAN NATIONAL STANDARDS INSTITUTE (ANSI) (1986). The Database Language SQL, Document ANSI X3.135.

AMERICAN NATIONAL STANDARDS INSTITUTE (ANSI) (1991). Object Oriented Database Task Group Final Report, X3/SPARC/DBSSG OODBTG.

AMERICAN NATIONAL STANDARDS INSTITUTE (ANSI) (1994). Database Language SQL 3, Melton, J. (Ed.), Database Committee (X3H2).

ANDERSON, J. T. and STONEBRAKER, M. (1994). SEQUOIA 2000 metadata schema for satellite images, *Association for Computing Machinery SIGMOD Record*, **23** (4), 42–48.

ARIAV, G. (1986). A temporally oriented data model. *Association for Computing Machinery Transactions on Database Systems*, **11**, 499–527.

ARMSTRONG, M. A. (1979). *Basic Topology*, Maidenhead: McGraw-Hill.

ARNAUD, A. M., CRAGLIA, M., MASSER, I., SALGE, F. and SCHOLTEN, H. (1993). The research agenda of the European Science Foundation's GISDATA scientific programme, *International Journal of Geographical Information Systems*, **7** (5), 463–470.

BANCILHON, F., DELOBEL, C. and KANELLAKIS, P. (1992). *Building an Object-Oriented Database: The story of $O_2$*, San Mateo, CA: Morgan-Kaufmann.

BAUMGART, B. (1975). A polyhedron representation for computer vision, *Proceedings AFIPS National Conference*, **44**, 589–596.

BEKKER, J. H. TER (1992). *Semantic Data Modelling*, New York: Prentice-Hall.

BELL, D. and GRIMSON, J. (1992). *Distributed Database Systems*, Reading, MA: Addison-Wesley.

BENTLEY, J. L. (1977). Algorithms for Klee's Rectangle Problems. Unpublished notes, Department of Computer Science, Carnegie-Mellon University, Pittsburgh, PA.

BENTLEY, J. L. (1979). Multidimensional binary search trees in database applications. *IEEE Transactions on Software Engineering*, **5**, 333–340.

BENTLEY, J. L. and FRIEDMAN, J. H. (1979). Data structures for range searching. *Association for Computing Machinery Computing Surveys*, **11** (4), 397–409.

BERTINO, E. and MARTINO, L. (1993). *Object-oriented Database Systems*, Reading, MA: Addison-Wesley.

BERTOLOTTO, M., DE FLORIANI, L. and MARZANO, P. (1994). An efficient representation for pyramidal terrain models, in Pissinou, N. and Makki, K. (Eds) *Proceedings of the Second ACM Workshop on Advances in Geographic Information Systems, Gaithersburg, MD:* National Institute for Standards and Technology, pp. 129–136.

BÉZIER, P. (1970). *Emploi des Machines á Commande Numérique*, Paris: Masson et Cie. Translated by Forrest, A. R. and Pankhurst, A. F. as Bézier, P. (1972) *Numerical Control – Mathematics and Applications*, London: Wiley.

BLANKENAGEL, G. and GÜTING, R. H. (1994). External segment trees, *Algorithmica*, **12**, 498–532.

BOOCH, G. (1994). *Object-Oriented Analysis and Design with Applications*, 2nd Edn, Redwood City, CA: Benjamin/Cummings.

BOWYER, A. (1981). Computing Dirichlet tessellations, *Computer Journal*, **24**, 162–166.

BRACKEN, I. and WEBSTER, C. (1989). Towards a typology of geographical information systems, *International Journal of Geographical Information Systems*, **3** (2), 137–152.

BRINKHOFF, T., KRIEGEL, H.-P., SCHNEIDER, R. and BRAUN, A. (1995). A measure for the complexity of spatial objects. Submitted for publication.

BRODIE, M., MYLOPOULOS, J. and SCHMIDT, J. (Eds) (1984). *On Conceptual Modelling*, Berlin: Springer-Verlag.

BROOKSHEAR, J. G. (1989). *Theory of Computation: Formal Languages, Automata and Complexity*, Redwood City, CA: Benjamin/Cummings.

BROOME, F. R. (1986). Mapping from a topologically encoded database: The US Bureau of Census example, in Blakemore, M. J. (Ed.) *Proceedings of Auto-Carto 7*, London: Royal Institution of Chartered Surveyors, pp. 402–411.

BRYANT R. and SINGERMAN, D. (1985). Foundations of the theory of maps on surfaces with boundaries, *Quarterly Journal of Mathematics*, **2** (36), 17–41.

BURROUGH, P. A. (1986). *Principles of Geographical Information Systems for Land Resources Assessment*, Oxford: Clarendon Press.

BURROUGH, P. A. (1992). Development of intelligent geographical information systems, *International Journal of Geographical Information Systems*, **6** (1), 1–11.

BURROUGH, P. A. (1994). Accuracy and error in GIS, in Green, D. R. and Rix, D. (Eds) *The AGI Source Book for Geographic Information Systems*, London: Association for Geographic Information, pp. 87–91.

BUTTENFIELD, B. P. (1991). Visualization, in Maguire, D. J., Goodchild, M. F. and Rhind, D. W. (Eds) *Geographical Information Systems*, Vol 1, Harlow: Longmans, pp. 427–443.

CATTELL, R. G. G. (Ed.) (1994). *The Object Database Standard ODMG-93*, San Mateo, CA: Morgan-Kaufmann.

CERI, S. GOTTLOB, G. and TANCA, L. (1990). *Logic Programming and Databases*, Berlin: Springer-Verlag.

CHAMBERLIN, D. D. and BOYCE, R. (1974). SEQUEL: A structured English query language, *Proceedings of the Association for Computing Machinery Workshop on Data Description, Access and Control*, Ann Arbor, Michigan.

CHAMBERLIN, D. D. et al. (1976). SEQUEL 2: a unified approach to data definition, manipulation and control, *IBM Journal of Research and Development*, **20** (6), 560–575.

CHANCE, A., NEWELL, R., THERIAULT, D. (1990). An object-oriented GIS: Issues and solutions, *Proceedings European Geographical Information Systems (EGIS) Annual Conference*, Utrecht: EGIS Foundation, pp. 179–188.

CHAZELLE, B. (1991). Triangulating a simple polygon in linear time, *Discrete Computational Geometry*, **6**, 485–524.

CHEN, P. P-S. (1976). The entity-relationship model – toward a unified view of data. *Association for Computing Machinery Transactions on Database Systems*, **1** (1), 9–36.

CHEYLAN, J.-P. and LARDON, S. (1993). Towards a conceptual data model for the analysis of spatio-temporal processes: the example of the search for optimal grazing strategies, in Frank, A. U. and Campari, I. U. (Eds) *Spatial Information Theory, Lecture Notes in Computer Science 716*, Berlin: Springer-Verlag, pp. 113–138.

CHOI, A. and LUK, W. S. (1992). Using an object-oriented database system to construct a spatial database kernel for GIS applications, *Computer System Science and Engineering*, **7**, 100–121.

CHOU, H. and KIM, W. (1986). A unifying framework for version control in a CAD environment, *Proceedings of the Conference on Very Large Databases*, Los Altos, CA: Morgan-Kaufmann, pp. 336–346.

CHRISMAN, N. R. (1975). Topological information systems for geographic representation. *Proceedings of Second International Symposium on Computer-Assisted Cartography (Auto-Carto 2)*, Falls Church: ASPRS/ACSM, pp. 346–351.

CHRISMAN, N. R. (1978). Concepts of space as a guide to cartographic data structures, in Dutton, G. (Ed.) *Proceedings of the First International Advanced Study Symposium on Topological Data Structures for Geographic Information Systems*, Cambridge, MA: Harvard Laboratory for Computer Graphics and Spatial Analysis, pp. 1–19.

CHRISMAN, N. R. (1991). The error component in spatial data, in Maguire, D. J., Goodchild, M. F. and Rhind, D. W. (Eds) *Geographical Information Systems*, Vol 1, Harlow: Longmans, pp. 165–174.

CHRISTENSEN, A. H. J. (1987). Fitting a triangulation to contour lines, *Proceedings of Auto-Carto*, **8**, 57–67.

CLARKE, B. L. (1981). A calculus of individuals based on connection, *Notre Dame Journal of Formal Logic*, **2** (3).

CLARKE, B. L. (1985). Individuals and points, *Notre Dame Journal of Formal Logic*, **26** (1).

CLIFF, A. D. and ORD, J. K. (1981). *Spatial Processes, Models and Applications*, London: Pion.

CLIFFORD, J. and WARREN, D. S. (1983). Formal semantics for time in databases, *Association for Computing Machinery Transactions on Database Systems*, **8**, 214–254.

CLIFFORD, J. and TANSEL, A. U. (1985). On an algebra for historical relational databases: Two views, *Proceedings of the 14th Association for Computing Machinery SIGMOD Conference*, Austin, Texas, New York: ACM Press, pp. 247–265.

CLIFFORD, J. and ISAKOWITZ, T. (1994). On the semantics of (bi)temporal variable databases, in Jarke, M., Bubenko, J. and Jeffery, K. (Eds) *Advances in Database Technology, EDBT'94, Lecture Notes in Computer-Science 779*, Berlin: Springer-Verlag, pp. 215–230.

CODD, E. (1970). A relational model for large shared data banks. *Communications of the Association for Computing Machinery*, **13** (6), 377–387.

CODD, E. (1979). Extending the relational database model to capture more meaning, *Association for Computing Machinery Transactions on Database Systems*, **4** (4), 397–434.

COHN, A. G. (1987). A more expressive formulation of many sorted logic, *Journal of Automated Reasoning*, **3** (2), 113–200.

COMER, D. (1979). The ubiquitous B-tree, *ACM Computing Surveys*, **11** (2), 121–137.

COPPOCK, J. T. and RHIND, D. W. (1992). The history of GIS, in Maguire, D. J., Goodchild, M. F. and Rhind, D. W. (Eds) *Geographical Information Systems*, Vol 1, Harlow: Longmans, pp. 21–43.

CORI, R. and VAUQUELIN, B. (1981). Planar maps are well labelled trees, *Canadian Journal of Mathematics*, **33** (5), 1023–1042.

COUCLELIS, H. (1992). Beyond the raster-vector debate in GIS, in Frank, A. U., Campari, I. and Formentini, U. (Eds) *Theories of Spatio-Temporal Reasoning in Geographic Space, Lecture Notes in Computer Science 639*, Berlin: Springer-Verlag, pp. 65–77.

COULOURIS, G., DOLLIMORE, J. and KINDBERG, T. (1994). *Distributed Systems: Concepts and Design*, 2nd Edn, Reading, MA: Addison-Wesley.

COXETER, H. S. M. (1961). *Introduction to Geometry*, New York: Wiley.

CUI, Z., COHN, A. G. and RANDELL, D. A. (1993). Qualitative and topological relationships in spatial databases, in Abel, D. and Ooi, B. C. (Eds) *Advances in Spatial Databases, Proceedings of SSD'93, Singapore, Lecture Notes in Computer Science 692*, Berlin: Springer-Verlag pp. 296–315.

DANGERMOND, J. (1983). A classification of software components commonly used in geographic information systems, in Peuquet, D. and O'Callaghan, J. (Eds) *Design and Implementation of Computer-based Geographic Information Systems*. New York: IGU, pp. 70–91. Reprinted in Peuquet, D. J. and Marble, D. F. (Eds) (1990). *Introductory Readings in Geographic Information Systems*, London: Taylor & Francis, pp. 30–51.

DATE, C. J. (1995). *An Introduction to Database Systems*, 6th Edn, Reading, MA: Addison-Wesley.

DATE, C. J. and DARWEN, H. (1993). *A Guide to the SQL Standard*, 3rd Edn, Reading, MA: Addison-Wesley.

DAVID, B., RAYNAL, L., SCHORTER, G. and MANSART, V. (1993). GeO$_2$: Why objects in a geographical DBMS?, in Abel, D. and Ooi, B. C. (Eds) *Advances in Spatial Databases, Proceedings of SSD'93, Singapore, Lecture Notes in Computer Science 692*, Berlin: Springer-Verlag, pp. 264–276.

DE FLORIANI, L. (1989). A pyramidal data structure for triangle-based surface description, *IEEE Computer Graphics and Applications*, New York: IEEE, pp. 67–78.

DE FLORIANI, L. and PUPPO, E. (1992). A hierarchical triangle-based model for terrain description, in Frank, A. U., Campari, I. and Formentini, U. (Eds) *Theories of Spatio-Temporal Reasoning in Geographic Space, Lecture Notes in Computer Science 639*, Berlin: Springer-Verlag, pp. 236–251.

DE FLORIANI, L., FALCIDIENO, B., PIENOVI, C., ALLEN, D. and NAGY, G. (1986). A visibility-based model for terrain features, *Proceedings of the 2nd International Symposium on Spatial Data Handling*, Seattle, Columbus, OH: International Geographical Union, pp. 600–610.

DE FLORIANI, L. MARZANO, P. and PUPPO, E. (1993). Spatial queries and data models, in Frank, A. U. and Campari, I. U. (Eds) *Spatial Information Theory, Lecture Notes in Computer Science 716*. Berlin: Springer-Verlag, pp. 113–138.

DE FLORIANI, L. MARZANO, P. and PUPPO, E. (1994). Hierarchical terrain models: survey and formalization, *Proceedings SCA'94*, Phoenix, AR, pp. 323–327.

DICKINSON, H. J. and CALKINS, H. W. (1988). The economic evaluation of implementing a GIS, *International Journal of Geographical Information Systems*, **2** (4), 307–328.

DIGITAL CARTOGRAPHIC DATA STANDARDS TASK FORCE (DCDSTF) (1988). The proposed standard for digital cartographic data, *The American Geographer*, **15** (1), 9–140.

DIJKSTRA, E. W. (1959). A note on two problems in connexion with graphs, *Numerische Mathematik*, **1**, 269–271.

DIRICHLET, G. L. (1850). Über die reduction der positeven quadratischen formen mit drei unbestimmten ganzen zahlen, *Journal für die Reine und Angewandte Mathematik*, **40**, 209–227.

DOUGLAS, D. (1974). It makes me so CROSS. From an unpublished manuscript, Harvard Laboratory for Computer Graphics and Spatial Analysis. Reprinted in Peuquet, D. J. and Marble, D. F. (Eds) (1990). *Introductory Readings in Geographic Information Systems*, London: Taylor & Francis, pp. 303–307.

DÜPPE, R. D. and GOTTSCHALK, H. J. (1970). Automatische Interpolation von Isolinien bei willkürlich verteilten Stützpunkten, *Allgemeinen Vermessungnachrichten*, **77**, 423–426.

DUTTON, G. H. (1984). Geodesic modelling of planetary relief, *Cartographica*, **21** (2 & 3), 188–207.

DUTTON, G. H. (1989). Planetary modelling via hierarchical tessellation, in *Proceedings of Autocarto 9*, Baltimore, MD: ACSM-ASPRS, pp. 462–471.

DUTTON, G. H. (1990). Locational properties of quaternary triangular meshes, in Brassel, K. and Kishimoto, H. (Eds) *Proceedings of the 4th International Symposium on Spatial Data Handling*, Zurich, Columbus, OH: International Geographical Union, pp. 901–910.

DWYER, R. A. (1987). A fast divide-and-conquer algorithm for constructing Delaunay triangulations, *Algorithmica*, **2**, 137–151.

EASTERFIELD, M. E., NEWELL, R. G. and THERIAULT, D. G. (1990). Version management in GIS: Applications and techniques, *Proceedings European Geographical Information Systems (EGIS) Annual Conference*, Utrecht: EGIS Foundation.

EDDY, C. A. and LOONEY, B. (1993). Three-dimensional digital imaging of environmental data: selection of gridding parameters, *International Journal of Geographical Information Systems*, **7** (2), 165–172.

EDELSBRUNNER, H. (1987). *Algorithms in Combinatorial Geometry*, Berlin: Springer-Verlag.

EDMUNDS, J. (1960). Combinatorial representation for polyhedral surfaces, *Notices of the American Mathematical Society*, **7**, 646–646.

EGENHOFER, M. J. (1989a). Spatial Query Languages. PhD Thesis, University of Maine, Orono, ME, USA.

EGENHOFER, M. J. (1989b). A formal definition of binary topological relationships, In Litwin, W. and Schek, H.-J. (Eds) *Proceedings of the Third International Conference on Foundations of Data Organization and Algorithms (FODO)*, Paris, Lecture Notes in Computer Science 367, Berlin: Springer-Verlag, pp. 457–472.

EGENHOFER, M. J. (1990). Interactions with Geographic Information Systems via Spatial Queries, *Journal of Visual Languages and Computing*, **1**, 389–413.

EGENHOFER, M. J. (1991a). Deficiencies of SQL as a GIS query language, in Mark, D. M. and Frank, A. U. (Eds) *Cognitive and Linguistic Aspects of Geographic Space*, Nato ASI Series, Dordrecht: Kluwer Academic, pp. 477–491.

EGENHOFER, M. J. (1991b). Reasoning about binary topological relations, in Günther, O. and Schek, H.-J. (Eds) *Advances in Spatial Databases, Proceedings of SSD'91, Zurich, Switzerland, Lecture Notes in Computer Science, No. 525*, Berlin: Springer-Verlag, pp. 143–160.

EGENHOFER, M. J. (1991c). Categorizing topological relationships – Quantitative refinement of qualitative spatial information, Technical Report, Department of Surveying Engineering, University of Maine, Orono, ME, USA.

EGENHOFER, M. J. (1992). Why not SQL!, *International Journal of Geographical Information Systems*, **6** (2), 71–85.

EGENHOFER, M. J. (1994). Spatial SQL: A query and presentation language, *IEEE Transactions on Knowledge and Data Engineering*, **6** (1), 86–95.

EGENHOFER, M. J. and FRANK, A. (1987). Object-oriented databases: Database requirements for GIS, in *Proceedings of the International GIS Symposium: The Research Agenda*, Vol 2, Washington DC: US Government Printing Office, pp. 189–211.

EGENHOFER, M. J. and HERRING, J. R. (1990). A mathematical framework for the definition of topological relationships, in Brassel, K. and Kishimoto, H. (Eds) *Proceedings of the 4th International Symposium on Spatial Data Handling*, Zurich, Columbus, OH: International Geographical Union, pp. 803–813.

EGENHOFER, M. J. and HERRING, J. R. (1991). Categorizing binary topological relationships between regions, lines and points in geographic databases, Technical Report, Department of Surveying Engineering, University of Maine, Orono, ME, USA.

EGENHOFER, M. J. and FRANK, A. (1992). Object-oriented modeling for GIS, *Journal of the Urban and Regional Information Systems Association*, **4**, 3–19.

EGENHOFER, M. J. and RICHARDS, J. R. (1993). Exploratory access to geographic data based on the map-overlay metaphor, *Journal of Visual Languages and Computing*, **4**, 105–125.

EGENHOFER, M. J. and HERRING, J. R. (1994). Querying a geographical information system, in Medyckyj-Scott, D. and Hearnshaw, H. M. (Eds) *Human Factors in Geographical Information Systems*, London: Belhaven Press, pp. 124–135.

EGENHOFER, M. J. and FRANZOSA, R. D. (1995). On the equivalence of topological relations, *International Journal of Geographical Information Systems*, **9** (2), 133–152.

EGENHOFER, M. J., FRANK, A. and JACKSON, J. (1989). A topological data model for spatial databases, in Günther, O. and Smith, T. (Eds) *Design and Implementation of Large Spatial Databases, Proceedings of SSD'89*, Santa Barbara, CA, *Lecture Notes in Computer Science, No. 409*, Berlin: Springer-Verlag, pp. 189–211.

ELMASRI, R. and NAVATHE, S. B. (1994). *Fundamentals of Database Systems*, 2nd Edn, Redwood City, CA: Benjamin/Cummings.

FAGIN, R., NIEVERGELT, J., PIPPENGER, N. and STRONG, H. (1979). Extendible hashing – A fast access method for dynamic files, *Association for Computing Machinery Transactions on Database Systems*, **4** (3), 315–344.

FALOUTSOS, C. (1988). Gray codes for partial match and range queries, *IEEE Transactions on Software Engineering*, **14** (10), 1381–1393.

FALOUTSOS, C., SELLIS, T. and ROUSSOPOULOS, N. (1987). Analysis of object-oriented spatial access methods, *Proceedings of the 16th Association for Computing Machinery SIGMOD Conference*, San Francisco, New York: ACM Press, pp. 426–439.

FEKETE, G. and DAVIS, L. S. (1984). Property spheres: a new representation for 3-D object recognition, *Proceedings of the Workshop on Computer Vision: Representation and Control*, Annapolis, MD, pp. 176–186.

FERNÁNDEZ, R. N. and RUSINKIEWICZ, M. (1993). A conceptual design of a soil database for a geographical information system, *International Journal of Geographical Information Systems*, **7** (6), 525–539.

FEUCHTWANGER, M. (1989). Geographical logical database requirements, *Proceedings of Autocarto 9*, Baltimore, MD: ACSM-ASPRS, pp. 599–609.

FINKEL, R. A. and BENTLEY, J. L. (1974). Quad trees: a data structure for retrieval on composite keys, *Acta Informatica*, **4** (1), 1–9.

FISHER, P. F. (1994a). Probable and fuzzy models of the viewshed operation, in Worboys, M. F. (Ed.) *Innovations in GIS I*, London: Taylor & Francis, pp. 161–176.

FISHER, P. F. (1994b). Stretching the viewshed operation, in Waugh, T. C. and Healey, R. G. (Eds) *Proceedings of the 6th International Symposium on Spatial Data Handling*, Edinburgh, London: Taylor & Francis, pp. 725–738.

FLANAGAN, N., JENNINGS, C. and FLANAGAN, C. (1994). Automatic GIS data capture and conversion, in Worboys, M. F. (Ed.) *Innovations in GIS I*, London: Taylor & Francis, pp. 25–38.

FLOYD, R. W. (1962). Algorithm 97: Shortest path, *Communications of the ACM*, **5**, 345.

FOLEY, J. D., VAN DAM. A., FEINER, S. K. and HUGHES, J. F. (1990). *Computer Graphics: Principles and Practice*, Reading, MA: Addison-Wesley.

FORTUNE, S. (1987). A sweepline algorithm for Voronoi diagrams, *Algorithmica*, **2**, 153–174.

FOURNIER, A. and MONTUNO, D. Y. (1984). Triangulating simple polygons and equivalent problems, *Association for Computing Machinery Transactions on Graphics*, **3** (2), 153–174.

FRANK, A. (1983). Storage methods for space related data: The field tree, Tech. rep. Bericht no. 71, Eidgenössische Technische Hochschule, Zürich.

FRANK, A and BARRERA, R. (1989). The field-tree: A data structure for geographic information systems, in Günther, O. and Smith, T. (Eds) *Design and Implementation of Large Spatial Databases, Proceedings of SSD'89, Santa Barbara, CA, Lecture Notes in Computer Science, No. 409*. Berlin: Springer-Verlag, pp. 29–44.

FRANK, A. and MARK, D. M. (1991). Language issues for GIS, in Maguire, D. J., Goodchild, M. F. and RHIND, D. W. (Eds) *Geographical Information Systems*, Vol 1, Harlow: Longmans, pp. 147–163.

FRANKLIN, WM. R. and RAY, C. K. (1994). Higher isn't necessarily better: Visibility algorithms and experiments, in Waugh, T. C. and Healey, R. G. (Eds) *Proceedings of the 6th International Symposium on Spatial Data Handling*, Edinburgh, London: Taylor & Francis, pp. 751–763.

FREEMAN, H. (1961). On the encoding of arbitrary geometric configurations, *IEEE Transactions Elec. Computers*, **10**, 260–268.

FREEMAN, J. (1975). The modelling of spatial relations, *Computer Graphics and Image Processing*, **4**, 156–171.

FREESTON, M. (1987). The BANG file: a new kind of grid file, *Proceedings of the Association for Computing Machinery SIGMOD Conference*, San Francisco, New York: ACM Press, pp. 260–269.

FREESTON, M. (1989). A well-behaved file structure for the storage of spatial objects, in Günther, O. and Smith, T. (Eds) *Design and Implementation of Large Spatial Databases, Proceedings of SSD'89, Santa Barbara, CA, Lecture Notes in Computer Science, No. 409*, Berlin: Springer-Verlag, pp. 287–300.

FREESTON, M. (1993). Begriffsverzeichnis: a concept index, in Worboys, M. F. and Grundy, A. F. (Eds) *Advances in Databases, Proceedings of the 11th British National Conference on Databases, BNCOD11, Keele, UK, Lecture Notes in Computer Science, No. 696*, Berlin: Springer-Verlag, pp. 1–22.

FREKSA, C. (1991). Qualitative spatial reasoning, in Mark, D. M. and Frank, A. U. (Eds) *Cognitive and Linguistic Aspects of Geographic Space*, Nato ASI Series, Dordrecht: Kluwer Academic, pp. 361–372.

FRIESEN, P., KUCERA, H. and SONDHEIM, M. (1993). SAIF profiles: the missing link, *Proceedings of GIS'93*, Vancouver, British Columbia, Canada, pp. 1113–1120.

FUCHS, H., ABRAM, G. D. and GRANT, E. D. (1983). Near real-time shaded display of rigid objects, *Computer Graphics*, **17** (3), 65–72.

FUCHS, H., KEDEM, Z. M. and NAYLOR, B. F. (1980). On visible surface generation by *a priori* tree structures, *Computer Graphics*, **14** (3), 124–133.

GARDARIN, G. and VALDURIEZ, P. (1989). *Relational Databases and Knowledge Bases*, Reading, MA: Addison-Wesley.

GAREY, M. R., JOHNSON, D. S., PREPARATA, F. P. and TARJAN, R. E. (1978). Triangulating a simple polygon, *Information Processing Letters*, **7**, 175–179.

GATRELL, A. C. (1991). Concepts of space and geographical data, in Maguire, D. J., Goodchild, M. F. and Rhind, D. W. (Eds) *Geographical Information Systems*, Vol 1, Harlow: Longmans, pp. 119–134.

GIBLIN, P. J. (1977). *Graphs, Surfaces and Homology – An Introduction to Algebraic Topology*, London: Chapman and Hall.

GILBERT, P. N. (1979). New results on planar triangulations. Technical Report ACT-15, Coord. Sci. Lab., University of Illinois at Urbana, USA.

GITTINGS, B. M., SLOAN, T. M., HEALEY, R. G., DOWERS, S. and WAUGH, T. C.

(1994). Meeting expectations: a review of GIS performance issues, in Mather, P. M. (Ed.) *Geographical Information Handling – Research and Applications*, Chichester: Wiley, pp. 33–45.

GOLD. C. (1994). Three approaches to automated topology, and how computational geometry helps, in Waugh, T. and Healey, R. G. (Eds) *Advances in GIS research, Proceedings SDH'94*, London: Taylor & Francis, pp. 145–158.

GOLD, C. and CORMACK, S. (1987). Spatially ordered networks and topographic reconstructions, *International Journal of Geographical Information Systems*, **1**, 137–148.

GONZALES, R. C. and WINTZ, P. (1987). *Digital Image Processing*, Reading, MA: Addison-Wesley.

GOODCHILD, M. F. (1989). Tiling large geographical databases, in Günther, O. and Smith, T. (Eds) *Design and Implementation of Large Spatial Databases, Proceedings of SSD'89, Santa Barbara, CA, Lecture Notes in Computer Science, No. 409*, Berlin: Springer-Verlag, pp. 137–146.

GOODCHILD, M. F. (1991). Geographical data modeling, *Computers and Geosciences*, **18**, 400–408.

GOODCHILD, M. F. (1992). Geographic information science, *International Journal of Geographical Information Systems*, **6** (1), 31–45.

GOODCHILD, M. F. (1993). The state of GIS for environmental problem-solving, in Goodchild, M. F., Parks, B. O. and Steyaert, L. T. (Eds) *Environmental Modelling with GIS*, Oxford: Oxford University Press, pp. 8–15.

GOODCHILD, M. F. (1994). Information highways, in Green, D. R. and Rix, D. (Eds) *The AGI Source Book For Geographic Information Systems*, London: Association for Geographic Information, pp. 107–120.

GOODCHILD, M. F. and SHIREN, Y. (1992). A hierarchical data structure for global geographical information systems, *Graphical Models and Image Processing*, **54** (1), 31–44.

GRAY, P. M. D., KULKARNI, K. G. and PATON, N. W. (1992). *Object-oriented Databases: A Semantic Data Model Approach*, New York: Prentice Hall.

GREENE, D and YAO, F. (1986). Finite resolution computational geometry, *Proceedings of the 27th IEEE Symposium on the Foundations of Computer Science*, New York: IEEE, pp. 143–152.

GUIBAS, L. and STOLFI, J. (1985). Primitives for the manipulation of general subdivisions and the computation of Voronoi diagrams, *Association for Computing Machinery Transactions on Graphics*, **4** (2), 74–123.

GÜNTHER, O. (1987). Efficient structures for geometric data management, PhD dissertation, University of California at Berkeley, CA. Published as *Lecture Notes in Computer Science 337*, (1988), Berlin: Springer-Verlag.

GÜNTHER, O. and BILMES, J. (1988). The implementation of the cell tree: design alternatives and performance evaluation, TRCS88-23, University of California at Santa Barbara, CA.

GÜNTHER, O. and BILMES, J. (1991). Tree-based access methods for spatial databases: implementation and performance evaluation, *IEEE Transactions on Knowledge and Data Engineering*, **3** (3), 342–356.

GÜNTHER, O. and BUCHMANN, A. (1990). Research issues in spatial databases, *Association for Computing Machinery SIGMOD Record*, **19**, 61–68.

GUPTILL, S. (1991). Spatial Data Exchange and Standardization, in Maguire, D. J., Goodchild, M. F. and Rhind, D. W. (Eds) *Geographical Information Systems*, Vol 1, Harmondsworth: Longmans, pp. 515–530.

GUPTILL, S. and STONEBRAKER, M. (1992). The Sequoia 2000 approach to managing large spatial object databases, in Corwin, E. and Cowen, D. (Eds) *Proceedings of the 5th International Symposium on Spatial Data Handling*, Columbus, OH: International Geographical Union, pp. 642–651.

GÜTING, R. H. and SCHNEIDER, M. (1993a). Realms: A foundation for spatial data types in database systems, in Abel, D. and Ooi, B. C. (Eds) *Advances in Spatial Databases, Proceedings of SSD'93, Singapore, Lecture Notes in Computer Science 692*, Berlin: Springer-Verlag, pp. 14–35.

GÜTING, R. H. and SCHNEIDER, M. (1993b). Realm-based spatial data types: The ROSE algebra. Fachbereich Informatik, FernUniversität Hagen, Report 141.

GUTTMAN, A. (1984). R-trees: a dynamic index structure for spatial searching, *Proceedings of the 13th Association for Computing Machinery SIGMOD Conference*, Boston, New York: ACM Press, pp. 47–57.

HAAS, L. M., FREYTAG, J. C., LOHMAN, G. M. and PIRAHESH, H. (1989). Extensible query processing in Starburst, *Proceedings of the Association for Computing Machinery SIGMOD Conference*, Portland, Oregon, New York: ACM Press, pp. 377–388.

HAAS, L. M., CHANG, W., LOHMAN, G. M., MCPHERSON, J., WILMS, P. F., LAPIS, G., LINDSAY, B., PIRANSESH, H., CAREY, M. J. and SHEKITA, E. (1990). Starburst mid-flight: as the dust clears, *IEEE Transactions on Knowledge and Data Engineering*, **2** (1), 143–160.

HAMMER, M. M. and MCLEOD, D. J. (1981). Database description with SDM, *Association for Computing Machinery Transactions on Database Systems* **6** (3), 351–386.

HARARY, F. (1969). *Graph Theory*, Reading, MA: Addison-Wesley.

HASKIN, R. L. and LORIE, R. A. (1982). On extending the functions of a relational database system, *Association for Computing Machinery Proceedings of SIGMOD*, New York: ACM Press, pp. 207–212.

HAZELTON, N. W. J. (1991). Integrating time, dynamic modelling and geographical information systems: Development of four-dimensional GIS, PhD Thesis, University of Melbourne, Australia.

HEALEY, R. G. (1991). Database management systems, in Maguire, D. J., Goodchild, M. F. and Rhind, D. W. (Eds) *Geographical Information Systems*, Vol 1, Harlow: Longmans, pp. 251–267.

HEARNSHAW, H. M. and UNWIN, D. J. (Eds) (1994). *Visualization in Geographical Information Systems*, Chichester: Wiley.

HENLE, M. (1979). *A Combinatorial Introduction to Topology*, San Franscisco, CA: Freeman.

HERRING, J. (1991a). TIGRIS: A data model for an object-oriented geographic information system, *Computers and Geosciences*, **18**, 443–452.

HERRING, J. (1991b). The mathematical modelling of spatial and non-spatial information in geographic information systems, in Mark, D. M. and Frank, A. U. (Eds) *Cognitive and Linguistic Aspects of Geographic Space*, Nato ASI Series, Dordrecht: Kluwer Academic, pp. 313–350.

HERZNER, W. and KAPPE, F. (Eds) (1994). *Multimedia/Hypermedia in Open Distributed Environments*, Berlin: Springer-Verlag.

HOFFMANN, C. M. (1989). *Geometric and Solid Modeling – An Introduction*, San Mateo, CA: Morgan-Kaufmann.

HOPKINS, S. HEALEY, R. G. and WAUGH, T. (1992). Algorithm scalability for line intersection detection in parallel polygon overlay, in Corwin, E. and Cowen, D. (Eds) *Proceedings of the 5th International Symposium on Spatial Data Handling*, Columbus, Ohio,: International Geographical Union, pp. 210–229.

HOWE, D. R. (1989). *Data Analysis for Data Base Design*, London: Edward Arnold.

HUGHES, G. E. and CRESSWELL, M. J. (1968). *An Introduction to Modal Logic*, London: Methuen (reprinted with corrections 1972).

HULL, R. and KING, R. (1987). Semantic data modelling: Survey, applications and research issues, *Association for Computing Machinery Computer Surveys*, **19**, 201–260.

HUNTER, G. M. (1978). Efficient computation and data structures for graphics, PhD Thesis, Princeton University, Princeton, NJ, USA.

INTERNATIONAL ORGANIZATION FOR STANDARDIZATION (ISO) (1992). The Database Language SQL. Document ISO/IEC 9075:1992.

JAYAWARDENA, D. P. W. (1994). The role of triangulation in spatial data handling, PhD Thesis, Keele University, Keele, Staffordshire, UK.

JAYAWARDENA, D. P. W. and WORBOYS, M. F. (1995). The role of triangulation in spatial data handling, in Fisher, P. (Ed.) *Innovations in GIS*, London: Taylor & Francis, pp. 7–17.

JEANSOULIN, R. and MATHIEU, C. (1994). A modal logic for spatial knowledge and hypothesis. Unpublished paper. Laboratoire d'Informatique de Marseille, URA CNRS 1787, Université de Provence, 13331 Marseille, France.

JENSEN, C. S. (1994). A consensus glossary of temporal database concepts, *Association for Computing Machinery SIGMOD Record*, **23** (1), 52–64.

JENSEN, C. S., SNODGRASS, R. T. and SOO, M. D. (1992). Extending normal forms to temporal relations. Technical Report 92–17, Department of Computer Science, University of Arizona, Tucson, AZ, USA.

JOHNSON, M. (1987). *The Body in the Mind*, Chicago: University of Chicago Press.

JONES, C. B. (1989). Data structures for 3-D spatial information systems, *International Journal of Geographical Information Systems*, **3**, 15–32.

KÄMPKE, T. (1994). Storing and retrieving changes in a sequence of polygons, *International Journal of Geographical Information Systems*, **8** (6), 493–514.

KATZ, R. H. and CHANG, E. (1988). Managing change in a computer-aided design database, *Proceedings of the Conference on Very Large Databases*, Los Altos, CA: Morgan-Kaufmann, pp. 455–462.

KEMP, K. K. (1992). Environmental modelling with GIS: A strategy for dealing with spatial continuity, *Proceeding GIS/LIS Annual Conference*, Bethseda, MD: ASPRS and ACSM, pp. 397–406.

KHOSHAFIAN, S. (1993). *Object-Oriented Databases*, New York: Wiley.

KHOSHAFIAN, S. and ABNOUS, R. (1990). *Object Orientation: Concepts, Languages, Databases and Interfaces*, New York: Wiley.

KIM, W. (Ed.) (1995). *Modern Database Systems – The Object Model, Interoperability and Beyond*, Reading, MA: Addison-Wesley.

KIM, W. and LOCHOVSKY, F. H. (Eds) (1989) *Object-oriented Concepts, Databases and Applications*, New York: ACM Press.

KIRBY, G. H., VISVALINGAM, M. and WADE, P. (1989). Recognition and representation of a hierarchy of polygons with holes, *Computer Journal*, **32** (6), 554–562.

KIRKPATRICK, D. G., KLAWE, M. M. and TARJAN, R. E. (1992). Polygon triangulation in O($n$ log log $n$) time with simple data structures, *Discrete Computational Geometry*, **7**, 329–346.

KLINE, N. (1993). An update of the temporal database bibliography, *Association for Computing Machinery SIGMOD Record*, **22** (4), 66–80.

KNUTH, D. (1973). *The Art of Computer Programming, Volume 3: Sorting and Searching*, Reading, MA: Addison-Wesley.

KRIEGEL, H. P., BRINKHOFF, T. and SCHNEIDER, R. (1993). Efficient spatial query processing in geographic database systems, *IEEE Data Engineering Bulletin*, **6** (3), 10–15.

KRIPKE, S. (1963). Semantical analysis of modal logic 1: Normal propositional calculi, *Zeit. Math. Logik. Grund.*, **9**, 67–96.

KUHN, W. and FRANK, A. U. (1991). A formalization of metaphors and image-schemas in user interfaces, in Mark, D. M. and Frank, A. U. (Eds) *Cognitive and Linguistic Aspects of Geographic Space*, Nato ASI Series, Dordrecht: Kluwer Academic, pp. 419–434.

LANGRAN, G. (1989). A review of temporal database research and its use in GIS applications, *International Journal of Geographical Information Systems*, **3** (3) 215–232.

LANGRAN, G. (1991). Dilemmas of implementing a temporal GIS, *Proceedings of the International Cartographic Association*, Bournemouth, UK, pp. 547–555.

LANGRAN, G. (1992). *Time in Geographic Information Systems*, London: Taylor & Francis.

LANGRAN, G. (1993). Issues of implementing a spatiotemporal system, *International Journal of Geographical Information Systems*, **7** (4), 293–304.

LANGRAN, G. and CHRISMAN, N. R. (1988). A framework for temporal geographic information, *Cartographica*, **25** (3), 1–14.

LAURINI, R. (1994). Distributed geographic databases, in Green, D. R. and Rix, D. (Eds) *The AGI Source Book for Geographic Information Systems*, London: Association for Geographic Information, pp. 45–55.

LAURINI, R. and THOMPSON, D. (1992). *Fundamentals of Spatial Information Systems*, London: Academic Press.

LAWSON, C. L. (1977). Software for $C^1$ surface interpolation, in Rice, J. (Ed.) *Mathematical Software III*, New York: Academic Press, pp. 161–194.

LEE, D-T. and SCHACHTER, B. J. (1980). Two algorithms for constructing the Delaunay triangulation, *International Journal of Computer and Information Sciences*, **9** (3), 219–242.

LEE, J. (1991). Analyses of visibility sites on topographic surfaces. *International Journal of Geographical Information Systems*, **1**, 413–429.

LEE, Y. C. and CHIN, F. L. (1995). An iconic query language for topological relationships in GIS, *International Journal of Geographical Information Systems*, **9** (1), 25–46.

LEUNG, Y. and LEUNG, K. S. (1993). An intelligent expert system shell for knowledge-based GIS: 1 and 2, *International Journal of Geographical Information Systems*, **7** (3), 189–214.

LIENHARDT, P. (1991). Topological models for boundary representation: a comparison with $n$-dimensional generalized maps, *Computer-aided Design*, **23** (1), 59–82.

LOHMAN, G., LINDSAY, B., PIRAHESH, H. and SCHIEFER, K. B. (1991). Extensions to Starburst: Objects, types, functions and rules, *Communications of the Association for Computing Machinery*, **34**, 94–109.

LUSE, M. (1993). *Bitmapped Graphics Programming in C++*, Reading, MA: Addison-Wesley.

MACKANESS, W. A. and BUTTENFIELD, B. P. (1991). Incorporating time into geographic process: A framework for analysing process in GIS, *Proceedings of the International Cartographic Association, Bournemouth, UK*, pp. 565–574.

MADSEN, K. H. (1994). A guide to metaphorical design, *Communications of the Association for Computing Machinery*, **37** (12), 57–62.

MAGUIRE, D. J. and DANGERMOND, J. (1991). The functionality of GIS, in Maguire, D. J., Goodchild, M. F. and Rhind, D. W. (Eds) *Geographical Information Systems*, Vol 1, Harlow: Longmans, pp. 319–335.

MAGUIRE, D. J. and DANGERMOND, J. (1994). Future GIS technology, in Green, D. R. and Rix, D. (Eds) *The AGI Source Book for Geographic Information Systems*, London: Association for Geographic Information, pp. 113–120.

MAGUIRE, D. J., GOODCHILD, M. F. and RHIND, D. W. (1991). *Geographical Information Systems*, Harlow: Longman.

MAHONEY, M. S. (1979). *René Descartes. Le Monde, ou Traité de la Lumière*, New York: Abaris Books.

MALING, D. H. (1991). Coordinate systems and map projections for GIS, in Maguire, D. J., Goodchild, M. F. and Rhind, D. W. (Eds) *Geographical Information Systems*, Vol 1, Harlow: Longmans, pp. 135–146.

MÄNTYLÄ, M. (1988). *An Introduction to Solid Modeling*, Rockville, MD: Computer Science Press.

MARBLE, D. F. (1994). An introduction to the structured design of geographic information systems, in Green, D. R. and Rix, D. (Eds) *The AGI Source Book for Geographic Information Systems*, London: Association for Geographic Information, pp. 31–38.

MARK, D. M. (1979). Phenomenon-based data structuring and digital terrain modeling, *Geo-Processing*, **1**, 27–36.

MARK, D. M. (1987). Recursive algorithms for the analysis and display of digital elevation data, *Proceedings of the First Latin American Conference On Computers in Cartography*, San Jose, Costa Rica, pp. 562–571.

MARK, D. M. (1989). Cognitive image-schemata for geographic information: relations to users' views and GIS interfaces, *Proceedings of GIS/LIS '89*, Vol 2, Falls Church: ASPRS/ACSM, pp. 551–560.

MARK, D. M. and FRANK, A. U. (Eds) (1991). *Cognitive and Linguistic Aspects of Geographic Space*, Nato ASI Series, Dordrecht: Kluwer Academic.

MARX, R. W. (1986). The TIGER system: automating the geographic structure of the United States Census, *Government Publications Review*, **13**, 181–201.

MASON, D. C., O'CONAILL, M. A. and BELL, S. B. M. (1994). Handling four-dimensional geo-referenced data in environmental GIS, *International Journal of Geographical Information Systems*, **8** (2), 191–216.

MATTOS, N. and DEMICHIEL, L. G. (1994). Recent design trade-offs in SQL3, *Association for Computing Machinery Association for Computing Machinery SIGMOD Record*, **23** (4), 84–89.

MATTOS, N. M., MEYER-WEGENER, K. and MITSCHANG, B. (1993). Grand tour of concepts for object-orientation from a database point of view. *Data and Knowledge Engineering*, **9**, 321–352.

MAYO, T. (1994). Computer-assisted tools for cartographic data capture, in Worboys, M. F. (Ed.) *Innovations in GIS I*, London: Taylor & Francis, pp. 39–52.

MCGREGOR, J. D. and SYKES, D. A. (1992). *Object-oriented Software Development: Engineering Software for Reuse*, New York: Van Nostrand Reinhold.

MCMASTER, R. B. and SHEA, K. S. (1992). *Generalization in Digital Cartography*, Washington DC: Association of American Geographers.

MEDYCKYJ-SCOTT, D. and HEARNSHAW, H. M. (Eds) (1994). *Human Factors in Geographical Information Systems*, London: Belhaven Press.

MEHLHORN, K. and NÄHER, S. (1995). LEDA: A platform for combinatorial and geometric computing, *Communications of the Association for Computing Machinery*, **38** (1), 96–102.

MILENKOVIC, V. (1993). Robust construction of the Voronoi diagram of the polyhedron, *Proceedings of the 5th Canadian Conference on Computational Geometry*, Waterloo, Ontario: University of Waterloo, pp. 473–478.

MILLER, C. L. and LAFLAMME, R. A. (1958). The digital terrain model – theory and application, *Photogrammetric Engineering*, **24** (3), 433–442.

MILNE, P., MILTON, S. and SMITH, J. L. (1993). Geographical object-oriented databases: A case study, *International Journal of Geographical Information Systems*, **7**, 39–56.

MONTANARI, U. (1969). Continuous skeletons from digitized images, *Journal of the Association for Computing Machinery*, **16** (4), 534–549.

MOREHOUSE, S. (1990). The role of semantics in geographic data modelling, in Brassel, K. and Kishimoto, H. (Eds), *Proceedings of the 4th International Symposium on Spatial Data Handling, Zurich*, pp. 689–700.

MORRIS, R. (1968). Scatter storage techniques, *Communications of the Association for Computing Machinery*, **11** (1).

MULLER, D. E. and PREPARATA, F. P. (1978). Finding the intersection of two convex polyhedra, *Theoretical Computer Science*, **7** (2), 217–236.

MULLER, J.-C. (1991). Generalization of spatial databases, in Maguire, D. J., Goodchild, M. F. and Rhind, D. W. (Eds) *Geographical Information Systems*, Vol 1, Harlow: Longmans, pp. 457–475.

MULMULEY, K. (1994). *Computational Geometry: An Introduction through Randomized Algorithms*, Englewood Cliffs, NJ: Prentice-Hall.

MUSAVI, M. T., SHIRVAIKAR, M. V., RAMANATHAN, E. and NEKOREI, A. R. (1988). A vision-based method to automate map processing, *International Journal of Pattern Recognition*, **21** (4), 319–326.

NATIONAL CENTER FOR GEOGRAPHIC INFORMATION AND ANALYSIS (1989). The research plan of the National Center for Geographic Information and Analysis, *International Journal of Geographical Information Systems*, **3**, 117–136.

NATIONAL CENTER FOR GEOGRAPHIC INFORMATION AND ANALYSIS (1992). A research agenda for Geographic Information and Analysis. Technical Report 92-7, National Center for Geographic Information and Analysis, University of California, Phelps Hall, Santa Barbara, CA, USA.

NELSON, R. C. and SAMET, H. (1986). A consistent hierarchical representation for vector data, *Computer Graphics*, **20** (4), 197–206.

NEWELL, R. (1992). Practical experience of using object-orientation to implement a GIS, *Proceeding GIS/LIS Annual Conference*, Bethseda, MD: ASPRS and ACSM, pp. 624–629.

NIEVERGELT, J. (1989). 7 ± 2 criteria for assessing and comparing spatial data structures, in Günther, O. and Schek, H.-J. (Eds) *Advances in Spatial Databases, Proceedings of SSD'91, Zurich, Switzerland, Lecture Notes in Computer Science, No. 525*, Berlin: Springer-Verlag, pp. 3–27.

NIEVERGELT, J., HINTERBERGER, H. and SEVCIK, K. C. (1984). The grid file: An adaptable, symmetric, multikey file structure, *ACM Transactions on Database Systems*, **9** (1), 38–71.

NIJSSEN, G. (Ed.) (1976). *Modelling in Data Base Management Systems*, Amsterdam: North-Holland.

NUNES, J. (1991). Geographic space as a set of concrete geographical entities, in Mark, D. M. and Frank, A. U. (Eds) *Cognitive and Linguistic Aspects of Geographic Space.*, Nato ASI Series, Dordrecht: Kluwer Academic, pp. 9–34.

O'CONAILL, M. A., MASON, D. C. and BELL, S. B. M. (1994). Spatiotemporal GIS techniques for environmental modelling, in Mather, P. M. (Ed.) *Geographical Information Handling – Research and Applications*, Chichester: Wiley, pp. 103–112.

O'ROURKE, J. (1994). *Computational Geometry in C*, Cambridge: Cambridge University Press.

OHYA, T., IRI, M. and MUROTA, K. (1984a). Improvements of the incremental method for the Voronoi diagram with computational comparisons of various algorithms, *Journal of the Operations Research Society of Japan*, **27**, 306–336.

OHYA, T., IRI, M. and MUROTA, K. (1984b). A fast Voronoi-diagram algorithm with quaternary tree bucketing, *Information Processing Letters*, **18** (4), 227–231.

OKABE, A., BOOTS, B. and SUGIHARA, K. (1992). *Spatial Tessellations: Concepts and Applications of Voronoi Diagrams*, Chichester: Wiley.

OPENSHAW, S. (1991). Developing appropriate spatial analysis methods for GIS, in Maguire, D. J., Goodchild, M. F. and Rhind, D. W. (Eds) *Geographical Information Systems*, Vol 1, Harlow: Longmans, pp. 389–402.

OPENSHAW, S. (1994). A concepts-rich approach to spatial analysis, theory generation, and scientific discovery in GIS using massively parallel computing, in Worboys, M. F. (Ed.) *Innovations in GIS I*, London: Taylor & Francis, pp. 123–137.

OTOO, E. J. and ZHU, H. (1993). Indexing on spherical surfaces using semi-quadcodes, in Abel, D. and Ooi, B. C. (Eds) *Advances in Spatial Databases, Proceedings of SSD'93, Singapore, Lecture Notes in Computer Science 692*, Berlin: Springer-Verlag, pp. 510–529.

OZSU, T. and VALDURIEZ, P. (1991). *Principles of Distributed Database Systems*, Englewood Cliffs, NJ: Prentice-Hall.

PEUCKER, T. K. (1978). Data structures for digital terrain models: discussion and comparison, *Harvard Papers on Geographic Information Systems, Number 5*, Harvard: Harvard University Press.

PEUCKER, T. K. and CHRISMAN, N. R. (1975). Cartographic data structures, *American Cartographer*, **2**, 55–69.

PEUCKER, T. K., FOWLER, R. J., LITTLE, J. J. and MARK, D. M. (1978). The triangulated irregular network, *Proceedings of the ASP Digital Terrain Models (DTM) Symposium*, Falls Church, VA: American Society of Photogrammetry, pp. 516–540.

PEUQUET, D. J. (1984). A conceptual framework and comparison of spatial data models, *Cartographica*, **21** (4), 66–113. (Also in Peuquet, D. J. and Marble, D. F. (Eds) *Introductory Readings in Geographic Information Systems*, London: Taylor & Francis, pp. 250–285.)

PEUQUET, D. J. (1986). The use of spatial relationships to aid spatial database retrieval, *Proceedings of the 2nd International Symposium on Spatial Data Handling*, Seattle, Columbus, OH: International Geographical Union, pp. 459–471.

PEUQUET, D. J. (1988). Towards the definition and use of complex spatial relationships, *Proceedings of the 3rd International Symposium on Spatial Data Handling*, Sydney, Columbus, OH: International Geographical Union, pp. 211–223.

PEUQUET, D. J. and ZHAN, C.-X. (1987). An algorithm to determine the directional relationship between arbitrarily-shaped polygons in the plane, *Pattern Recognition*, **20**, 65–74.

PEUQUET, D. J. and MARBLE, D. F. (1990). *Introductory Readings in Geographic Information Systems*, London: Taylor & Francis.

PHILLIPS, A. D. M. (1993). *The Potteries: Continuity and Change in the Staffordshire Conurbation*, Gloucestershire: Alan Sutton.

PIGOT, S. (1991). Topological models for 3-D spatial information systems, in *Proceedings of Autocarto 10* (Falls Church, Virginia, USA), Baltimore, MD: ACSM-ASPRS, pp. 368–392.

PIGOT, S. (1994). Generalized singular 3-cell complexes, in Waugh, T. C. and Healey, R. G. (Eds) *Proceedings of the 6th International Symposium on Spatial Data Handling*, Edinburgh, London: Taylor & Francis, pp. 89–111.

PISSINOU, N., SNODGRASS, R. T., ELMASRI, R., *et al.* (1994). Towards an infrastructure for temporal databases: Report of an invitational ARPA/NSF workshop, *Association for Computing Machinery SIGMOD Record*, **23** (1), 35–51.

PRATT, I. (1991). Path finding in free space using sinusoidal transforms: III, in Mark, D. M. and Frank, A. U. (Eds) *Cognitive and Linguistic Aspects of Geographic Space*, Nato ASI Series, Dordrecht: Kluwer Academic, pp. 219–233.

PREECE, J., ROGERS, Y., SHARP, H., BENYON, D., HOLLAND, S. and CAREY, T. (1994). *Human-Computer Interaction*, Reading, MA: Addison-Wesley.

PREPARATA, F. P. and SHAMOS, M. I. (1985). *Computational Geometry: An Introduction*, Berlin: Springer-Verlag.

PULLAR, D. and EGENHOFER, M. J. (1988). Towards formal definitions of spatial relationships among spatial objects, *Proceedings of the 3rd International Symposium on Spatial Data Handling*, Sydney, Columbus, OH: International Geographical Union, pp. 225–242.

PUPPO, E., DAVIS, L., DE MENTHON, D. and TENG, Y. A. (1994). Parallel terrain triangulation, *International Journal of Geographical Information Systems*, **8** (2), 105–128.

RAAFAT, H., YANG, Z. and GAUTHIER, D. (1994). Relational spatial topologies for historical geographical information, *International Journal of Geographical Information Systems*, **8** (2), 163–173.

RAPER, J. F. (Ed.) (1989). *Three Dimensional Applications in Geographical Information Systems*, London: Taylor & Francis.

RAPER, J. F. (1990). An atlas of 3-D functions, *Proceedings of the Symposium on Three Dimensional Computer Graphics in Modelling Geologic Structures and Simulating Processes*, Freiburger Geowissenschafliche Beitrage, **2** 74–75.

RAPER, J. F. and BUNDOCK, M. S. (1991). UGIX. A layer based model for a GIS user interface, in Mark, D. M. and Frank, A. U. (Eds) *Cognitive and Linguistic Aspects of Geographic Space*, Nato ASI Series, Dordrecht: Kluwer Academic, pp. 449–475.

RAPER. J. F. and KELK, B. (1991). Three-dimensional GIS, in Maguire, D. J., Goodchild,

M. F. and Rhind, D. W. (Eds) *Geographical Information Systems*, Vol 1, Harlow: Longmans, pp. 299–317.

REDDY, D. R. and RUBIN, S. (1978). Representation of three-dimensional objects, Technical Report CMU-CS-78-113, Department of Computer Science, Carnegie-Mellon University, Pittsburgh, USA.

REQUICHA, A. A. G. (1980). Representations for rigid solids: theory, methods and systems, *Association for Computing Machinery Computing Surveys*, **12** (4), 437–464.

REQUICHA, A. A. G. and VOELCKER, H. B. (1985). Boolean operations in solid modeling: Boundary evaluation and merging algorithms, *Proceedings of IEEE*, **73** (1), 30–44.

RESCHER, N. and URQUEHART, A. (1971). *Temporal Logic*, Berlin: Springer-Verlag.

RIKKERS, R., MOLENAAR, M. and STUIVER, J. (1994). A query oriented implementation of a topologic data structure for 3-dimensional vector maps, *International Journal of Geographical Information Systems*, **8** (3), 243–260.

RITTER, G. X., WILSON, J. and DAVIDSON, J. (1990). Image algebra: An overview, *Computer Vision, Graphics and Image Processing*, **49**, 297–331.

ROBINSON, J. T. (1981). The K-D-B-tree: A search structure for large multidimensional dynamic indexes, *Association for Computing Machinery SIGMOD*, **10**, 10–18.

ROSENFELD, A. (1980). Tree structures for region representation, in Freeman, H. and Pieroni, G. G. (Eds) *Map Data Processing*, New York: Academic Press, pp. 137–150.

ROSENFELD, A. and PFALTZ, J. L. (1966). Sequential operations in digital image processing, *Journal of the ACM*, **13** (4), 471–494.

ROTH, M. A., KORTH, H. F. and SILBERSCHATZ, A. (1988). Extended algebra and calculus for nested relational databases, *Association for Computing Machinery Transactions on Database Systems*, **13** (4), 389–417.

ROUSSOPOULOS, N. and LIEFKER, D. (1985). Direct spatial search on pictorial databases using packed R-trees, *Association for Computing Machinery SIGMOD*, **14**, 17–31.

ROWE, L. A. and STONEBRAKER, M. (1987). The POSTGRES next generation database management system, *Proceedings of the 13th Conference on Very Large Databases*, Brighton, pp. 83–96.

RUMBAUGH, J., BLAHA, M., PREMERLANI, W., EDDY, F. and LORENSEN, W. (1991). *Object-Oriented Modeling and Design*, Englewood Cliffs, NJ: Prentice-Hall.

SAMET, H. (1989a). *The Design and Analysis of Spatial Data Structures*, Reading, MA: Addison-Wesley.

SAMET, H. (1989b). *Applications of Spatial Data Structures: Computer Graphics, Image Processing and GIS*, Reading, MA: Addison-Wesley.

SAMET, H. and WEBBER, R. E. (1985). Storing a collection of polygons using quadtrees, *ACM Transactions on Graphics*, **4** (3), 182–222.

SCHLEIFLER, R. W. and GETTYS, J. (1992). *X Window System*, Bedford, MA: Digital Press.

SCHOLL, M. and VOISARD, A. (1992a). Geographic applications: An experience with $O_2$, in Bancilhon, F., Delobel, C. and Kanellakis, P. (Eds) *Building an Object-Oriented Database: The story of $O_2$*, San Mateo, CA: Morgan-Kaufmann, pp. 585–618.

SCHOLL, M. and VOISARD, A. (1992b). Object-oriented databases for geographic applications: An experiment with $O_2$, in Gambosi, G., Scholl, M. and Six, H.-W. (Eds) *Geographic Database Management Systems*, Berlin: Springer-Verlag, pp. 103–137.

SCHWEIZER, R. H. (1973). *Mapping Urban America with Automated Cartography*, Suitland, MD: Bureau of the Census, US Department of Commerce.

SEDGEWICK, R. (1983). *Algorithms*, Reading, MA: Addison-Wesley.

SELLIS, T., ROUSSOPOULOS, N. and FALOUTSOS, C. (1987). The R+-tree: a dynamic index for multi-dimensional objects. Computer Science TR-1795, University of Maryland, College Park, MD.

SHAMOS, M. I. (1975). Geometric complexity, *Proceedings of the 7th Annual ACM Symposium on Theory of Computing*, New York: ACM Press, pp. 224–233.

SHAMOS, M. I. and HOEY, D. (1975). Closest point problems, *Proceedings of the 16th Annual IEEE Symposium on Foundations of Computer Science*, New York: IEEE, pp. 151–162.

SHU, N. C. (1988). *Visual Programming*, New York: Van Nostrand Reinhold.

SMITH, J. M. and SMITH, D. C. P. (1977). Database abstractions: aggregation and generalization, *Association for Computing Machinery Transactions on Database Systems*, **2** (2), 105–133.

SMITH, P. and BARNES, G. (1987). *Files and Databases: An Introduction*, Reading, MA: Addison-Wesley.

SMITH, T. R. and YIANG, J. E. (1991). Knowledge-based approaches in GIS, in Maguire, D. J., Goodchild, M. F. and Rhind, D. W. (Eds) *Geographical Information Systems*, Vol 1, Harlow: Longmans, pp. 413–425.

SNODGRASS, R. T. (1987). The temporal language TQuel, *Association of Computing Machinery Transactions on Database Systems*, **12**, 247–298.

SNODGRASS, R. T. (1992). Temporal databases, in Frank, A. U., Campari, I. and Formentini, U. (Eds) *Theories of Spatio-Temporal Reasoning in Geographic Space, Lecture Notes in Computer Science 639*, Berlin: Springer-Verlag, pp. 22–64.

SNODGRASS, R. T., AHN, U., ARIAV, G. (1994). TSQL2 language specification, *Association for Computing Machinery SIGMOD Record*, **23** (1), 65–86.

SONDHEIM, M. (1993). Modelling the real world, *Proceedings of GIS'93*, Vancouver, British Columbia, Canada, pp. 1099–1111.

SOO, M. D. (1991). Bibliography on temporal databases, *Association for Computing Machinery SIGMOD Record*, **20**, 14–23.

STALLINGS, W. (1987). *Computer Organization and Architecture*, New York: Macmillan.

STALLINGS, W. and VAN SLYKE, R. (1994). *Business Data Communications*, New York: Macmillan College Publishing.

STELL, J. G. and WORBOYS, M. F. (1994). Towards a representation for spatial objects in diverse geometries, in Pissinou, N. and Makki, K. (Eds) *Proceedings of the Second ACM Workshop on Advances in Geographic Information Systems, National Institute for Standards and Technology*, Gaithersburg, Maryland, New York: ACM Press, pp. 28–33.

STEVENS, S. S. (1946). On the theory of scales and measurement, *Science*, **103**, 677–680.

STONEBRAKER, M. (1993). The MIRO DBMS, *Proceedings of the Association for Computing Machinery SIGMOD Conference*, New York: ACM Press.

STONEBRAKER, M. and ROWE, L. A. (1986). The design of POSTGRES, *Proceedings of the 15th Association for Computing Machinery SIGMOD Conference*, Washington DC, New York: ACM Press.

STONEBRAKER, M. and DOZIER, J. (1991). Large capacity object servers to support global change research, Sequoia 2000. Technical Report, Department of Computer Science, University of California at Berkeley, CA, USA.

STONEBRAKER, M., SELLIS, T. and HANSON, E. (1986). An analysis of rule indexing implementations in data base systems, *Proceedings of the First International Conference on Expert Database Systems*, Charleston, SC, pp. 353–364.

SURVEYS AND RESOURCE MAPPING BRANCH (1992). Spatial Archive and Interchange Format: Formal Definition, Release 2.0. British Columbia Ministry of Environment, Lands and Parks, Vancouver, British Columbia, Canada.

SUTHERLAND, W. A. (1975). *Introduction to Metric and Topological Spaces*, Oxford: Clarendon Press.

TAKEYAMA, M. (forthcoming). Mapping geographic models into Geo-Algebra.

TAKEYAMA, M. and COUCLELIS, H. (forthcoming). Map dynamics: integrating cellular automata and GIS through Geo-Algebra, *International Journal of Geographical Information Systems*.

TANSEL, A. U., CLIFFORD, J., GADIA, S., SUSHIL, J., SEGEV, A. and SNODGRASS, R. T. (Eds) (1993). *Temporal Databases: Theory, Design and Implementation*, Redwood City, CA: Benjamin/Cummings.

TARJAN, R. E. and WYK, C. J. V. (1988). An O($n$ log log $n$)-time algorithm for triangulating a simple polygon, *SIAM Journal of Computing*, **17**, 143–178.

TEOREY, T. J., YANG, D. and FRY, J. P. (1986). Logical design methodology for relational databases, *Association for Computing Machinery Computing Surveys*, **18** (2), 197–222.

THAYSE, A. (Ed.) (1989). *From Modal Logic to Deductive Databases*, Chichester: Wiley.

TOBLER, W. (1970). A computer movie simulating urban growth in the Detroit region, *Economic Geography*, **46** (2), 234–240.

TOBLER, W. (1993). Non-isotropic geographic modelling, in Three Presentations on Geographical Analysis and Modelling, Technical Report 93-1, National Center for Geographic Information and Analysis, University of California at Santa Barbara, CA, USA.

TOBLER, W. and CHEN, Z.-T. (1993). A quadtree for global information storage, *Geographical Analysis*, **14** (4), 360–371.

TOMLIN, C. D. (1990). *Geographic Information Systems and Cartographic Modelling*, Englewood Cliffs, NJ: Prentice-Hall.

TOMLIN, C. D. (1991). Cartographic modelling, in Maguire, D. J., Goodchild, M. F. and Rhind, D. W. (Eds) *Geographical Information Systems*, Vol 1, Harlow: Longmans, pp. 361–374.

TOUSSAINT, G. T. (1992). Scanning the issue: Computational geometry, *Proceedings of IEEE*, **80** (9), 1347–1363.

TSAI, V. J. D. (1993). Delaunay triangulations in TIN creation: an overview and a linear-time algorithm, *International Journal of Geographical Information Systems*, **7** (6), 501–524.

TSICHRITZIS, D. and KLUG, A. (Eds) (1978). *The ANSI/SPARC DBMS Framework*, AFIPS Press.

ULLMAN, J. D. (1988). *Principles of Database and Knowledge-Base Systems*, Rockville, MD: Computer Science Press.

UNWIN, D. J. (1981). *Introductory Spatial Analysis*, London: Methuen.

VAN KREVELD, M. (1994). On quality paths on polyhedral terrains, in Nievergelt, J., Ross, T., Schek, H.-J. and Widmayer, P. (Eds) *IGIS'94: Geographic Information Systems. Lecture Notes in Computer Science 884*, Berlin: Springer-Verlag, pp. 113–122.

VAN OOSTEROM, P. J. M. (1993). *Reactive Data Structures for Geographic Information Systems*, Oxford: Oxford University Press.

VAN OOSTEROM, P. J. M. and VAN DEN BOS, J. (1989). An object-oriented approach to the design of geographic information systems, *Computers and Graphics*, **13**, 409–418.

VETTER, R. J. and DU, D. H. C. (Eds) (1995). Issues and challenges in ATM Networks, *Communications of the Association for Computing Machinery*, **38** (2), 28–109.

VORONOI, G. (1908). Nouvelles applications des paramètres continus à la théorie des formes quadratiques, deuxième memoire, recherches sur les parallelloèdres primitifs, *Journal für die Reine und Angewandte Mathematik*, **134**, 198–287.

WARSHALL, S. (1962). A theorem on Boolean matrices, *Journal of the Association for Computing Machinery*, **9** (1), 11–12.

WAUGH, T. C. and HEALEY, R. (1987). The GEOVIEW design. A relational database approach to geographic data handling, *International Journal of GIS*, **1**, 101–118.

WAUGH, T. C. and HOPKINS, S. (1992). An algorithm for polygon overlay using cooperative parallel processing, *International Journal of Geographical Information Systems*, **6** (6), 457–468.

WEBBER, R. E. (1984). Analysis of quadtree algorithms, PhD Thesis, University of Maryland, College Park, MD, USA.

WEIBEL, R. and HELLER, M. (1991). Digital terrain modelling, in Maguire, D. J., Goodchild, M. F. and Rhind, D. W. (Eds) *Geographical Information Systems*, Vol 1, Harlow: Longmans, pp. 269–297.

WEILER, K. (1985). Edge-based data structures for solid modeling in curved-surface environments, *Computer Graphics Applications*, **5** (1), 21–40.

WEIMAN, C. F. R. (1980). Continuous anti-aliased rotation and zoom of raster images, *ACM SIGGRAPH*, **80**, 286–293.

WILLIAMS, G. J. (1995). Templates for spatial reasoning in responsive geographical information systems, *International Journal of Geographical Information Systems*, **9** (2), 117–131.

WILSON. A. A., REES, J. G., CROFTS, R. G., HOWARD, A. S., BUCHANAN, J. G. and WAINE, P. J. (1992). Stoke-on-Trent: A geological background for planning and development. Technical Report WA/91/01, British Geological Survey, Keyworth, Nottingham, England.

WORBOYS, M. F. (1991). The role of modal logics in the description of a geographical information system, in Mark, D. M. and Frank, A. U. (Eds) *Cognitive and Linguistic Aspects of Geographic Space*, Nato ASI Series, Dordrecht: Kluwer Academic, pp. 403–413.

WORBOYS, M. F. (1992a). A generic model for planar geographic objects, *International Journal of Geographical Information Systems*, **6**, 353–372.

WORBOYS, M. F. (1992b). A model for spatio-temporal information, in Corwin, E. and Cowen, D. (Eds) *Proceedings of the 5th International Symposium on Spatial Data Handling*, Columbus, OH: International Geographical Union, pp. 602–611.

WORBOYS, M. F. (1994a). A unified model of spatial and temporal information, *Computer Journal*, **37** (1), 26–34.

WORBOYS, M. F. (1994b). Object-oriented approaches to geo-referenced information, *International Journal of Geographical Information Systems*, **8** (4), 385–399.

WORBOYS, M. F. (1994c). Unifying the spatial and temporal components of geographical information, in Waugh, T. C. and Healey, R. G. (Eds) *Proceedings of the 6th International Symposium on Spatial Data Handling*, Edinburgh, London: Taylor & Francis, pp. 505–517.

WORBOYS, M. F. and BOFAKOS, P. (1993). A canonical model for a class of areal spatial objects, in Abel, D. and Ooi, B. C. (Eds) *Advances in Spatial Databases, Proceedings of SSD'93, Singapore, Lecture Notes in Computer Science 692*, Berlin: Springer-Verlag, pp. 36–52.

WORBOYS, M. F., HEARNSHAW, H. M. and MAGUIRE, D. J. (1990). Object-oriented data modelling for spatial databases, *International Journal of Geographical Information Systems*, **4**, 369–383.

WORBOYS, M. F., MASON, K. T. and DAWSON, B. R. P. (1993). The object-based paradigm for a geographical database system: modelling, design and implementation issues, in Mather, P. M. (Ed.) *Geographical Information Handling – Research and Applications*, Chichester: Wiley, pp. 91–102.

YOELI, P. (1985). The making of intervisibility maps with computer and plotter, *Cartographica*, **22** (3), 88–103.

ZADEH, L. A. (1965). Fuzzy sets, *Information and Control*, **8**, 338–353.

ZADEH, L. A. (1988). Fuzzy logic, *IEEE Computer*, **21**, 83–93.

ZHANG, T. Y. and SUEN, C. Y. (1984). A fast parallel algorithm for thinning digital patterns, *Communications of the Association for Computing Machinery*, **27** (3), 236–239.

# Index